PENGUIN BOOKS

Waste

Tristram Stuart has been a freelance writer for Indian newspapers, a project manager in Kosovo and a prominent critic of the food industry. He has made regular contributions to television documentaries and radio and newspaper debates on the social and environmental aspects of food. His first book, *The Bloodless Revolution*, 'a genuinely revelatory contribution to the history of human ideas' (*Daily Telegraph*), was published in 2006. He lives in the UK, where he rears pigs, chickens and bees.

D0335687

Waste

Uncovering the Global Food Scandal

TRISTRAM STUART

PENGUIN BOOKS

PENGUIN BOOKS

Published by the Penguin Group

Penguin Books Ltd, 80 Strand, London WC2R ORL, England

Penguin Group (USA) Inc., 375 Hudson Street, New York, New York 10014, USA

Penguin Group (Canada), 90 Eglinton Avenue East, Suite 700, Toronto, Ontario, Canada M4P 2Y3
(a division of Pearson Penguin Canada Inc.)

Penguin Ireland, 25 St Stephen's Green, Dublin 2, Ireland (a division of Penguin Books Ltd)

Penguin Group (Australia), 250 Camberwell Road, Camberwell, Victoria 3124, Australia
(a division of Pearson Australia Group Pty Ltd)

Penguin Books India Pvt Ltd, 11 Community Centre, Panchsheel Park, New Delhi – 110 017, India

Penguin Group (NZ), 67 Apollo Drive, Rosedale, North Shore 0632, New Zealand
(a division of Pearson New Zealand Ltd)

Penguin Books (South Africa) (Pty) Ltd, 24 Sturdee Avenue,
Rosebank, Johannesburg 2196, South Africa

Penguin Books Ltd, Registered Offices: 80 Strand, London WC2R ORL, England

www.penguin.com

First published 2009

1

'The Monkey', words and music by Dave Bartholomew and Pearl King,
copyright © 1957, reproduced by permission of EMI United Partnership Ltd,
London W8 5SW

Set in 11/13 pt Monotype Bembo
Typeset by Rowland Phototypesetting Ltd, Bury St Edmunds, Suffolk
Printed in England by Clays Ltd, St Ives plc

ISBN: 978–0–141–03634–2

www.greenpenguin.co.uk

In Memory of Gudrun

Contents

List of Illustrations

Unless otherwise stated, all photographs are by the author.

Acknowledgements

Thanks to my editor at Penguin, Stuart Proffitt, for his astounding meticulousness and enthusiasm, and to Phillip Birch, who made innumerable astute contributions and to Mark Handsley; also to Bob Weil at Norton and my agent, David Godwin, ever irrepressible and supportive. Andrew Parry, Mark Barthel and Keith James at the Waste & Resources Action Programme (WRAP) gave me access to data and insight and kindly took up much of their precious time reading an early draft of the manuscript, saving me from mistakes and sharpening insights and policy recommendations. Tim Lang and Tara Garnett generously agreed to read drafts and provided me with invaluable feedback. I am extremely grateful to all of them, but equally, any errors are mine alone, and the stated opinions are not necessarily shared by them.

I was taken aback by the willingness of some individuals within the food industry to co-operate, including Alison Austin, Gus Atri, James Cherry, Julian Walker Palin, Kev Stevenson, Mark Bartlett, Rowland Hill, Maureen Raphael, Takeshi Tanami, Toshiro Miyoshi, Kouichi Takahashi and Hiroyuki Yakou. A special thanks to Lord Chris Haskins who has been feeding me ideas and honing my understanding of the global food industry for years.

Technical assistance with my calculations and questions of scientific method was selflessly and unstintingly provided by Jannick Schmidt, Jo Howes, Randi Dalgaard and Vaclav Smil. Many other valued suggestions came from academics, friends and other individuals. I am particularly grateful to Adrien Assous, Robyn Kimber, Peter Jones, Marion Nestle, Akifumi Ogino, Tomoyuki Kawashima, Karoline Schacht, Eric Evans, Michael

Mann, Eric Audsley, Hyunook Kim, James Parsons, Cinzia Cerri, Sung-Heon Chung, David Jukes, Keith Waldron, Faqir Mohammad, Belinda Fletcher, Jean Buzby, Itisaki Shimadu, Frank Filardo, Thomas Stuart, Deborah Stuart, Alex Evans, Jack Mathers, Ricardo Sibrían, Daniel Wilson, Michael Chesshire, Ozunimi Iti, Nicola Kohn, Timothy Jones, Catherine Gaillochet and Shahin Rahimifard.

I was blessed with dedicated research assistants, Laura Yates and Simon Inglethorpe, who among many other things crunched thousands of statistics from various sources: the result of much of their work is concentrated in the appendix, where raw data has been converted into stunning maps, charts and tables through the work of Alan Gilliland. In South Korea, Japan and India I received diligent help from Tehion Kim, Haruka Ezaki and Arshinder Kaur.

Crucial doses of hospitality while researching the book came from Declan Walsh, John, Louise and Rose Dargue, Zainab Dar, Usman and Hussain Qazi, and the Kawasaki family. Above all I have to thank my wife, Alice Albinia, whose careful editing, drafting and unending patience made the book possible.

But if they perished, in his possession, without their due use; if the fruits rotted, or the venison putrified, before he could spend it, he offended against the common law of nature, and was liable to be punished . . . if either the grass of his enclosure rotted on the ground, or the fruit of his planting perished without gathering, and laying up, this part of the earth, notwithstanding his enclosure, was still to be looked on as waste, and might be the possession of any other . . . He was only to look, that he used them before they spoiled, else he took more than his share, and robbed others. And indeed it was a foolish thing, as well as dishonest, to hoard up more than he could make use of. If he gave away a part to any body else, so that it perished not uselesly in his possession, these he also made use of. And if he also bartered away plums, that would have rotted in a week, for nuts that would last good for his eating a whole year, he did no injury; he wasted not the common stock; destroyed no part of the portion of goods that belonged to others, so long as nothing perished uselesly in his hands . . . the exceeding of the bounds of his just property not lying in the largeness of his possession, but the perishing of any thing uselesly in it.

John Locke, *Second Treatise of Government* (1690), v.37–38, 46

Introduction

The Effluent of Affluence

Picture an aerial view of the earth. Over the last ten millennia, it has changed beyond recognition. By far the biggest invaders of the natural landscape are not tarmac and concrete, but fields – those green, furrowed units in the countryside. Where forests once cloaked the ground, fields now stretch across continents, transforming the land into a food factory. Cities, roads and industries are mere spots and veins on the body of the earth compared to the changes brought about by cultivation. Since the advent of agriculture, humans have replaced the world's diverse ecosystems with a handful of domesticated species designed to harness the sun and soil exclusively for human food.[1] Civilization is based on the yields that result. But the advance of agriculture now threatens the life it was designed to support.

In a globalized food industry, almost everything we eat – from bananas to locally grown beef – is connected to the system of world agriculture. Demand for food in one part of the world indirectly stimulates the creation of fields thousands of miles away. The onward march of agriculture into natural forests is currently most visible in Latin America and south-east Asia. On one side of the frontier there are pristine forests; on the other are monocultures of soybeans, oil palms and grass; at the margin there is a strip of fire, and an army of loggers.[2] Unknown species of plant and animal are lost and billions of trees vaporized into as many tonnes of greenhouse gases to satisfy our hunger. This process is also upsetting the climate, hydrological cycle and soil to such an extent that the United Nations now estimates that the world's agricultural land may decline in productivity by up to 25 per cent this century, which could undermine humanity's future ability to grow enough food at all.[3]

The encroachment of agriculture into forests may provide a short-term abundance of food and profits, but such environmental devastation is not essential in order to grow enough to feed the world's population. As this book will show, in the United States around 50 per cent of all food is wasted.[4] In Britain up to 20 million tonnes of food waste is created each year. The Japanese – with their love of sushi, caviar and imported luxuries – manage to dispose of food worth ¥11 trillion (US$101.6bn) annually. Throughout the developed world, food is treated as a disposable commodity, disconnected from the social and environmental impact of its production. Most people would not willingly consign tracts of Amazon rainforest to destruction – and yet that is happening every day. If affluent nations stopped throwing away so much food, pressure on the world's remaining natural ecosystems, and on the climate, would be lifted.

The environmental motive for tackling food waste at a global level is clear. The social imperative for finding a solution is, if anything, even starker. By buying more food than we are going to eat, the industrialized world devours land and resources that could otherwise be used to feed the world's poor. There are nearly a billion undernourished people in the world – but all of them could be fed with just a fraction of the food that rich countries currently throw away.

The connection between food profligacy in rich countries and food poverty elsewhere in the world is neither simple nor direct, but as this book will argue, it is nevertheless real, and we all have the power to relieve it. Obviously, the solution is not for rich countries to send old tomatoes or stale bread over to poor countries after saving them from the rubbish bin. Instead, I shall argue that in a global food market in which rich and poor countries buy food from a pool of internationally traded commodities, we are all essentially buying food from the same common source. If rich countries buy hundreds of millions of tonnes of food and end up throwing these into the bin, they are gratuitously removing food from the market which could have remained there for other people to buy.

Wasting food also uses up the world's limited available agricultural land. If rich countries wasted less this could liberate agricultural land for other uses, including growing food that the world's hungry could buy in the normal ways. This sequence of causes and effects is most clearly evident for internationally traded commodities such as wheat, and less clearly for fresh produce grown and purchased within individual nations – but it does still apply even there. If that food wasn't being bought and wasted, the land and other resources could be used to grow something else, including food such as cereals that could contribute to much-needed global supplies.

We are beginning to accept that we have personal responsibility for the impact that our greenhouse gas emissions have on people worst affected by global warming. Similarly, we need to recognize that the world is affected by our consumption and waste of food. There has been enormous attention paid to the 'waste' of food by the US and European decision to use food grains and pulses to make biofuels. Critics of the biofuels industry point out that we are filling fuel tanks instead of filling hungry stomachs, and in 2008 the UN special rapporteur on the right to food, Jean Ziegler, called biofuels 'a crime against humanity'.[5] Partly in response to this, both the UK government and the European Commission (EC) have now revised their biofuels targets – while the US continues to promote biofuels, especially ethanol made from maize. It is true that under current circumstances, and in some instances, the biofuel industry is causing hunger, damaging natural ecosystems and doing more harm than good to the climate.[6] But what about filling wastebins in preference to feeding people? This is surely worse.

In the year from 2007 to 2008, up to 95 million tonnes of cereals were manufactured into biofuels, and the World Bank and International Monetary Fund estimated that this was responsible for most of the spike in global food prices which pushed tens of millions more people into poverty and malnourishment.[7] And yet, the amount going to biofuels is less than half of the food wasted unnecessarily in the world.[8] The public found it easy to believe that diverting food into fuel contributed to the starvation of millions. It may be harder for us to accept that the same applies to the food we divert into our bins.

During a BBC radio debate about the food crisis in March 2008, the former British member of parliament Michael Portillo – one of Margaret Thatcher's old acolytes – asked sardonically what possible connection there was between our waste of food and the want of the world's poor.[9] Do people wish to repudiate the connection because their own behaviour is implicated? People either misconstrue the basic connections between themselves and far-off people, or they focus on the real barriers intervening in the confusing sequence of causes and effects. These barriers do complicate the argument, but they do not undermine it.

There are legitimate objections – for example, that rich countries' demand for food can stimulate production and contribute to the economies of poor nations and therefore throwing food away merely increases demand which raises some farmers' revenues. It is also true that in some circumstances – as I shall discuss in chapter 11 – growing surplus can be a necessary and desirable measure to prevent food shortages. But creating surplus food involves a trade-off in terms of land use, resource depletion and stretching supplies, and, therefore, when ecological or production limits are reached, the costs of waste outweigh the potential benefits. It is true, too, that if rich countries stopped wasting so much, the food that would be liberated might merely be bought by other relatively affluent people, for example to fatten more livestock, rather than being eaten by the poorest families. But, overall, pressure on world food supplies would decrease, helping to stabilize prices and improving the condition of the vast majority of poor people who depend on these markets for their food.[10] It is difficult to predict exactly what the economic result of reducing demand by decreasing waste would be, but if the increased demand for biofuels caused up to 70 per cent of the price rises that led to the food crisis in 2008,[11] then decreasing demand by reducing waste could relieve pressure on supplies and prices by that much at least. The bottom line is that the present and growing demand for food is not sustainable, by which I mean that if it continues unabated, the equilibrium in the world's biosphere could well be irremediably altered to the detriment of many species on Earth, including our own. Indeed, in some parts of the world this moment has already arrived.

Hunger and malnutrition are not exclusively foreign concerns either; millions in the developed world also do not have enough to eat. In Britain alone, 4 million people are unable to access a decent diet.[12] In the United States around 35 million live in households that do not have secure access to food,[13] and in the European Union (EU) an estimated 43 million are at risk of food poverty.[14] This situation persists even while supermarkets throw away millions of tonnes of quality food. Here one potential solution *is* for surplus food to be redistributed and given to people who need it while it is still fresh and good to eat.

This book argues that the world's mountain of surplus food is currently an environmental liability – but it is also a great opportunity. There is a fantastic amount of slack in the world's food supply, where efficiency measures could create enormous savings, help the fight against hunger and guarantee food for future generations. As well as potentially feeding hungry people, salvaging food would help tackle global warming. More than 30 per cent of Europe's greenhouse gas emissions come from food production.[15] If food waste was halved, emissions could be slashed by 5 per cent or more. In a hypothetical scenario, if we planted trees on the land currently used to grow unnecessary surplus and wasted food, we could theoretically offset a maximum of 50–100 per cent of the world's greenhouse gas emissions. And, unlike most environmental measures, reducing food waste has minimal negative consequences. It is not like the tough dilemma of whether to forgo car or aeroplane travel or give up eating so much meat. We can help reduce our impact on the planet and improve the lives of the world's poor just by using food instead of throwing it away.

Once food has been discarded it creates another environmental problem of its own. In most countries throughout the developed world, organic waste is buried in landfill sites, where it decomposes into toxic effluent and methane, a greenhouse gas twenty-one times more potent than carbon dioxide.[16] Some developed nations are running out of space to bury their rubbish.[17] In South Korea and Taiwan, sending food to landfill is already banned, and it is only a matter of time before other countries will have to do the same. As these countries are learning, the very concept of waste has to be redefined. Even when unfit for human consumption, food waste is a valuable resource that can be used for animal feed, power production and fertile compost.[18]

Sorting out the food waste problem would be good for society and the environment, but it would also be good for business. As I show in the chapters that follow, where waste has been cut, profit margins consistently soar. Farmers have doubled their income by marketing previously discarded produce; manufacturers have saved 20 per cent of their costs just by tackling waste; and retailers have defeated their rivals by tightening up on efficiency. Those who say we should grow our own food and use local farmers' markets are portraying an ideal solution. But even if we acknowledge that the supermarkets are here to stay, and that consumer culture will not relinquish its infinite choices until necessity demands it – still, there is immense scope for cutting down on waste.

In researching this book, I took trains and buses from London into Europe, through Russia and central Asia, through Pakistan and India, across China and by ferry to South Korea and Japan. Between nations there are striking differences in attitudes to food, and particularly to waste. In some cultures, such as in South Korea, wasting food is a sign of profusion and hospitality. In others – such as among the Uighurs of western China – wasting food is taboo. The Uighurs' frugality is not merely a reaction to hardship, but stems from a profound appreciation of farming, cooking and eating – of knowing that food is just too good to waste. It would not be beyond any of us to realize that food ownership carries with it a weight of responsibility – an understanding that we are its custodians, bound to treat it carefully.

The United Nations has backed a call for food waste to be halved by 2025, but the target could be achieved even faster with the co-operation of businesses, governments and the public.[19] If industries have been slow to tackle waste, most governments have been sloth-like. Those in Europe and America have been unwilling even to find out how much the food industry throws away. The British government has financed the examination of private household bins, but so far it has not dared to take a peek into the skipfuls of food thrown away by supermarkets. In the developed world as a whole, there have been few attempts even to calculate how much food is wasted, or how many acres are needlessly ploughed to fill rubbish bins. Glimpsing the totality of this waste, and understanding its grotesque scale and gratuitous causes, is the first step towards reducing it.

Poor countries can certainly benefit from technologies used in the industrialized world. But in countries where resources are scarce and waste is avoided wherever possible, people's parsimony also holds lessons for the West. In the typical middle-class neighbourhood where I lived in Delhi from 1999 to 2001, many restaurants and food shops ensure that leftovers are distributed to beggars. Garbage collectors take household organic waste to concrete bunkers at the bottom of the street, and there cows and pigs gather at their leisure to pick through a meal of mango skins, potato peelings and coriander stalks, converting them

into meat, milk and manure. Paper and plastic are collected on
bicycles by small-time entrepreneurs who take it to nearby cottage industries for recycling (though the emissions from these
unregulated industries can be injurious to their health).[20] Gleaning in the developing world may be enforced by poverty; but
that is the point. Industrialized nations need to learn what it
means to live in scarcity – because the appearance of infinite
abundance is an illusion.

However, the sad truth is that even in nations filled with
hungry people, there are phenomenal levels of waste. In affluent
countries, food is discarded deliberately; but in poor ones it is
inadvertently lost owing to lack of resources and technology.[21]
India alone wastes more than Rs 580bn ($14bn) of agricultural
produce every year, largely because the infrastructure to bring
harvests to market before they are spoiled is lacking. In Sri
Lanka, 30–40 per cent of fruit and vegetables are lost, even while
daily vegetable consumption by Sri Lankans is as low as 100g
per person. Simple measures can be taken to cut these losses
down to a fifth of current levels. Reducing waste is not just a
way of improving efficiency, but also of lifting some of the
world's poorest people out of malnutrition.

The starving poor often elicit guilt about Western wastefulness. But if Western governments want to reclaim resources,
they could look to places such as India, where systems for
recycling and re-using have never died. In the Indian state of
Maharasthra, for example, more than a hundred domestic biogas
plants are being fitted every month, using organic waste to
power kitchen stoves.

The West's thoughtlessness about waste could be overcome by
learning from the developing world's frugality. Poorer nations,
meanwhile, could dramatically increase the amount of food
available to them with the efficient technologies that are taken
for granted in the West. At the moment we have the worst of
both worlds. This book will show that the best of both are not
beyond our reach.

PART I

Perishing Possessions

1. Liber–ate

When they had all had enough to eat, he said to his disciples,
'Gather the pieces that are left over. Let nothing be wasted.'[1]

Feeding of the five thousand, John 6:12

I first caught sight of the world's food waste mountain when I
started rearing pigs in my mid-teens. I had bought a Gloucester
Old Spot sow to breed from and intended to raise the piglets in
the most environmentally friendly and traditional way – by
feeding them on waste food. The staff in my school kitchens
were only too happy to set aside uneaten dishes and peelings
in exchange for a few cuts of pork at the end of the season.
Although they were not allowed to give me plate scrapings or
anything containing meat, even so, the waste regularly amounted
to 25 gallons of nutritious leftovers each day. My pigs learnt to
live on mixed salads, roast potatoes, macaroni cheese, sponge
cake, rice pudding and a hundred other delights of canteen
cuisine. They used to squeal in anticipation when I arrived back
from school carrying the bright yellow buckets. One litter of
ten earnt themselves the name 'the decibels' for the power of
their collective cry as I tipped out the food my school-fellows
had turned their noses up at.

In addition to the school swill, I was soon collecting a weekly
van-load of cauliflower leaves from the town market and several
sacks of organic bread from the local baker, which helped feed
my flock of laying hens. As payment, I used to take windfall
apples and eggs down to the baker, who turned them into cakes
and pies. Out-sized potatoes would come in from a deer farmer

down the road who picked them up by the tonne for next to nothing. He would swap a few sackloads with me for a dozen eggs whenever I needed them. I used to cross my sow, Gudrun, with a high-quality Landrace boar; and her piglets, which came in litters as large as seventeen, had lovely long bacon-backs. The meat was indescribably succulent, and I gave the rich manure to my dad in exchange for the use of his farm buildings. One year a swaying field of tomatoes grew where the pigs dunged in the run, and I remembered seeing crate-loads of tomato salad in the swill-buckets months earlier. I harvested them all and made gallons of green-tomato chutney.

As a schoolboy chasing hard-earnt pocket money and delicious home-grown pork, these free sources of food were a boon. Commercial pig feed is expensive at the best of times, and is almost always produced in environmentally unsustainable ways. It was only when I reflected that much of the 'waste' I was picking up was still fit for human consumption that I began to see it as a problem as well. Before I started collecting it, most of that food was dumped into landfill sites along with other forms of garbage, and I knew my efforts were insignificant beside the mounds of food being thrown away by numerous local food outlets. Many shopkeepers had turned me away, either because they were anxious that I was not a licensed waste collector, or because they did not have the time or inclination to separate their food waste. I had seen sackloads of food being locked into bins behind supermarkets, whose managers were not willing even to discuss how much they threw away or what they did with it. I tried to imagine the quantities of food being lost all over the country by every different kind of shop and kitchen. The waste did not bear thinking about. Surely some of that food could have been eaten by people, and anything that couldn't could be fed to livestock?

As it was, the discarded organic bread I fed my pigs was better quality than the supermarket loaves my dad and I were eating at home. There was one particularly fragrant sun-dried tomato bread which I used to see in my pigs' trough every now and then. One morning before school, I decided to try it. As my pigs munched through their breakfast, I tore open the rich loaf and ate mouthfuls of tomato-flavoured dough. I had always been keen on foraging for wild mushrooms, nuts and berries; this seemed like hunter-gathering in a new context. The bread was salty and soft, still fresh enough to eat and enjoy. But mixed with the pleasure was a strong sense of unease that such good food, all over the world, could be allowed to go to waste. Eating that bread was my first act of what I later learnt to call 'freeganism' – the consumption of free discarded food. Eating it grew into a kind of protest – a way of demonstrating that food like this should not be thrown away.

It was only years later that I read the words of John Locke, one of England's pre-eminent philosophers, expounding the idea that if people let food perish in their possession they lose the right to own it: 'if the fruits rotted, or the venison putrified, before he could spend it, he offended against the common law of nature, and was liable to be punished . . .'[2] Here was the compelling idea that it was not I who was stealing from those who had discarded the food, but they who had stolen it from the world's common resources by hoarding it, allowing it to go to waste, and depriving the people who needed it most.

The couple who ran the local bakery were at least doing what they could to reduce waste by letting me take their leftovers. By contrast, all those supermarkets and others who wasted good edible food and refused to change would henceforth be legitimate targets of protest. It was then that I decided to find out more about what was being wasted up and down the food chain, from farms to manufacturers, retailers and consumers. I wanted to find out what level of waste was inevitable, and what was avoidable, how much food in total we were really wasting, what happened to the food we did throw out, and what impact this had on us and our environment. Above all, I wanted to know what could be done to change it.

By the time I left school, I had learnt that I could live off the food being thrown away by supermarkets and other retailers. For as long as I could remember I had been an environmentalist and during the year after school I visited protest camps erected in the path of the Newbury bypass, where I found that it was common for people to live on food that had been scavenged for free. Freeganism became part of the wider environmental protest movement I was part of, as we aimed simultaneously to minimize our impact on the environment and to raise awareness of the food waste problem. If someone like me, who was easily able to buy food, could still live off waste, then surely something must be wrong with the system. I never regarded rummaging in bins as a viable solution to the problem itself: the logical telos of freeganism was to bring an end to itself by encouraging businesses to do something more sensible with their surplus than throwing it in the bin.

During my first term at university, the vegetable market was the source of much of my food. But I soon started to concentrate on what the supermarkets were throwing away. I was introduced to the subterranean loading-bay beneath Sainsbury's by a homeless man called Spider – so called for the web of tattoos on his face. When he encouraged me to start raiding their bins, I told him that I did not want to take food which homeless people were relying on. They actually needed the food, while I was merely making a point about waste. 'You don't understand, mate,' came Spider's reply, 'if all the homeless in the country got their food out of this bin, there'd still be enough left for you.'

It was true: the quantity of good food that Sainsbury's placed in its six wheelie bins defied explanation. Admittedly, my standards were not so high as a student, and so my then girlfriend (now wife) and I used to eat a lot of white sliced bread which was thrown out by the bin-load on most days (though we tended to ignore the sugary doughnuts and chocolaty snacks which were the preferred sugar rush of the town's cold and hungry homeless population). Mixed up with all this, the staff would regularly deliver sacks of fresh vegetables, cheese and yoghurt and heaps of chilled ready-meals, which I took home to cook up in my student halls. Ever since university I have continued to dine on food thrown away by high street shops in cities and towns throughout the developed world – from New York to Paris, London to Stockholm. Each time, I feel the hunter-gatherer in me grunting with satisfaction over another successful forage. But this atavistic reaction is tempered by outrage. There are simple methods of using surplus food for its proper purpose – eating it – and there are no good reasons why these should not be practised universally.

The waste that goes into their bins is a scar on supermarkets' reputations. In one day, a store can easily throw out enough to feed over a hundred people. Last week, for example, I visited the bins of a branch of Waitrose and found the following, by no means exceptional, shopping list of items: 28 chilled high-quality ready-meals (including lasagne, prawn linguine, beef pie, chicken korma with rice, chicken tikka with rice, chicken with Madeira wine and porcini mushrooms); 16 Cornish pasties; 83 yoghurts, chocolate mousses and other desserts; 18 loaves of bread and 23 rolls; one chocolate cake; 5 pasta salads; 6 large melons; 223 individual items of other assorted fruits and vegetables including nectarines, oranges, fair-trade organic bananas, papaya, organic carrots, organic leeks and avocados, 7 punnets of soft fruit, one pack of mushrooms, 6 bags of potatoes; a bag of onions and two thriving potted herbs (chives and parsley – now planted in my herb garden); one almost full box of serving-size pots of margarine; a box of serving-size UHT milk cartons; several bunches of flowers, and a potted orchid. Not one of these items was unfit to eat (apart from the flowers) and I have been living off the vegetables all week.

As an undergraduate, I wanted to help the local supermarket do something more constructive with its surplus. I knew a number of people who worked in the Sainsbury's store and talked to them about the food being thrown out; they asked the managers if staff could take home any surplus that was still in date. But the most the supermarket was willing to do was sell it to them at a 10 per cent discount. Neither would the managers give food to homeless shelters and other charities. They claimed they did not want to be sued if anyone fell ill – though other supermarkets like Marks & Spencer had been donating their surplus for years without any problem.

In those days, most supermarket managers were not interested in the issue of what to do with unsold food. Their activities were constrained by company policies which determined that surplus should be sent to landfill. They were in the business of selling food, so many bosses believed that giving it away could undermine their sales. It made more sense for the supermarkets to lock the food in bins and send it off to be buried in the ground, regardless of the social and environmental costs. Although there are signs of change within the industry which I discuss in the following chapters, this remains the default position today. Food corporations are profit-making machines. They may try to reduce waste as far as it affects profits, but once surplus has arisen most of them will not do anything other than throw it away – unless they can save significant amounts of money on disposal costs, or if, as is now increasingly becoming the case, they believe that adopting better policies will give them good publicity.

A supermarket's profit margins may allow tonnes of food to be overstocked and wasted, but for the staff actually responsible for putting good food in the bin, this can be bad for morale. Numerous interviews with supermarket staff have shown that people find wasting food on this scale wrong – there appears to be something instinctive in human beings that tells us so. One supermarket worker who called in to a BBC Radio 2 show I was on described how he had resigned in protest at the amount of food he was being forced to bin.[3] I have met many people who worked in shops, restaurants and canteens who protested bravely and vociferously against the waste in the industry. Their efforts are just beginning to yield results.[4]

Soon after graduating from university, frustrated by the super-markets' unwillingness to engage with the food waste issue, I started to work on a media campaign. The method was easy: take newspaper, radio and television crews round the back of supermarkets and show them just how much was being thrown away. The level of interest in this underbelly of the food industry was overwhelming, and scores of requests came in from broad-casters and newspapers from England, Scotland, Wales, Northern Ireland and Ireland, and even further afield, from Germany and Italy to Brazil, Australia and the United States. There was clearly enormous appetite in all parts of the world for discovering the truth behind the industrial scale of food waste, and the public response showed that many people had long been troubled by it.

The first occasion was when a journalist twisted my arm into making a feature for the BBC *Politics Show* in 2003.[5] Taking the camera team to one of my favourite spots in central London, I could see their jaws dropping as I selected six bin-bags of fresh food from the waste that one small sandwich shop had taken off its shelves only minutes before. The BBC set up an interview with Lord Haskins, then one of the chief advisors to the govern-ment on food and farming, and the ex-chairman of one of Britain's largest food-processing firms, Northern Foods. I was just preparing my tirade when Lord Haskins launched into his own: sell-by dates were absurdly strict and by his estimation an incredible 70 per cent of all food produced was wasted. I nearly fell off the park bench we were sitting on. This was the most extreme figure I had heard, and it was coming not from an environmentally minded campaigner but a very senior member of the industry itself. My hunch about the scale of the problem was confirmed.

In 2007 Sky News asked me to put together a breakfast of scavenged food for representatives of the food industry and the

Director of Fareshare, Tony Lowe. The big five supermarkets
were asked along, but they declined. Instead we spoke with
an employee of the British Retail Consortium (BRC), a trade
organization that represents the retailers. The aim was to ask
why supermarkets chose to waste food rather than donating it
to charities like Fareshare for redistribution to the disadvantaged,
and why the BRC in particular had not promoted this cause.
As we sat there munching on discarded croissants and bananas,
the BRC representative tried to point the finger of blame at
consumers, saying that 'most of the wasted food that we have
actually comes from domestic waste, so it comes out of homes
rather than out of the back of the supermarkets.' In fact, I had
just been examining the latest figures and was able to prevent
this apparent attempt to mislead the public by observing that
consumers were responsible for only about one third of all food
wasted in the UK, while the rest came from the industry. 'We
accept that,' said the man from the BRC.[6] Some months later,
BBC Radio 4's *Today* programme invited me on air, again with
the BRC. Despite having neglected Fareshare hitherto, on this
occasion the Director-General of the BRC, Kevin Hawkins,
applauded their work and highlighted the examples where supermarkets had donated them food.[7] But he eventually lost his cool,
spluttering angrily that if I carried on reclaiming discarded food
from bins I would get food poisoning. He had made my point
for me: if he thought I, who was healthy and reasonably well-
educated, was at risk, how much more vulnerable were all those
homeless and hungry who currently rummage in skips for their
meals? I wished I could have introduced him to the old south
Asian homeless man I had met the week before, who was looking
for food in a bin of grimy high street waste, mucus running
down his face, apparently sick with pneumonia. Why not give
people like him food in a dignified and hygienic manner through
established redistribution charities?

At last, in recent years governments and corporations have
begun to give the problem more attention. A handful of

countries have funded research and campaigns to address food waste, and most of the major supermarkets in Britain have begun to donate a very small proportion of their surplus to charity. But, even now, food corporations keep their waste statistics locked away from the public eye. Waste is their dirty secret. As for solving the problem thoroughly, we have barely begun.

Although it may seem counter-intuitive, it is the supermarkets who have developed some of the most efficient systems for food management and processing in the whole food industry. There have been improvements, but there is still a very long way to go before any major chain anywhere in the world can claim to have done everything in its power to reduce their colossal waste of food. These huge businesses, with vast human, technological and financial resources at their disposal, can allocate personnel and other facilities to focus on the issue – and they are at last beginning to do so. Before anybody trolleys off happily to the biggest supermarket they can find under the impression that they waste less as a proportion of their sales than other shops, it is worth noting that supermarkets conceal much of the waste they cause by pushing it further up and down the supply chain, as I shall discuss in more detail later. It may not end up in the bins at the back of their stores, but they are still largely responsible for the waste produced by their suppliers and even, partially, their customers.

In terms of back-of-store waste, however, smaller shops and chains often perform worse than the large supermarkets. Dr Timothy Jones, a specialist on food waste in the US, claims that the best-managed supermarket chains in North America have 'leaned down' their waste to less than 1 per cent of the food coming into the stores. Convenience stores as a sub-sector compare unfavourably, with average wastage levels of 26 per cent.[8] This is partly because the drop-in and top-up shopping for which people use convenience stores is inherently more unpredictable than the regular weekly shop they do in the big supermarkets. Non-supermarket stores also sell a higher proportion of Class II fruit and vegetables, and other produce past its peak, which are therefore more susceptible to spoilage.[9] But the higher waste levels in these smaller stores are also because they simply do not have, or fail to allocate, the time and expertise required to reduce waste. Smaller shops often employ people with little formal expertise, and although they may be great entrepreneurs in many ways, they simply do not have as much knowledge about food ordering and store management as the supermarket giants.

I realized this phenomenon first-hand (or second-hand perhaps) when living out of the bins in the well-heeled Primrose Hill area of north London from 2002. This was perhaps the heyday of my freeganism. The relatively small shops in the area were targeting an increasingly affluent clientele, and they were phenomenally bad at managing stock. Every single evening, on the street in front of one of these smaller stores, there would be two or three sackloads of luxury food – from mature Stilton cheese to smart jars of apricots preserved in Cognac, packaged organic vegetables and fruit yoghurts. I would regularly find one or two hundred pounds' worth of food dumped on the street as garbage – far too much even for our collective household to consume. Once my housemates found a sack of fifty pots of Ben & Jerry's, Häagen Dazs and Green & Black's organic ice cream, straight out of the freezer. It took the best part of six months to work our way through it all.

Cream also appeared by the gallon, and rather than block my arteries by using it all up in puddings and homemade ice cream, I used to churn it with an electric whisk to make big pats of golden butter. This was a joy in itself, and through it I discovered buttermilk, a delicious by-product of butter-making which seeps out as a milky juice just as the cream coagulates into butter, with the consistency of milk but the flavour of cream. It took me some time to find out why this is seldom marketed as a product in its own right – because in the 1980s inventive food engineers found a way of solidifying it along with vegetable oils into butter substitutes – a magnificent instance of how waste avoidance and industrial interests are often allied.[10]

More dismaying than the practice of ordinary convenience stores is the level of waste emanating from organic wholefood shops. Indeed, in the various places I have lived – London, Edinburgh, Sussex – organic shops have always produced surprising quantities of waste. This is ironic when most of their customers undoubtedly believe they are buying into environmentally sustainable businesses. While this is no doubt true in respect of the provenance of the food, waste management has remained below the radar of consumer consciousness and thus also of many shopkeepers. In the bins of a local organic fruit and vegetable shop near my current home, I used to find the equivalent of ten sacks of produce – amounting to around a fifth of a tonne of waste – disposed of on Friday evenings. This was bemusing since the shop specializes in stock from local growers who would presumably benefit from the waste as feed for their livestock, or at least for compost. In fact, when I offered to collect their surplus to feed to my pigs, they told me they had just been approached by another pig farmer who was planning on collecting all their waste, and that they would call me if they ever had anything extra. For a couple of months the content of their bins was reduced. But soon they were back to filling them again with tonnes of edible food. It seemed like an unfortunate confirmation of the stereotype of organic shop owners being individuals with great ideals but no business sense. Even though supermarkets waste culpable amounts of food, a supermarket manager would lose their job if they wasted anything like the proportion of turnover reached by some of these smaller independent shops.

Occasionally, however, it is smaller enterprises – and in particular stall-holders at farmers' markets – who are most keen to cut down on waste. Those that operate most effectively have direct control over stocking and a good knowledge of likely sales. They can make instant price reductions in response to poor take-up of individual items and often make sophisticated calculations on the likely shelf-life of their produce. They also often have relatively short supply chains, thus cutting out waste at intermediary stages. Most importantly, they are not bound by the stringent cosmetic criteria that supermarkets impose on produce.

In 2004 I lived a short cycle ride from Spitalfields market in east London. There, each Sunday and Wednesday, the organic farmers who came to sell their fruit and vegetables would throw any over-ripe, blemished or surplus produce into a big heap of boxes in the middle of the market. Squatters, freegans and others came from miles around for this foraging bonanza. The stall-holders tolerated the situation to the point of actively encouraging it, and anything that really was not edible was sorted into separate bins to be composted at the local community garden where Bangladeshi mothers grew gourds and greens for their families.

Being farmers, these stall-holders knew the value of the food they grew and the importance of recycling organic matter to go back into the soil. Taking leftovers from them was not a protest; it was participation in a cycle of maximum utilization, and some of the foragers showed their gratitude for the farmers' complicity by volunteering to help carry boxes and load up vans when the market closed. The foragers would barter among themselves, sharing out fruit and vegetables, and at the end of each Sunday I would cycle home with a couple of boxes of organic produce. Most of what was being thrown away at the market was good food, but, being ripe, it needed to be used up quickly.

At home I would sort through the produce and decide what needed processing immediately, and what could be eaten later in the week. Boxes of tomatoes – redder and juicier than the pale watery specimens in supermarkets – were whisked up into tomato soup with any herbs, or pickled into tomato relish. I would make plums or cherries into jam or chutney, or bottle them in Kilner jars. I planted chard, kale and lettuces in vases of water to keep them fresh. Soaking greens in a bowl of cold water was enough to bring back crispness even if they had started to wilt.

A box-load of bananas was a regular staple – as it is in every market or supermarket I have ever known. Farmers from Africa, Latin America or the West Indies had toiled hard to send them over the seas to arrive in our markets. Supermarkets in particular used to throw them out while they were still green, either because they had received a new delivery and did not have enough space to store them all, or because individual bananas or pairs were separated from larger bunches. (Supermarkets prefer customers to buy big bunches, so the spares often go straight in the bin, regardless of their state.) To deal with this constant stream of fruit I made it into banana chutney until I stumbled on a version of the smoothie which remains my favourite break-fast to this day. When the bananas are at their sweetest, with black spots appearing through the yellow, I peel them, bag them up and pack them into the freezer. Then on summer mornings I whiz a few of them up with a splurge of milk into the smoothest ice-cream-textured slush. Any other fruit like strawberries, rasp-berries, blackcurrants, nectarines or peaches can be added to the bananas or crushed separately to make beautifully coloured fragrant desserts. Being 'bin-ripened', they are as delicious as any fruit available.

One summer, my wife's cousin asked us to make mango lassi for their wedding, and by luck, on a Sunday at Spitalfields, I stumbled upon a stack of twenty-five box-loads of the ripest, loveliest organic mangoes I had ever seen in Britain. Twenty-four hours more in store and they would have been past it, but on that day they were at the height of perfection. We peeled and stoned them all, put the pulp in bags and froze it. Just before the wedding we churned the pulp up with milk and yoghurt and added a sprig of mint to each serving. The result was sensational. (That mango find was also the night I converted my mother to freeganism. Hitherto she had been sceptical about my eating habits but when she dropped in during the great mango-processing stint, even she could not resist taking home a few boxes to make a batch of chutney.)

Such salvaging efforts may seem peripheral to the main problem of food wastage, but, magnified many times, these concepts can go mainstream. The Spitalfields collaboration I have described is rare in Britain, but in food-obsessed France it is common to see hordes of foragers gathering at the end of a market, picking over discarded vegetables and fruit; and when I lived in Paris my friends and I regularly fed an extended house-hold of fifteen people on box-loads of reclaimed fruit and vegetables.

There are already some areas where these two problems – the difficulty of finding ready-to-eat fruit in the shops, and the waste of ripe produce by sellers – have been connected into business-friendly solutions. One practice, which is not as common as it ought to be, was explained to me by the owner of a Greek restaurant in south London, who – long fed-up with buying unripe tomatoes – would liaise with the local Cypriot grocery store and take its stock at closing time. With a modicum of lateral thinking, this model could be cost-effective enough to replicate it on a larger scale. Using surplus from retail in catering should be encouraged for economic, gastronomic and environmental reasons.

But the waste of locally grown fruit and vegetables is nothing compared to the waste of more valuable and resource-intensive produce, like meat and fish, which occurs on an unbelievable scale across the food supply chain. When I was writing my last book about the history of vegetarianism, my daily commute from the British Library took me past the doors of the K10 sushi outlet, which serves the affluent office workers of central London. I had never eaten sushi before, and I could see the staff emptying shelves at the end of the day. What I found when I opened the bin-bags was among the most offensive instances of waste I have encountered.

It was not the several kilos of rice, or the stodgy mass of teryaki chicken and curried pork I found tipped into bin bags which most upset me; nor was it the luxury pots of chocolate mousse, tiramisu or fruit salads. It was the trays and trays of neatly prepared prawn, salmon, swordfish and tuna sushi or sashimi which they disposed of every single day of the week. In an interview, a senior representative of the chain told me that K10 only wasted food on Fridays. The discrepancy between this claim and the undeniable evidence I have long witnessed is perhaps not that surprising: K10 said that if businesses were asked to declare how much they waste, 'everyone will lie.'

Under pressure, K10 had the honesty to explain that it was indeed company policy deliberately to prepare more fish than anyone wanted, and that if there was zero waste from a store, it would be seen as a weakness in the business. K10 explained that if a store runs out of any product, it means it is probably losing out on sales. The retail price of any item is two to three times the cost price, which means that it is better to waste two of each product than lose even just one sale by selling out of it. This is the critical equation that all food retailers juggle with. The size of profit margins and the low cost of food waste disposal influences the amount of waste retailers create as an 'affordable' by-product of their marketing policies.

Besides, said K10, 'empty shelves put customers off.' If stocks are low they enter the shop and receive the impression that they are getting the remainder that no-one else wants. Furthermore, K10 does not give price reductions at the end of the day because it is believed that customers will wait until then, rather than paying the full price earlier. It also refuses to let staff take home surplus because it cannot trust its chefs not to deliberately over-produce with that motive in mind.

K10 does have a policy which, if practised, would obviate the need for this waste. If the store accidentally runs out of a product towards the end of the day, rather than making do with a bare shelf, K10 says that the staff are required to fill the space with photographs of the product and instruct customers that it can be made fresh to order within five to ten minutes. Any whole fish which has not been prepared at the end of the day can be cooked up the following day. This means that preparing sushi to order, rather than pre-preparing it, cuts out the waste. If K10 had faith in the system, it could aim to sell out of ready-prepared sushi before closing time, and then for the last hour or so make up the rest of the products to order, rather than deliberately producing trays of fish that no-one is likely to buy. But, like most other stores, K10 chooses to sustain the pretence that food, and all the resources that go into producing it, are infinitely available and expendable.

K10 says that each of its stores makes daily records of every item staff throw in the bin but, like most businesses, it would not reveal these figures on the grounds that competitors might use the information to their advantage. (Neither would it reveal the source and sustainability of its fish, and claimed not to know where its prawns came from or even whether they were farmed or caught.)[11] To assess their waste, therefore, you will have to make do with my estimate: I once calculated that the retail value of the ninety-five items I saw in their bins one evening was £474.05.

K10 is alas not exceptional: rather, it is representative of how food is condensed by businesses into numbers on a profit sheet, and all other considerations evaporate. But monetary cost is a minor issue compared to the cost to the planet. The ability of tuna fish, particularly the bluefin species, to withstand the onslaught of the world's booming demand is far from certain as they are fished at about four times the sustainable level, and an irrevocable collapse in their population may now be unavoidable.[12] Swordfish, likewise, is fished unsustainably – North Atlantic stocks halved twenty years ago; it is on the International Union for Conservation of Nature Redlist of threatened species, and US chefs have boycotted it sporadically since the late 1990s. Both fishing for and farming prawns can be among the most unsustainable ways of producing food known to man. The first step in making way for prawn-rearing lagoons is generally to demolish coastal mangrove swamps which are vital nurseries for wild fish stocks and serve as natural coastal defence systems, as the south-east Asians learnt with appalling results in the 2004 tsunami. The global shrimp-farming industry has already destroyed 30 per cent of the world's mangroves, in the process also releasing enormous quantities of greenhouse gases, particularly methane. Wild fish stocks on which locals depend for their sustenance are then caught to feed the prawns, and it takes at least 2 kg of wild fish to produce a kilo of prawns.[13] To exacerbate their already huge carbon footprint, the prawns are then flown from hungry parts of the world to satisfy the palates of the affluent.[14] If only it was satisfying their palates; but purveyors of seafood across the world dump a proportion of them straight in the bin – as if all the environmental and social damage incurred in producing them counted for nothing.

I invite anyone who enjoys eating fish – and particularly those who wish the next generation to enjoy it too – to ask their local store manager what they do with their surplus at the end of the day, and whether they might consider buying just a little less of each of these valuable animals, so that everything they source is actually sold, rather than sent to landfill. It is up to us, the customers, to convince corporations and local shops alike that at – literally – the end of the day we would rather see sparse shelves and empty bins than shelves heaving with food that will be thrown away. In France, where this element of the food culture is also different, customers patronize particular artisanal bakers and butchers in their neighbourhood, arriving early in the day before the *pain de campagne* or house pâté sells out.

One evening at the K10 bins I found to my dismay that staff were opening the trays of sushi before throwing them in the bin-bags apparently in a bid to spoil the food and prevent anyone reclaiming it. I asked the staff about this policy, and they told me they had been instructed to collect the little bottles of soy sauce in each tray. I have no idea whether they were telling the truth, but it certainly casts a shadow over K10's claim that it cannot give away surplus because of food safety. Those bottles are in packets with raw fish all day, so if they are reused the next day . . . I leave the multiplication of bacteria to your imagination.

Whereas good-natured staff in some food outlets leave food bagged up clearly in the knowledge that people (often low-paid cleaners) will come and take it away, spoiling food deliberately before disposing of it is becoming a perverse standard practice on the high street. In the chain store Eat in central London, lunching professionals buy from a range of ready-to-eat packaged sandwiches, soups, pies, noodles, chicken salads, granola yoghurts and creamy desserts. Until recently, there was a half-hour-long window, known to a handful of homeless people in the city, between the stores' closing time at around 5.30pm and the rubbish collection at about 6pm, during which any down-and-out could pass by the store, quickly determine which of the ten to twenty green rubbish bags left on the street were full of bounty, untie the knot and help themselves to a relatively nutritious evening meal.

However, at various times during the past few years, one or two of these outlets began the practice of opening each of the sandwiches and salads and tipping the food into a dirty mess before putting the sacks out on the street. This would be distressing enough if it were an effective way of preventing homeless people from getting their meals out of bins. But, on the contrary, I still see people, clearly desperate for a meal, extracting sandwiches from this disgusting jumble and eating them. All the staff achieve is to increase dramatically the chances that these vulnerable people will contract food poisoning. The last time I inspected a selection of Eat branches in central London in August 2008, I was frustrated to find that every one of them was now mucking up their surplus food. I chatted to a couple of the homeless people whom I knew had previously relied on their sandwiches: they showed me what they had managed to salvage, and despite repeated refusals, tried to give me whatever I wanted for myself.

An Eat branch manager I spoke to about this denied that they put food out on the streets any-more (which certainly was not the case ten days earlier). In fact a couple of Eat stores do now give away some food – but at the time of writing, they still waste most of their surplus. I have tried asking branch managers about the business policy behind this decision, but other than saying that the instruction comes from the head office, none of them are willing to explain their decision to spoil food.[15] I received no replies to my repeated enquiries by telephone and in writing to customer services and the Eat head office.

So why do they do it? If a sandwich store overstocks by, say 100 sandwiches and salads, and it knows people are scavenging for them, managers may believe they will lose out on sales. However, in the past fifteen years I have never encountered a bin-rummager (apart from myself) who fits the likely demographic of their customer base. In Islington, I often used to meet an old woman, her glasses stuck together with tape, gleaning from the stalls at Chapel Market, who would quote the words of Jesus after the feeding of the five thousand from the Gospel according to St John, thanks to her the epigraph of this chapter: 'he said to his disciples, "Gather the pieces that are left over. Let nothing be wasted."' All across London, from Borough Market to Primrose Hill, I met several old women and men like her gleaning food to supplement their meagre pensions, homeless people who do not have any money at all, or immigrants who are working for less than a living wage. None of these are potential customers at Eat.

Another fear managers may have is that some wily entrepreneur will lift the whole sackload of waste food off the street and sell it on for a tidy profit. As far as I know, Eat has never been affected by such a scam, and, in any case, breaking the seal on the packages without tipping out the contents or spraying them with harmless dye would make such a move extremely unlikely and would not endanger the health of the hungry.

The only other motive – besides sheer bloody-mindedness – that I can think of is more difficult to dismiss. It is a sad truth that a small minority of particularly forlorn scavengers are either too out of their minds or too frustrated to care if they tear holes in bin-bags and strew the contents on the ground. This can make an unsightly mess, complicates the lives of rubbish collectors, and attracts vermin. It raises the question of whether food shops are legally obliged to prevent scavenging of this sort. In the UK, the Environmental Protection Act (1990) Waste Management Code Of Practice specifies that all waste holders 'must act to keep waste safe against scavenging of waste by vandals, thieves, children, trespassers or animals'. However, leaving tied bags on the street for collection is one of the procedures the Code of Practice actually requires them to follow: 'Waste left for collection outside premises should be in containers that are strong and secure enough to resist not only wind and rain but also animal disturbance, especially for food waste. All containers left outside for collection will therefore need to be secured or sealed. For example, drums with lids, bags tied up, skips covered. To minimize the risks, waste should not be left outside for collection longer than is necessary. Waste should only be put out for collection on or near the advertised collection times.'[16] All the stores under discussion follow this specification. But it is the shops' own policy of unpacking their waste food before throwing it away that attracts dogs and rats.

I have very rarely seen scavengers strew the contents of bins like this on the pavement, but I have seen it occasionally round the back of supermarkets. When I lived in Camberwell in south London, for example, I used to frequent the Safeway skips, and there I sometimes bumped into a man who spoke very little English. He was methodical in his rummaging – standing on a crate so he could reach further into the bin, and throwing anything he did not want over his shoulder into the loading bay.

I tried to convince him that if he continued in this way, the supermarket would start locking their bins, as many other stores already did. Sure enough, that supermarket evidently got so tired of clearing up after this man that they invested hundreds of pounds in an eight-foot-high padlocked cage to enclose their four wheelie bins. Locking skips has, over the years, regrettably become the normal practice of supermarkets, particularly in large cities regardless of whether they have been affected by inconsiderate litterbugs. Instead of redistributing surplus, they lock it away to prevent anyone from eating it.

If only retailers did not throw away so much good food, they would not be bothered by scavengers. If food outlets either effectively avoided overstocking, or tried to donate their surplus food to charities, they would eliminate the problem of wastage and scavengers and alleviate the hunger of the poor.

2. Supermarkets

The monkey speaks his mind: . . .
'That man descended from our noble race –
The very idea is a big disgrace.

           . . . you will never see –
A monkey build a fence around a coconut tree,
And let all the coconuts go to waste
Forbidding all other monkeys to come and taste.
Why, if I put a fence around this tree
Starvation will force you to steal from me.'[1]

'The Monkey', lyrics by Dave Bartholomew
and Pearl King, New Orleans, 1957

In 1997 the US government estimated that retailers waste around 2.5 million tonnes of food each year, or just under 2 per cent of the country's total food supply.[2] In Japan, the official figure is 2.6 million tonnes.[3] In the UK, data obtained by the government-funded, not-for-profit company Waste & Resources Action Programme (WRAP) and by the Environment Agency suggests that British retailers are proportionally more than three times more wasteful than their American counterparts (who themselves have plenty of room for improvement).[4] Although other estimates of UK retailer waste are much lower, and it is difficult to compare figures from different studies, the apparent comparative wastefulness of British retailers has been monitored by academics working in this area.[5]

The discrepancy can be attributed to the British supermarkets' general ineptitude and sloppy store management, their historic neglect of environmental and social responsibilities, their slow-ness at adopting technologies that increase efficiency, their legacy of refusing to donate surplus (in contrast to America, where the practice is universal), and the British government's failure to deliver sufficiently powerful inducements to bring about change. According to WRAP, each year retailers in the UK produce 1.6 million tonnes of food waste.[6]

Enormous though this figure is, individuals within WRAP suggest that it may even be an underestimate, because it has been calculated by extrapolating from the waste statistics pro-vided by some of the most efficient businesses in the entire industry and thus ignores much more wasteful companies. Evi-dence in the US suggests that the worst category of retailers can throw away proportionally thirty-five times more food than the most efficient companies.[7] Worse still, the WRAP figures were based on data presented voluntarily by retailers themselves. This data generally lacked sufficient detail, and was not comparable – as some supermarkets only provided information on their back-of-store food waste, while others also included food waste in distribution. All of them did so on the condition of total secrecy, so the truth remains hidden from the public.

However, other estimates of retail food waste are dramatically lower than that issued by WRAP. Some studies suggest that retailers may only produce 455,000 tonnes of food waste annually,[8] and Biffa, the waste company, estimates that retailers waste 500,000 tonnes of food a year.[9] Some of this discrepancy could be attributable to the varying definitions of what is included under the term 'retailer' and whether this does or does not include, for example, prepared-meal convenience stores. The other main cause of the discrepancy would be the variable definition of 'food'.

It is difficult not to be cynical about the veracity of the available information on retailers. As the general manager of one food retail business told me, 'figures based on self-reported data are not that meaningful because retailers will always understate their waste. Supermarket chains hide their profits: it is even easier to hide waste, and managers will do so to improve their environmental credentials and disguise mistakes.'[10]

The available data on food companies is in marked contrast to the information that governments have gathered on the wastefulness of the general public. Researchers discovered that when householders are asked how much food they waste, they always underestimate, sometimes by thirty times.[11] Self-reporting is thus regarded as completely unreliable, and the only way of checking is actually to examine the bins. The British and American people have had their bins raked through with forensic precision to determine exactly what they have thrown away, down to the last crust of bread and half-eaten apple.

And yet, when it comes to the waste of the biggest, most wily businesses in the world, all anyone has to go on is that which the supermarkets choose to publish; there is no official mechanism for cross-checking their data. They are allowed to bury the world's valuable food supplies without giving any account of their activities, despite the fact that it would be logistically far simpler to spot-check supermarket skips than to trawl through thousands of filthy domestic dustbins. There have been numerous attempts to guess what the food industry throws away, but, as far as I know, in Britain there has only been one investigation where researchers actually picked through the food a supermarket threw away – and that was only over a two-week period in a single store. Eric Evans, director of the company Bio-Recycle and author of the report, points out that the supermarkets do not want people to know how much food they waste: 'just imagine the headlines in the *Daily Mail* – a lot of the food is wasted and a lot of it is still edible.' He agrees that all available data is based on completely unreliable self-reporting and insists that if you want to know the answers 'you have to get your hands dirty.'[12]

The supermarkets' excuse for hiding the facts is that waste levels are 'commercially sensitive'. They claim that if rivals know how much others waste, they can gain an advantage by working out how to make their own system more efficient. If anything, this is an additional reason for governments to make the publication of waste data mandatory – because it would speed up the race to increase efficiency.

At first glance it appears a mystery why so many shops order more than they can sell and end up having to throw it away. Wasting more than necessary affects profit margins, so it is in the business's interest to avoid it. The real reasons for this can be broken down into various strands of economic logic, similar to those exhibited by the sushi company in the previous chapter, but on a vaster scale. Firstly, supermarkets feel they have to ensure that their customers' favourite products are always available for fear of losing dissatisfied clientele. As one supermarket manager from Asda explained in an interview, this generally means that supermarkets will 'always put more stock on there rather than less' even if it means ending up with wasted food.[13]

But beyond this requirement to meet actual demand, supermarkets will deliberately overstock because they believe that shoppers like to see full shelves. Full shelves give the impression of infinite abundance – an illusion which remains central to expectations of choice in today's consumer culture. Even if supermarkets know they are overstocking certain products, they believe that the losses that result will be recouped by attracting customers to their store. If shelves are empty, managers believe that customers might turn round and go to another shop where the illusion of cornucopian choice is maintained more gaudily.[14] According to some individuals in the industry, this assumption has become prevalent, despite counter-suggestions that consumers like to see empty shelves because it assures them that the food they purchase has not been sitting around for too long.[15]

The second factor is that profit margins gained by retailers determine that overstocking can still be profitable. For example, if the retail price of a sandwich is twice the cost price, it will be

more profitable to overstock than to forgo sales by understocking. Costs of the disposal of surplus must be subtracted
from the potential profits of sales, but since binning a sandwich
weighing 200 g costs less than 1 pence whereas the profit of a
sale can be up to a hundred times that, this does not often have
a significant impact on the profitability of overstocking.[16]

In addition to these waste-creating policies, a great deal of
overstocking occurs simply because people who make food
orders predict sales badly − day after day. This has long been
evident to me because of the variability in the contents of the
skips of different supermarket stores. When I was last living
in London, every time I cycled past the Sainsbury's store in
Knightsbridge in west London, for example, I would find at
least two sacks of bakery products in the bins on the street −
scores of croissants and pains-au-chocolat, doughnuts and bread
rolls. Who is making the bakery orders at Sainsbury's, Knightsbridge? Another supermarket will have a problem with its fruit
and vegetable section or with its chilled ready-meals, regularly
reflected in the disproportionate representation of those foods
in its bins. I am not alone in using bin inspections to assess
efficient business practice. Supermarket bosses are often known
to test the competence of the store managers by occasionally
rummaging at the back of the store.

In addition to surplus stock, a great deal of food is thrown
away because its packaging gets damaged. This often consists of
a minor tear or mark on the outer packaging, which doesn't
affect the food inside, but the supermarkets still choose to throw
it away. In addition, if one item within a larger pack is blemished,
the whole pack will generally be chucked out. I regularly find
packs of fresh apples, vegetables and whole crates of egg-boxes
(containing over a hundred eggs) that have been wasted on such
slight grounds.

If you care about these things and want to put pressure on
supermarkets to do something about them, choosing between
chains on the basis of their relative wastefulness is complicated

by the fact that they do not reveal the relevant figures. This lack of transparency makes it difficult for those outside the retail sector, including the general public, to ascertain whether these companies are meeting any of their public waste reduction commitments. A government-instigated group of industry figures and other stakeholders called the 'Champions' Group on Waste' has itself recommended that 'The major retailers should explore whether it is possible to develop and agree a voluntary framework for recording waste arisings from stores with a view to using this to identify opportunities for waste reduction.' But there are no moves to effect this because the supermarkets prefer to keep the information a secret.[17] For the time being, estimating how much food each supermarket really wastes requires certain extrapolations and assumptions. It should be stressed that such an analysis will be inherently weak because of the lack of reliable data.

The Co-op chain is unique in reporting straightforwardly in its Sustainability Report 2007–8 that its stores threw away 10,100 tonnes of food in the year. None of its competitors publish such a clear statement. However, its figure has no independent verification and it is not made clear whether this includes any waste further up the supply chain. It is the only figure available, but it may not be comparable to statistics available for other supermarkets. Co-op did not agree to be interviewed on the subject, and therefore no further details are known.

Marks & Spencer does not publish figures, but in an interview its representative, Rowland Hill, gave a very rough estimate that it threw away 20,000 tonnes of food per year.[18] He has also estimated that around 4 per cent of food displayed on shelves · ends up unsold.[19] Alison Austin at Sainsbury's acceded to an estimate that the food waste it sends to landfill may be around 60,000 tonnes per year.[20]

For the other chains, a figure must be arrived at by taking the total waste they send to landfill and estimating what proportion of this is food. Waitrose states on its website that 60 per cent of all its waste is food; Sainsbury's food waste total amounts to 70 per cent of its landfilled waste and Asda says that 69 per cent of its landfilled waste is biodegradable. The proportion of landfilled waste that consists of food is largely a factor of the level of recycling of other materials (cardboard, paper, plastic, glass). Sainsbury's claims to be recycling around 63 per cent of its waste, Tesco says it recycles 70 per cent of store waste, Asda 65 per cent of store waste and Morrisons 72 per cent.[21]

Tesco and Morrisons have not reported how much of their landfilled waste is food, but since they recycle more of their general waste than Sainsbury's, it is reasonable, though clearly tentative, to make calculations on the basis that the proportion of residual landfilled waste that is food is similar to Sainsbury's, i.e. 70 per cent. The figures for waste sent to landfill necessarily omit tonnages of raw meat and other animal products diverted for specialist treatment such as rendering. One survey that examined supermarket waste in 2004 suggested that these raw-meat products comprised around 19 per cent of supermarket food waste,[22] and there are also fluids such as milk which are often poured away. To arrive at a total amount of food wasted, therefore, it is necessary to add an approximate value of 23 per cent to the amount of food waste going to landfill. Using this to work out approximate tonnages of food waste per year, based on reported tonnages of total waste sent to landfill, it would appear that Tesco wastes around 125,000 tonnes of food each year and Morrisons 46,000 tonnes. Asda's representative, Julian Walker-Palin, casually estimated in an interview that in the space of a week its stores typically only generate a tonne or so of 'biodegradable waste', which in the case of a supermarket will clearly be mainly food. Extrapolating this figure to the number of stores nationwide, Asda's food waste from stores would amount to just a little over 17,000 tonnes per year. However, this estimate is less than a quarter of the total that can be inferred from the company's published waste statistics. Of the 88,000 tonnes of waste sent to landfill, it says 69 per cent is biodegradable, or 61,000 tonnes, to which must be added 23 per cent diverted food, yielding a total of 75,000 tonnes.[23]

Altogether, these figures would suggest that the seven super-markets which I surveyed (Tesco, Asda, Morrisons, Sainsbury's, Co-op, Marks & Spencer and Waitrose) waste 367,000 tonnes of food annually. If anything, this figure merely illustrates that supermarkets may be under-reporting their waste figures. For example, this represents only 23 per cent of WRAP's estimated 1.6 million tonnes of food waste per year. WRAP's figure is for the retail sector as a whole and should therefore include waste from smaller high street grocers, convenience stores, sandwich shops and so forth. However, even taking this into consideration, the food waste generated by the supermarkets should be higher considering the fact that six of the supermarkets under consider-ation (not including Marks & Spencer) represented 83 per cent of the total UK grocery market share in 2008.[24]

In any case, either WRAP has got it badly wrong, and is including in its estimate a great deal of packaging and other non-food waste, or the supermarkets' published waste statistics are wholly unreliable, or both.[25] WRAP is currently doing more concerted work on calculating food waste within the industry, and it remains to be seen how its estimates will change. Rowland Hill from Marks & Spencer – who also chairs the BRC's Environment Policy team – regards the WRAP figure as an overestimate and suggests that the true figure should be less than 1 million tonnes of food waste and probably 'towards the shallower end of 500,000 tonnes'. This huge range – from 367,000 to 1.6 million tonnes – shows how little people really know, and how expert the supermarkets are at hiding the truth.

Making comparisons between different supermarket chains on the basis of such incomplete data will necessarily be flawed. Even for those companies that are relatively open about their waste statistics there is insufficient detail to draw confident conclusions. Nevertheless, bearing these caveats in mind, some rough indications can be deduced. If any supermarket manager feels aggrieved by the negative conclusions drawn here, it will hopefully encourage them to publish the true figures and have them independently verified.

In order to assess the relative efficiency of the various supermarkets, it is necessary to take into account their relative size and compare that to how much waste they produce. To make this assessment I have devised a score system based on the estimated total mass of food wasted as a function of gross takings. In other words, it is a rating based on the amount of food waste per pound earnt at the till in each supermarket chain. I have condensed the number into a point-system where 100 represents the average performance. A score above average indicates that the chain is relatively wasteful; a lower figure suggests it is more efficient. The table showing these calculations is in the appendix at the end of the book. In summary, the Co-op is the best-performing supermarket, 27 per cent more efficient than average. Sainsbury's is the worst, apparently 14 per cent more wasteful than the average and 55 per cent worse than the Co-op. Morrisons and Tesco are about average and Waitrose and Asda are slightly more wasteful than average and around 47 per cent worse than the Co-op. I should reiterate that this assessment is based on extrapolations from the scant available data.

Some of the differences between chains revealed here will be due to the different proportions of perishable and non-perishable goods sold in each chain, which affects how much the supermarket wastes. So Waitrose could be worse than average partly because it sells a relatively large proportion of fresh fruit and vegetables. On the other hand, one would expect Waitrose to perform better than it has because, as an upmarket chain, its monetary takings per tonne of food should be higher than in downmarket chains. Probably the biggest factor is that Waitrose does not reduce its products near the sell-by date as much as other supermarkets. Tesco may be less wasteful than suggested here because it sells a higher proportion of non-food items than supermarkets such as Sainsbury's, so its bins are occasionally filled with clothes, CDs and other items (trust me, I know), which means that the proportion of its landfilled waste that is food could actually be lower than calculated here.[26] However, this distortion should be taken care of to some extent by the fact that these calculations were made on the basis of total takings in the grocery chains, including sales of those non-food items, so these distortions should cancel each other out.

Another feature worthy of note is that while the Co-op appears, according to these figures, to be far and away the best in terms of the mass of food it wastes relative to its takings, it does have the very poorest record in terms of recycling its non-food waste, such as glass, cardboard, paper and plastic. The Co-op claims to recycle just 47 per cent of its waste in comparison to Tesco, Morrisons, Asda and Sainsbury's, who recycle 63–72 per cent. The Co-op's pledge to 'ensure that less than 50 per cent of total waste arisings are landfilled by 2013' is one of the weakest targets of the seven supermarkets. This is probably reprehensible negligence on the Co-op's part, but one possible extenuation is that rather than focusing on headline-grabbing initiatives about diverting waste from landfill, it may have been focusing on the arguably more important issue of not wasting food. I know from my examination of its bins, for example, that when Co-op store managers dispose of food, they account for the waste and label each bin-bag with an identity code. This at least implies they are measuring losses, which is a good sign.

It seems astonishing that no attempt has been made to compare supermarket performance in this way, despite many government-led round table discussions on food waste reduction. It is elementary that the worst-performing companies should be held to account and encouraged to emulate the best-performing companies in terms of both transparency and efficiency. Retailers could be encouraged to improve by giving an award for the winners in a league table.[27] And yet there is currently no benchmark and no official assessment of performance. Instead, supermarket PR teams have been focusing on quick-win policy packages and impressive-sounding pledges to stop sending their waste to landfill. But the environmental and social gains of their initiatives have not been quantified and may in the end turn out to be relatively small. Not wasting food, by contrast, has very clear and substantial benefits.

Supposing Tesco did achieve the waste avoidance efficiency apparently exhibited by the Co-op: according to these figures it would save around 33,000 tonnes of food every year. If the other four (Morrisons, Waitrose, Asda, Sainsbury's) also came into line, the total saved would be 100,000 tonnes. This is undoubtedly an underestimate of the potential savings: firstly because the supermarkets may be under-reporting their waste figures, and secondly because the Co-op itself is very far from doing everything it can to reduce waste.

Another way to compare the various supermarkets is to look at their published food waste management policies. The two methods of comparison do not always match up. On this count, Sainsbury's and Marks & Spencer are among the best in their class, having some relatively good practices such as a willingness to donate surplus to charities. After years of resistance, Sainsbury's now co-operates extensively with the food redistribution charities Fareshare, the Salvation Army and Food For All. Tribute should be made to it for its relatively progressive attitude in this respect. Indeed, according to my calculations on its published data, it donates around 400 kg of food for every million pounds it takes at the till, which may not sound very much, but it would appear that this is the same rate of food donation as Kroger, one of the more 'generous' supermarkets in the US, though Kroger also makes many other contributions both in kind and in cash, and Kroger's US rival Safeway appears to donate more than four times as much.[28] This bucks the general trend of much higher rates of food donation as a whole in America than in Britain (see chapter 14). Having said this, Sainsbury's still donates only a small proportion of the food it wastes and could therefore still do much better. In 2004 Sainsbury's reportedly sent 70,000 tonnes of food waste to landfill;[29] in 2006–7 it claims to have given away a total of 6,680 tonnes of food – which means it was still apparently discarding around 90 per cent of its unwanted food, most of it to landfill.[30]

Sainsbury's has now committed to divert all of this remaining food waste to other disposal methods such as turning it into pet food or sending it for anaerobic digestion – whereby it will be decomposed and the gas collected for power generation. They claim they can do this by the end of 2009 – though others in the industry are sceptical about whether this can be achieved.[31] Commendable though this pledge may be, as I shall discuss in greater detail below, anaerobic digestion is still basically turning food into gas and compost (which is sometimes contaminated by packaging). Composting crates of bananas, which could have fed hungry human beings, wastes almost all the energy and other resources that have gone into growing them. It is an ecologically favourable disposal method compared to landfill, but it is still a tragic waste of edible food. Furthermore, as I shall discuss in chapter 15, the supermarkets have made such pledges only now that landfill taxes have risen to a level where it is actually cheaper or around the same price to find alternatives such as anaerobic digestion. Essentially they have been forced to make these decisions by government tax regimes rather than through an 'ethical' decision on their part. And yet the supermarkets and other food corporations have been very successful in getting a great deal of environmental credit for these policies.

Marks & Spencer has a long history of donating surplus to hospitals, homeless shelters, old-people's homes and other charitable organizations and it deserves credit for providing a model that others are slowly beginning to follow. However, until very recently it has resisted reducing prices to sell off stock nearing the use-by date. It has now finally committed to doing so and estimates that this could reduce its waste of food by 10 per cent.[32] As an upmarket store, Marks & Spencer also has a tendency to impose excessively stringent quality controls on its products, meaning that perfectly edible food is discarded on very slight grounds. The produce that upmarket stores reject before it arrives in store does sometimes find its way back into lower-cost outlets, but this only highlights the fact that their 'quality' requirements often necessitate the rejection of good food.[33]

Waitrose did conduct a food redistribution trial with Fare-share, but it says that 'unfortunately these trials did not prove viable for Fareshare.' Waitrose has also supported composting trials; fourteen Waitrose branches have their waste processed via an anaerobic digestion plant in Bedford and the company plans to extend this to more branches in 2009. It also pioneered the marketing of imperfect 'ugly' fruit and vegetables such as apples, peas and rhubarb, but the overall impact of these endeavours is unclear and tends to be restricted to individual crops affected by extreme weather conditions, rather than a wholesale reappraisal of the waste caused by imposing arduous cosmetic standards on all produce. One Waitrose initiative with banana growers in the Windward Islands apparently reduced waste from a massive 40 per cent of the crop in 2002 to less than 3 per cent in 2008 – which, if true, demonstrates that enormous savings can be made by taking these issues seriously. Waitrose also announces that it no longer wants to 'waste' male calves born on dairy farms. Instead of discarding these calves as 'by-products' it now aims to 'ensure that every surplus male dairy calf is channelled into our beef supply chain and reared as meat'.[34] One suggestion I particularly liked was in a Waitrose magazine, where an article recommended feeding kitchen scraps to pigs and goats – a laudable exercise which, as I discuss in later chapters, also happens to be illegal.[35] Meaningful though these various projects may be, they still address only the fringes of the food waste problem. The fact is, Waitrose, like most other stores, throws away unnecessary quantities of good food. It claims that 'the majority of our destroyed wastage is related to items that are rendered inedible for a variety of reasons.'[36] But having lived on food from its bins for over a year I can confidently vouch for the fact that the majority of the food I find is edible, and often of a higher quality than I see on the shelves of most other supermarkets.

In the past, Morrisons and Asda have been among the worst in terms of their historic refusal to donate their own-brand surplus food to charities. Morrisons is the only one of the larger supermarkets to persist in its refusal to donate surplus to Fareshare. The company turned down the opportunity to tell me why. Having said this, Morrisons cuts down on food waste by proactively reducing prices to sell off stock before it goes out of date and it claims to have 'one of the best records in the supermarket sector for minimizing food waste', though it has produced no figures to support this claim, despite my enquiries.[37] There is anecdotal evidence that it takes measures such as removing raw chickens that are approaching their sell-by date off the shelves and cooking them to serve at the deli counters or in their 'Prepared for You' range, and preparing food in store can in itself help to reduce waste.[38] As of March 2009, Morrisons' commitment to reduce by 50 per cent the amount of waste it sends to landfill by 2010 is well behind the other major chains such as Sainsbury's and Asda.[39]

When I sent written questions to Asda two years ago asking why it insisted on destroying surplus food rather than allowing charities to take it, its response was that 'our preference is to sell it.'[40] Thankfully, in July 2008, Asda finally yielded to pressure and has lifted the ban imposed on its suppliers against donating unwanted produce, for example, if the packaging is defective. However, it still claims in interviews that it doesn't 'have the space' to allow donations of surplus food from its stores – even those with more than 70,000 square feet of floor space. Fareshare, however, does hold out hope that it will relent and start donating food from its depots.

To its credit, Asda has arguably been a pioneer in the retail sector for trying to 'back-haul' its store waste to regional recycling centres – a practice being adopted by Tesco and Sainsbury's. This makes it more economical for other companies to collect its waste for recycling. It also claims to be vigorously tackling the problem of food waste by examining what ends up in compactors outside stores and finding how waste can be minimized through supply chain improvements and in-store measures to ensure that products end up being sold.[41] Asda was also the first supermarket to make a 'zero waste' pledge, in October 2006, saying that it would send no waste to landfill by 2010.[42] But this does not mean it will cease to waste food. Instead, it is currently having the unsold food incinerated and is looking at ways of treating it by anaerobic digestion, composting or other similar processes. Asda does keep accurate accounts of how much food waste it produces, but it refuses to publish them because, as its representative explained in an interview, it does not want to help its competitor, Sainsbury's, to achieve its waste reduction pledge by releasing information on waste that could help it with logistics.[43] This is another confirmation of the fact that if supermarket chains were required by governments to publish their waste data, all supermarkets would be able to learn from each other about how to reduce the impact of their waste.

Tesco is beginning to participate in charitable food redistribution schemes, but this is still a relatively small-scale affair. Tesco is by far the biggest UK supermarket chain, so it is safe to say that it wastes more food than any other retailer. It has now committed to diverting 80 per cent of its waste from landfill to disposal methods such as anaerobic digestion by 2009. However, with recycling rates of 70 per cent in 2007, Tesco fell significantly short of its previous target of recycling 75 per cent in that year.[44]

Of course, none of these much-vaunted targets carry any penalties when the supermarkets fail to meet them.

If the British supermarkets' unwillingness to reveal their waste data seems suspicious, those in the US and Continental Europe are even worse. No specific figures on their food waste or landfilled waste could be gathered. Direct enquiries in France, Germany and the US, and several emails and telephone calls to the likes of Aldi, Carrefour, Lidl and the American supermarkets Target Corp, Kroger and Supervalu yielded little more than obfuscation, silence or referral to Corporate Responsibility reports that contain vague policy statements but no figures on tonnages of food waste or even the total amounts of waste sent to landfill or other disposal methods.[45]

American supermarkets can, however, be compared on their self-reported contributions to food distribution charities. Kroger is America's largest supermarket (though Wal-Mart – defined as a hypermarket largely owing to the typical size of its stores and range of non-food products – has supermarket-type sales of $134bn which are around twice those of Kroger's $70.2bn).[46] Kroger was a founding member of the food redistribution organization Second Harvest (henceforth referred to under its new name, Feeding America) and its executives serve on the board. In 2006, Kroger contributed 13,610 metric tonnes (30 million lb) of food valued at $45m and also helped to raise $6m in cash for Feeding America as well as giving grants for logistics and equipment. In 2007, Kroger was selected 'Retailer of the Year' by Feeding America – the fifth time in seven years it has received the award. According to Kroger's 2008 Sustainability Report, 'This program enables us to provide fresh, nutritious fruits, vegetables, meat, dairy and deli foods to thousands of hungry people and has the ancillary benefit of diverting food items from landfills nationwide.'[47]

Safeway personnel in the US sit on the boards of directors of food banks in various cities. In 2006, 'Safeway, its customers, employees and vendor partners' apparently donated $110m worth of 'merchandise' to local food banks and also helped to raise funds.[48] It does not say how many tonnes of food that came to, but by Kroger's valuation estimate it should be in the region of 33,244 tonnes. However, this figure may not be strictly comparable to Kroger's because it appears to include donations from customers and other companies.

Wal-Mart is reputed to have been among the least willing to

donate surplus food in the past, and in 2006 it announced a policy to cease donations of nearly expired food, claiming that it was concerned about liability if recipients got food poisoning, even though the law protects it from this (see chapter 14).[49] In a California court case, *Janes v. Wal-Mart* (2002), it was established that Wal-Mart had fired its employee, Jeffrey Janes, after he took expired meat from Wal-Mart's waste barrel and cooked it for lunch with several other employees in the store. Janes was dismissed for dishonesty, not stealing, but he was awarded $167,000 following a verdict of wrongful termination. The parties disputed whether the expired meat had a value to Wal-Mart.[50] In 2008, under pressure from the steeply rising food prices and levels of hunger in the US, Wal-Mart adopted a more progressive attitude and announced a new food donation programme, claiming that eventually it would be able to donate more than 31,750 tonnes (70 million lb) a year from its Sam's and Wal-Mart outlets.[51]

Supervalu donates millions of pounds of food, participates at board level in Feeding America and also helps to raise funds.[52] In 2006, Target donated more than 3,950 metric tonnes (8.7 million lb) of food.[53] The organic specialist in the US, Whole Foods Market, says that every one of its stores makes donations to food banks and shelters totalling millions of pounds of food each year; most stores back-haul any remaining spoiled food waste to regional centres for it to be composted and given to community gardens.[54]

Although figures for accurate comparisons between various supermarkets are unavailable, indicative figures can be calculated by taking the companies for which tonnages of donated food are available. This league table is extremely tentative as it is based on the companies' own published data. For each million dollars of sales in 2006, Safeway and its customers and partners appear to have donated the equivalent of 865 kg of food, though this may include some donations not included on the accounts of their rivals; Kroger gives 225 kg per million dollars of sales; and

Wal-Mart (grocery sales only) a negligible amount in 2006, though this may rise to 237 kg in the future.[55]

In conclusion, all of the three largest supermarkets in the US – Kroger Co., Safeway and Supervalu Inc., as well as the hypermarkets Wal-Mart and Target Corp. – have strong links with redistribution organizations and donate millions of kilos of food to them each year. However, there are enormous differences between the companies, which implies that a great deal more food could still be salvaged. If reliable and comparable data on food wasted and food donated were published by all supermarkets in Europe and the US, this would undoubtedly create an incentive for all companies to burnish their public images, with the result that a great deal of food could be saved.

3. Manufacturers

Classical economics imagined that primitive exchange occurred
in the form of barter . . . [but] a means of acquisition such
as exchange might have as its origin not the need to acquire
that it satisfies today, but the contrary need, the need to
destroy and lose.[1]

Georges Bataille, *The Notion of Expenditure* (1932)

In typical food-manufacturing plants across the Western world
the scene is the same: a pallet of packaged ready-meals – spaghetti
bolognese, noodles with crispy duck or chicken tikka masala –
sits in the docking bay of the manufacturer's warehouse. The
colourful cardboard wrappers advertise steaming, delicious fresh
food; fifty families could eat this food for a week. But instead of
loading the meals into delivery vans, a forklift truck reverses up,
lifts the pallet off the ground, and tips the entire load into
a crushing machine.

It was as a student protestor foraging for food round the back
of supermarkets that I first saw for myself the scale of our food
waste problem. But it was not until much later that I discovered
that what the supermarkets throw into bins at the back of their
stores is a mere fragment of the waste they cause across the food
chain.

This is reflected in the figures. According to a government
survey in 2002–3, conducted in order to comply with the EU
waste legislation, the food- and drink-manufacturing sector of
England, Wales and Scotland wallows in an annual yield of 4.6
million tonnes of food waste. A quarter of this consists of sludges

and some of the rest includes by-products, trimmings and other biodegradable matter which could never be eaten, and it is impossible to work out how much of it is actually food.[2] It is also a very rough estimate, based on surveys from just 3 per cent of the industry, which is an insufficient sample size from which to extrapolate confident estimates. For example, the Environment Agency data suggests that fish and seafood processing created around 10,000 tonnes of waste a year, while the Seafood Industry Association estimated that it ought to be more like 300,000 tonnes a year.[3]

One leading report that analysed the Environment Agency figures from a previous survey in 1998–9 calculated that the industry created 5.8 million tonnes of waste, or around 13 per cent of its total output of products.[4] Of this, 1.9 million tonnes was segregated biodegradable waste and 3.9 million tonnes was mixed general waste, which would typically include skipfuls of semi-processed or rejected animal and vegetable products mixed in with packaging and other factory wastes. There was an additional 3.4 million tonnes of organic by-products that were re-utilized within the sector, for example by being sent for animal feed. But this detailed study cannot be relied upon, not least because it draws on outdated material gathered before changes in the legal definition of waste,[5] and before the introduction of Europe's new laws on animal by-products which diverted hundreds of thousands more tonnes of food by-products into the waste stream.[6] According to figures from major European surveys, the manufacturing and processing industries in sixteen European countries produced 195 million tonnes of food waste (around a third of their output of final products).[7] There are no reliable figures for the United States but some studies suggest processors waste approximately 10 per cent of their food supply.[8] The truth is, no-one knows how much actual food is wasted by manufacturers. Apart from the lack of data collection, studies have highlighted that the methods manufacturers use to calculate their wastage levels 'always hide a proportion of the physical waste incurred', and when this is taken into account, waste levels can be nearly twice what has been estimated.[9] Until there is proper reporting across the industry, published figures and estimates are of limited utility. The government and industry official targets to reduce waste consequently have no satisfactory benchmarks against which to measure their success.[10]

A proportion of food-processing waste consists of plant and animal by-products which few of us would immediately recognize as food: whey from cheese-making, fruit and vegetable trimmings and animal by-products. These can be expensive to get rid of and involve a range of waste management options such as anaerobic digestion or livestock feed which I shall deal with in more detail later. More attractive than treating these by-products as waste, however, is to keep as much material as possible within the human food chain. One solution that food technologists around the world have been working on with increasing vivacity over the past decade is to find new ways of transforming by-products back into food ingredients.

Whey, the fluid remnant of cheese-making, used to be valued as a nutritious drink in its own right (as Little Miss Muffet knew).[11] But with the advent of industrial cheese-making there was insufficient demand, and dairies began to treat it as a waste product. In the past two decades, however, an increasing number of manufactured foods now contain processed whey – chocolate, cakes, cereals, diet drinks and sweets.[12] Likewise, citric acid, pectin and numerous other useful food additives are now extracted from citrus peels and pips that used to be wasted by juice manufacturers. Marmite was invented in 1902 as a way of using spent yeast from the brewing trade.[13] Such innovations have provided a model which many other food processors now seek to emulate.

Each year, for example, Europe disposes of 4 million tonnes of tomato by-products, consisting mainly of surplus crops and seeds or skins discarded when making juice and paste. Some of these by-products are fed to livestock, which is better than discarding them altogether, though it would be still more efficient if people ate whole tomatoes with all their goodness still in them. The seeds, in particular, are a rich source of nutrients such as carotenoids, proteins, sugars, fibre and oils. But without completely changing consumer preferences, the next best thing is to find new ways of extracting the useful components and using them as natural additives to other food.[14] Given our predilection for processed foods, we should not be surprised when in the future products contain an array of unfamiliar ingredients extracted from tomato seeds, apple pulp and orange pips.

Some waste of this kind is inevitable, so the question is how best to use it. But a great deal of food is currently wasted unnecessarily through the negligence of both the manufacturers and the retailers. At present, there is almost no public understanding of how the food supply chain operates, and thus grotesque instances of waste – far worse than what is thrown into skips at supermarket stores – go virtually unnoticed by either consumers or the media. This very invisibility gives industries a licence to waste food without besmirching their public image.

When supermarkets declare they will cut down on waste, they should include the food that they cause their suppliers to discard. In ordinary business relationships, this would be their responsibility; they are only able to shrug it off because their power allows them to. Take the Marks & Spencer waste reduction pledge. It claims on its website that it will work 'with our customers *and our suppliers* to combat climate change, reduce waste, safeguard natural resources'. But in fact its promises to send no waste to landfill by 2012 and to cut down on throwing away food do not cover its suppliers.[15] This is pretty relevant when its absurdly strict aesthetic requirements force one of its major sandwich suppliers to throw away four slices of bread for every loaf it uses – the crust *and the first slice at either end* – amounting to around 17 per cent of each loaf, or 13,000 slices from a single factory every single day (and see p. 258). As a

manager at the factory owned by Hain Celestial Group explained
in an interview, it is made to 'throw an awful lot of very, very
good food that is fit for human consumption out because it
doesn't meet that Marks & Spencer standard'.[16]

The most spectacular example of systematic wastage in the
manufacturing sector is what has been called 'over-production
waste'. This is exactly what it sounds like: when a manufacturing
company makes more of a product than the supermarkets can
actually sell. In the convenience food sector – supplying ready-
meals or sandwiches – over-production waste levels of 56 per
cent of a company's total output have been recorded (i.e. more
food wasted than sold), while a baseline of 5–7 per cent is
considered by many to be inevitable.[17] The industry-led Champions' Group on Waste estimated in 2007 that over-production
waste was often 10 per cent of production and that this level
could easily be brought down.[18] This does not include the
additional waste of unused ingredients, rejected items, or by-
products, nor the waste at retail or consumer levels. It simply
represents one bottleneck in the system – the waste largely
caused by the way supermarkets do business with their suppliers.

The most frustrating aspect of this type of waste is that it
occurs with finished products, so all the energy and resources
that go into making it have already been spent. Furthermore,
because it is already packaged, disposing of it becomes a further
problem. It is possible to put the waste through a de-packaging
machine to remove the plastic so the food can be sent for animal
feed, anaerobic digestion or composting – but with current
technology plastic often ends up in the organic material, so in
practice it is seldom used at the moment. The cheapest option
for the manufacturer is often just to send the food to landfill. Dr
Shahin Rahimifard from Loughborough University, who spent
years working with manufacturers on this problem, recalls his
horror when he first realized what was going on: 'I went into
these factories and saw whole sections of the warehouse full of
perfectly good packed ready-meals being written off as commer-

cial waste, and then a lorry would come and take them away to landfill; I couldn't understand it.'[19]

A sane outsider would assume that by discarding edible food, manufacturers are behaving inexplicably: they are losing profits, creating unnecessary environmental damage, and taking food out of the world markets which people elsewhere could have eaten. Few other industries regularly suffer from this level of over-production waste. There must be some extreme pressure exerted on them to elicit such inefficient behaviour.

It turns out that the principal reason why manufacturers waste so much finished produce is because of the policies of their gigantic customers, the supermarkets. Most food manufacturers are small or medium-sized enterprises, but even the larger ones nowadays are overpowered by the main supermarket chains. Typically, a food factory is designed to make a range of products which it supplies directly to individual supermarket chains. This means that the manufacturer often has only two or three customers, sometimes only one. It is thus entirely reliant on those few customers, or that single one, for the survival of its business, and any failure to satisfy them could result in bankruptcy. The supermarkets, by contrast, have innumerable suppliers and can painlessly switch between them, and this puts them in a position to bargain themselves into extremely advantageous business deals.[20] The supermarkets' disproportionate power allows them to manage contracts and sales in such a way that the losses incurred by their own failure to predict demand are met by the manufacturers. Ultimately, of course, the market cost of this wastage has to be added to the price of the product, and is thus paid for by consumers. Damage to the environment, meanwhile, is not generally financed by anyone – it is free.

Over-production waste typically occurs when a supermarket makes what is known as a forecast order – of say 100,000 assorted sandwiches. In other words, it tells the manufacturer a week in advance that on a certain day in August it will *probably* want 100,000 sandwiches. However, the supermarket will not confirm that order until the morning the sandwiches are to be delivered or, if the supplier is lucky, twenty-four hours in advance.[21] One day is not enough time for the manufacturer to make all those sandwiches from scratch – it takes maybe two or three days to order in the ingredients, set up the assembly line and put it into process, so manufacturers put together all those sandwiches on the strength of the forecast order before the supermarket has confirmed it. Very often, when the supermarket finally makes its confirmed order, it is less than it forecast – it hasn't been selling its bacon and egg sandwiches as fast as it hoped, so it doesn't want so many, or maybe it has started raining and fewer people are buying sandwiches of any kind, never mind that they have already been made.

The manufacturer then ends up with pallet-loads of fresh sandwiches and no-one to sell them to. You might think that it would be able to look for another customer, but the first problem it often faces is that many products nowadays carry the super-market's own brand name, so the packaging actually says 'Sainsbury's Taste the Difference', or 'Tesco's Finest'. It is possible to design packaging or stickers which can be removed and replaced with generic labels so products can be sold on to other retailers, a simple device which could reduce waste dramatically. But most supermarkets do not permit it, and insist that the food produced for them must be sold to them exclusively. Worse, they will often forbid the manufacturer from giving the surplus food to charities, because they suspect it might end up secretly being sold to informal market stall-holders or other members of the so-called 'grey market', thus causing damage to their brand. Often, the manufacturer has few options other than to throw the food away.

Even if the manufacturer is permitted to sell on surplus, it has another logistical problem to deal with. Fresh products like sandwiches and chilled ready-meals have short shelf-lives, and the clock is ticking rapidly from the moment they go into the packet. Most retailers insist on receiving the product with

70–90 per cent of the shelf-life still available: if the shelf-life of a ready-meal is twelve days, they want ten of them in which to try and sell it.[22] If the manufacturer has already held the food for three or four days, many retailers will not touch it: they would rather get produce from elsewhere with a longer shelf-life.

To exacerbate the situation still more, many food factories are poorly designed, lacking the refrigeration necessary to store a great deal of backlog. Dr Timothy Jones argues that in the US one of the simplest and most effective ways of cutting down on industrial food waste in the short term would be for factories to install more refrigeration capacity immediately.[23] However, as one study by Imperial College London pointed out, this could increase greenhouse gas emissions. In these terms, it might be better to reduce industrial refrigeration and cut down on over-production.[24]

So why do manufacturers not just make a little less than their forecast order, wait for the confirmed order, and then quickly make a top-up quantity to meet the difference? In fact, some of them already do this. But the downside is that frequently starting up and closing down production lines creates another kind of waste. Maybe a block of cheese – grated, ready to go into sandwiches – has not been finished before the end of the run. It may then be discarded rather than used in other products. Since most manufacturers of chilled foods buy in their ingredients ready-prepared, they only remain fresh for a couple of days, after which they are discarded.[25] Frequent changeovers also waste the manufacturer time and money in cleaning down and setting up the production line again.

If manufacturers tried to avoid this waste by making less than the supermarket's forecast order, they would risk failing to meet the contract. If they have a written contract, the supermarket may fine them for failing to meet their side of the bargain.[26] Much of the time, however, the supermarkets are careful to avoid making any sort of contract at all, preferring to suspend, reduce or cancel orders at the last minute. If a manufacturer fails to meet orders, the supermarket can simply take its custom elsewhere. It is for this reason that manufacturers will routinely over-produce just to avoid annoying the supermarkets.

Not all products incur such great magnitudes of waste. Those sold under manufacturers' brand names tend to be less wasteful than those marketed under the supermarket's own label. Take Ginsters, for example, whose Cornish pasty factory near Plymouth I visited in the summer of 2008 to witness their success in cutting down on waste. Here, over-production is not such a big problem. If Samworth Brothers, who own the Ginsters brand, are asked to produce 20,000 pasties for Asda, and then Asda decide they only want 15,000, Ginsters can sell the surplus 5,000 to one of the many other retailers who stock them. Furthermore, because the factory is dedicated to making a few similar products in large quantities, it can sometimes run all day without having to change production lines, and thus avoids the waste associated with changeovers.

Mark Bartlett, the environmental manager, claims that only 2–3.5 per cent of their finished products end up as either surplus or rejects. Most of these, Bartlett assured me, end up unwrapped in their seconds shop, where market traders and caterers buy them at cost price because they are not so fussy about whether the pie has a flake of pastry hanging off it, or whether it weighs 160 rather than 170 g. Even the Tesco pasties made in the same Ginsters factory can be liberated from their Tesco packaging and sold on unbranded – illustrating how easy it would be to avoid waste in other supply chains where re-selling branded products is currently forbidden.

Rows of quality controllers watch with glazed expressions, removing any items that are not visually perfect, as more than 3 million pies fly past them each week. The pace is so fast that one elderly worker I saw seemed in a constant state of panic trying to keep up with the machines. It was little surprise therefore to see that staff were throwing some rejected pies (which looked perfectly good to me) into the bin rather than taking the time or trouble to box them up for re-sale. To increase the speed of operations Ginsters have recently installed state-of-the-art visual sensor machines to do the job of the quality controllers – with robotic arms that take reject pies off the conveyor belts and drop them into boxes. But the sensors were working too fast for the staff to keep up with them and the boxes were overflowing, scattering pies onto the floor. Because Mark Bartlett said he 'couldn't find the key to the waste chute', I did not see for myself how many pies were actually being thrown away. But despite the informal negligence, I do believe that this wastage represents a small percentage of total production.

Owning a well-known brand like Ginsters puts Samworth Brothers in a much more powerful position: supermarkets and convenience stores cannot do without their popular products and thus cannot push them around so easily. Rather than pallets of surplus pies, the majority of the 29–37 tonnes of food waste produced by their factory every week consists primarily of churned up pastry, raw minced meat and vegetable peelings. Unlike many pie manufacturers, the Ginsters factory manages to re-use a lot of its pastry trimmings, which can comprise around 80 per cent of other factories' waste.[27] These used to go off for livestock feed, but new animal by-product legislation now makes this much more difficult for factories that handle meat.[28] As I watched coils of pastry pouring off the conveyor belt, it struck me that square-shaped pies eliminate this source of waste because they allow a whole of a sheet of pastry to be used, rather than discarding the spaces cut out between semi-circles.

Unfortunately, Ginsters also throw away whole pies that have been out-graded for visual or structural imperfections before reaching the ovens because it is cheaper to discard them at this early stage than cook and sell them on at cost price in their seconds shop. To reduce the impact of their waste, Samworth Brothers pay £40 a tonne to have it taken to the Holsworthy anaerobic digestion plant nearby, which turns it into gas to produce electricity. It is still a waste of resources to turn food into gas, but they have done more than most to mitigate the problem.

By contrast, the Tamar factory next door, owned by the same parent company, Samworth Brothers, makes ready-to-eat desserts and other meals exclusively for supermarket chains. It is much more wasteful as it is subject to the whim of supermarket ordering and short production runs. This is no doubt one reason why manufacturers producing their own branded products are on average over three times more profitable than firms producing supermarket own-label products.[29] Whereas the mass of food waste produced by the Ginsters factory represents 3–7 per cent of their total output, the Tamar factory figure is around 10–17 per cent (both figures supplied to me by the company). Taken together these factories are more efficient than the average waste in the chilled food sector which is 11.4 per cent of production, and Ginsters is much better than some poorer-performing pie manufacturers which waste around 24 per cent of production, mainly in the form of pastry trimmings which are often simply dumped rather than being rolled up and used again.[30] However, the discrepancy between the two Samworth Brothers factories confirms what many in the industry attest to on an anecdotal level. The figures suggest that the factory producing supermarket own-label products is between 1.4 and 5 times more wasteful than the manufacturing-brand factory.[31]

The contrast between these two factories under common ownership indicates that the causes of food waste may be less a matter of unavoidable logistical problems or the awareness of the management, and more to do with the supermarkets' arrangement of their supply systems. Supermarkets have found that they get higher profits by marketing their own-label products and this is why there has recently been such a boom in own-label ranges, both in the UK and in other European countries: in the five years between 2000 and 2005, the share of sales of consumer packaged goods held by supermarket own-label products in the UK rose from 35 to 45 per cent.[32] But these own-label products are among the most wasteful items in the entire food chain.

It would be relatively easy to knock out much of the current wastage without even affecting consumer choice, just by changing the way orders are made. Lord Haskins believes that the underlying problem is that manufacturers have simply failed to stand up to the bullying of the supermarkets. More or less his last act as chairman of Northern Foods was to have a 'punch-up' with Tesco in which he says he refused to agree to provide Tesco with products Northern Foods was making exclusively for other firms, despite Tesco threatening to withdraw business worth £90m. Standing up to Tesco improved Northern Foods' standing, but since his departure he says the company has been 'flattened' by faltering relations with other retailers, like Marks & Spencer. Haskins thinks supermarkets are only able to get away with this because manufacturing is a crowded industry: supermarkets can pick and choose.[33] Logically, this would mean that if many manufacturers collapsed, the remaining ones would be in a more powerful position to negotiate, which is probably true – but exceptionally costly.

Less dramatic than a total restructuring of the domestic market is the development of more-sophisticated forecasts and more-transparent communication with suppliers. Using new software and expertise, supermarkets are getting better at predicting demand themselves, but they have limited interest in passing the benefits on to suppliers, especially when suppliers are bearing the costs of over-production waste. They have been particularly reluctant to pass on their know-how to suppliers that also supply rival chains because they believe that this could indirectly assist their competitors.[34]

In contrast to the highly equipped supermarkets, food factories often operate with a minimum of staff, and managers work flat out 'just to get through the day', as one of them put it to me, 'without any cock-ups'. The personnel in charge very often do not have the time, inclination, resources or support to investigate

novel ways of reducing waste. Any spare energy goes into
designing and marketing an ever-increasing variety of new products, and keeping the profit margins in the black. In fact, one
survey in the UK found that 90 per cent of food and drink
companies felt that management did not provide sufficient
backing for improvements in waste policies, while less than
half of companies had a good awareness of relevant legislation.
Although there are a number of government-related initiatives
in the UK, Europe and America to reduce waste in the food
industry, manufacturers often have little idea which body to
consult.[35] This problem is particularly acute among the small
and medium-size enterprises. Even when presented with waste-
reducing innovations that will save them money and increase
profits, they often push them aside because they do not have the
time to implement them.

For example, there is one innovative way in which some
manufacturers could reduce the impact of inaccurate supermarket orders. When a particular product has been over-
produced, the problem is that those finished goods cannot be
converted into other things. There are, however, many ingredients that go into a wide variety of products – the same tomato
sauce, for example, which might be used in the chicken curry
meal of one supermarket or the ratatouille dish of another.
Instead of combining all the ingredients in separate product
streams, manufacturers could prepare them and then hold them
back until the supermarkets have made their confirmed orders,
and only then run the final production. Dr Shahin Rahimifard
and his colleagues designed a system along these lines and gave
it a trial in a real business situation. The waste reduction they
achieved saved the business a stunning £520,000–780,000 every
year as a result – up to one fifth of its costs.[36] This not only
gained the company a lot of money: it also reduced its impact
on the environment and cut out the waste of human and natural
resources. Another waste minimization project engaged in
jointly by thirteen food and drink companies in East Anglia

brought about annual savings of £1.1m; it also reduced raw materials use and solid waste production by 1,400 tonnes, carbon dioxide emissions by 670 tonnes and water use by 70,000 m³.[37] The government-funded body Envirowise has helped twenty-four different companies and over 600 of their suppliers to identify over £12m of potential savings.[38] These exceptional examples prove that much of the waste deemed inevitable by other companies is in fact unnecessary and can be eliminated.[39] But other companies have been slow to follow suit. Lord Haskins's verdict that this is 'just bad business' may be correct; it certainly illustrates that reducing waste has hitherto been a low priority for manufacturers.

Manufacturers should of course be held responsible for their own failure to improve efficiency, and pressure should be put on them to change. But the manufacturing sector is already under financial and legislative strain (even the biggest firms have issued profit warnings and pre-tax losses in recent years), and there would be a danger that manufacturing would simply move overseas if regulations became too demanding.[40] So it will probably be more equitable and effective to tackle the problem through the retailers. The large supermarket chains tend to be in a more powerful position to effect such changes, and have the resources to do so. Apart from anything else, supermarkets cannot so easily move overseas: they may do their financial dealings in offshore tax havens, but, as far as their operations are concerned, they are pinned to their respective national customers. Furthermore, if retailers imposed environmental standards they would have to apply wherever the manufacturer was based. National governments by contrast can effect change mainly only within their own borders.

Retailers are also more susceptible to their consumers' ethical concerns, and public campaigns influence their behaviour. They will only continue environmentally and socially unsustainable practices for as long as they believe the public are indifferent. At present, supermarkets are able to dismiss the waste they cause their suppliers to produce as none of their business. When the British Retail Consortium itself passed on an invitation to all the major retailers to respond to this issue, not one of them replied. And yet, a respected report on chilled-food manufacturing con-cluded that the supermarkets were to blame, after every one of the suppliers studied showed that the actions of retailers unnecessarily increased waste.[41] If retailers were held to account, they might decide that it was in their interests to take measures such as refining their forecast orders. Retailers could also be persuaded to give the manufacturers more warning. For products such as sandwiches the manufacturer would sometimes only need an extra twelve hours to be able to produce to demand, rather than to inaccurate estimates.[42] At the very least, super-markets could share their expertise to help manufacturers make their own predictions more accurate, rather than speculatively churning out food that nobody wants.

A great deal could be achieved merely by promoting aware-ness of the issues within the industry. As WRAP points out, business managers rarely have a complete idea of the cost of wasting food. For example, 'category buyers' (those retail man-agers responsible for buying food) often do not understand the extent to which over-ordering or significantly changing final orders can incur costs all the way up the supply chain. Most obviously, there is the cost of landfill tax, waste collection and treatment. If their supplier is forced to waste food as a result of inaccurate forecasts this will ultimately push up the suppliers' prices, as the cost of waste still has to be met somewhere. There are also hidden labour costs involved with sorting, segregating and distributing the food that ends up being wasted. Waste reductions could be achieved if these 'hidden costs' were more visible to those who are causing it. For example, if category buyers were given bonuses on the basis of performance in this area, they would be much more careful about when and how much they ordered.[43]

Recent developments in the UK indicate that there may be

another effective way to deal with this issue through sensitive regulation. In 2000, supermarkets in the UK were investigated by the Competition Commission which concluded that they engaged in inequitable business tactics, including obliging manufacturers and farmers to accept excessive financial liabilities for food that supermarkets had wasted or failed to sell. These observations referred to a number of different business practices. For example, supermarkets often agree a price for a product with their supplier, but when sales turn out to be less than predicted and it becomes necessary to put the products on price reduction offers, the supermarket will turn round and require the supplier to share the burden of the reduced revenue. There are also the notorious 'take-back' arrangements, by which supermarkets receive a delivery of produce and then return it to the manufacturer if they do not manage to sell it all. In the case of manufactured products, this often happens when around 75 per cent of a product's shelf-life has expired, which means that much of it could have still been sold during the final 25 per cent of the shelf-life, but the retailer prefers to send it back to the supplier to avoid being left with surplus. More often than not the supplier simply chucks it away at a great financial cost to itself.[44] The ability of the powerful supermarkets to shift the risk of financial losses of wastage onto suppliers removes the incentive for them to improve their forecast ordering. If the supermarket accepted liability for this wastage or at least shared the cost with the supplier, there would be incentives on both sides to solve the problem.[45]

In 2002 the Office of Fair Trading (OFT) tried to deal with the problems highlighted by the Competition Commission by establishing a voluntary Supermarket Code of Practice (SCOP) for the big four supermarkets, though it watered down many of the Competition Commission's recommendations.[46] Clause 7 of the SCOP stipulates that the issue of who accepts the financial cost of a supermarket's over-ordering should be dealt with in advance. However, in 2004 the OFT completed a review of the

SCOP and found that 'the Code is not working effectively', and that many of the unfair practices were still carrying on. As a report by Friends of the Earth in 2003 showed, many suppliers were still 'required to meet the cost of unsold or wasted products unrelated to a quality problem with the product'. Tesco, for example, was explicitly accused of giving 'verbal encouragement to overproduce and subsequent failure to buy'.[47]

Following public pressure, in 2006 the OFT once again referred the grocery market to the Competition Commission. This resulted in the Competition Commission's final report in 2008, which concluded that the supermarkets are still guilty of transferring unnecessary risks and excessive costs onto their suppliers, including the cost of unsold food. The wording of the Competition Commission's conclusions is surprisingly forthright. Firstly the Commission points out that it is inappropriate that supermarkets fail to make prior agreements with suppliers on who will bear the cost of unsold food. 'For example,' it says, 'if an erroneous sales forecast by the retailer results in significant overstocking of a product, it would not be appropriate for the retailer to then request or require that the supplier share the costs.' However, the Competition Commission goes further, saying that even when terms are agreed beforehand, the costs that the suppliers are currently asked to accept are still often 'excessive': 'We have particular concerns,' it says, 'regarding the transfer of risk from grocery retailers to suppliers in situations where this transfer creates a "moral hazard"; that is, where the retailer has the ability to affect the degree of risk incurred, but, as a result of the transfer, the retailer has less incentive to minimize that risk. In these situations, the transfer of risk increases the total risk borne by the parties, and also increases the costs to the supplier.'[48]

As one spokesperson from the Competition Commission elucidated in an interview: 'There is a problem where, for whatever reason, retailers end up with surplus produce and make the supplier liable for the cost of that, whether through heavy discounting to clear the product or refusing to pay for what has not sold. Naturally the supplier might think, "If you over-ordered why should we be responsible for the cost?"'[49]

The Competition Commission has called for, and (at the time of writing) is still in the process of drafting, a new, stricter Grocery Supply Code of Practice (GSCOP) and, crucially, arguing for the establishment of an Ombudsman to police the Code and resolve disputes.[50] The question remains whether this criticism of transferring excessive costs to suppliers will be translated into a clearly worded clause in the new Code, and whether it will be binding. Unfortunately, it seems probable that it will only be covered by a general 'fair dealing' provision. It would be up to the Ombudsman (if one is established) to judge whether there was unfair dealing when disputes arise.

But among the supermarkets there is currently resistance to the idea of an independent Ombudsman with the power to fine supermarkets for malpractice. The retailers are making what the Competition Commission calls 'exaggerated' claims about how expensive this would be, threatening that it would push up the price of food.[51] The Department for Business, Enterprise and Regulatory Reform, meanwhile, is sitting on the fence and hoping that the supermarkets agree voluntarily to sign up to the new Code before it indicates whether it will step in and legally oblige the retail sector to clear up its act. Despite the common perception that legislation is an unpopular way of instigating change, 80 per cent of suppliers surveyed by Friends of the Earth felt that 'new legislation to prohibit unfair trading practices of supermarkets' would have a positive effect.[52] Using competition law is a convoluted way of tackling the problem, but the drafting and enforcement of GSCOP does offer a way of making supermarkets accept liability for the food they waste. This is a fundamental step in creating an incentive for retailers to reduce food waste further up the supply chain, and similar measures could be taken internationally.

4. Selling the Sell-By Mythology

in such abundance lies our choice,
As leaves a greater store of Fruit untoucht

John Milton, *Paradise Lost*, Book IX.619–20

Food poisoning is the bogeyman of the food industry and companies with valuable brands live in fear of damaging headlines. A single case can cause sales to plummet and share prices to dive. Manufacturers and supermarkets who determine shelf-lives therefore often allow for huge margins of error when they calculate how quickly a food will go off. For some products they imagine a worst-case scenario in which customers leave food in a warm car for hours, expose it to bacteria from the air and their hands, and eventually leave it in a fridge which is a few degrees warmer than it should be. As a result, the date they set is often days earlier than when the food will go off if handled properly.

In the EU, food-labelling laws require most pre-packed products to carry either a 'use-by' or a 'best-before' date. The further category of 'sell-by' or 'display-until' used by many retailers is not mandatory; it is to help shop staff manage stock, and should be completely ignored by consumers.

The notorious best-before date is for lower-risk foods and indicates that the food would be at its best before this date, but not necessarily that it is unsafe to eat after it. This includes products that are stable at room temperature, like bread and cakes, or chilled foods which do not support the growth of food-poisoning organisms if stored properly, such as butter and margarines. The law explicitly states that fruit, vegetables and any bakery products designed to be eaten within a day do not have to be labelled at all – and yet supermarkets routinely package and date them. Arguably, these dates can help shops prevent waste by making sure things get sold in the right order, but they also mean that tonnes of good produce are thrown away unnecessarily. Anyone can tell when a piece of fruit has started to go wrinkly, and decide for themselves whether it is fit to eat. The legitimacy of the retail industry's application of best-before dates is therefore highly questionable, and the practice of putting dates on fruit and vegetables could be abandoned. (Arguably, one way of doing this would be to stop packaging fruit and vegetables. However, modern film packaging can extend shelf-life and can therefore help to reduce food waste. Getting rid of packaging might be an ideal scenario, but it would mean depending on shorter supply chains – for example, shopping at local farmers' markets.) The principal function of best-before dates on these products, as some in the industry admit, may merely be to increase the profit of the supermarket – primarily by giving consumers the impression of guaranteed freshness.[1] It may also be that they encourage us to throw away more food so that we come back for more.

Use-by dates are supposed to be reserved for those pre-packed foods which are 'highly perishable and are therefore likely after a short period to constitute an immediate danger to human health'.[2] The law does not stipulate how long the shelf-life of each product should be – that is determined by manufacturers according to their own assessments. Neither does it give a list of foods that fall into the category, but it is likely to include dairy products (not including butter), cooked or partly cooked products (including sandwiches), and cured meat or fish such as ham or smoked salmon which is not stable at room temperature. Although in practice raw meat and fish are almost always given a use-by date, by law they could be covered by best-before dates because the customer would be expected to refrigerate and then cook them, thus preventing exposure to dangerous bacteria.[3]

The exception to this rule is prepacked fresh poultry meat, which is covered by separate regulations and requires a use-by date.[4] Foods which would normally need a use-by date but which are sold to the consumer frozen do not need a use-by date either. Eggs are covered by separate EU legislation, which requires only a best-before date, though they cannot legally be sold and are not recommended to be eaten after this date.[5] The best-before date and the use-by date therefore have very different purposes under law: the first is merely a guarantee of quality; the second is a guide to food safety.

The UK guidelines on use-by date legislation state that 'The use-by date is the date up to and including which the food may be used safely' and that 'The incorrect use of a use-by date may create confusion in the mind of the consumer *and contravenes the legislation.*'[6] Problems arise when manufacturers of use-by category foods print a date that reflects when they think a product will be past its prime but is still a long way from being unsafe to eat. This is routine practice for some products, such as pork pies, where moisture from the filling can gradually soak through and change the texture of the crust. Pie manufacturers I spoke to admitted that the use-by date they put on their products did not reflect their food safety concerns – it was just a matter of how long the pastry would remain crisp.[7] In these cases, it might be less confusing to consumers if the products carried a best-before date reflecting these quality concerns, and then also a use-by date indicating when the food will actually be unsafe to eat. This way people can make up their own minds whether to throw away a pie just because it isn't as crisp as it once was, rather than being scared into chucking it out because they think it might kill them.

Despite the fact that the rudiments of this system have been in place for nearly thirty years, the public is still in utter confusion about their true meaning. According to some surveys, up to 80 per cent of the British public misinterpret the function of the various terms. Many consumers are even more cautious than the already exceptionally stringent system advises them to be. The biggest area of confusion that leads to food being wasted is when consumers treat a best-before (quality) date as if it were a use-by (food safety) date and throw away food because they believe it is unsafe to eat. According to one survey, more than a third of the population mistakenly believe that any product past its best-before date is liable to poison them and should never be eaten.[8] These misunderstandings mean we throw away a lot of food that is still good to eat.

Since the dating system is enshrined in law, governments should bear some responsibility for telling the public how it works. A simple message stating that food on or after its best-before or sell-by date is not necessarily unsafe to eat, and that food safety is as much to do with temperature, handling and cooking as it is to do with age, could help a great deal.[9] Food education, such as cooking lessons in schools, could play a significant role in teaching people to use eyes and noses in combination with their understanding of food safety, rather than relying solely on dates provided by businesses. Average room temperatures in the UK have risen by 30 per cent from 12 °C to 18 °C between 1970 and 2004, and this means that (unless people can be induced to turn their thermostats down) food once stored in naturally cool larders now has to be stored in energy-consuming fridges.[10] But some products are still best kept well away from fridges: lettuces, for example, are the products we waste most of out of all foods, with 45 per cent (by weight) or 60 per cent (by cost) of all purchased salad in the UK ending up in the bin.[11] Instead of storing whole lettuces in plastic bags in the fridge, they can last several times longer if they are kept like flowers, by cutting a thin slice off the base of the plant and standing them upright in water. Food storage instructions conveying this kind of message can have a significant impact. Fresh milk in Britain tends to carry a warning that it should be consumed within two to three days of opening, but one brand – Cravendale – says its milk stays fresh for seven days. WRAP asked it why, and Cravendale answered that it was because it used 'microfiltration', but WRAP discovered that other companies use microfiltration too. Those companies' storage instructions simply hadn't kept up with the technologies they were using.[12]

For dedicated technophiles, fridges have been designed to read the barcode of the food within them and alert their owner when it is approaching the end of its shelf-life. 'Smart labels' are also available that detect temperature changes, going red only when the food has been left out long enough for bacteria to develop. However, these devices are expensive and take even more responsibility away from consumers.

Before the advent of EU food-labelling legislation in 1979, the industry itself first introduced sell-by dates, following a chronic loss of public faith in supermarket food. Lord Haskins told me about his involvement in the invention of sell-by dates, following an *Evening Standard* story about Harrods cream being rancid. Back in the 1970s he was in discussion with Marks & Spencer, and between them they came up with the idea of coding dates to help staff bring old stock to the front of the display shelves. Products like biscuits, that had never been date-coded before, suddenly started carrying dates, despite them being safe to eat for an indefinite period. It was only later, according to Haskins, that the dates got mixed up with food safety. By then consumer culture had taken over, launching a marketing race between the supermarket chains – with Marks & Spencer running an advertising campaign along the lines of 'you know it's fresh by the sell-by date'.[13] Other manufacturers and retailers followed suit, though their competing brands for similar products often had shelf-lives twice as long. The dates used by manufacturers, according to Haskins, are now 'excessive, over-reactive and absurdly unnecessary', and he agrees that an overhaul of the system could help to reduce waste dramatically.

Products like chicken sandwiches which are eaten cold can certainly cause problems if eaten too late, but many argue that raw meat can be eaten after the date on the packet if it is cooked properly before eating. 'If meat is five or six days out of date,' said Lord Haskins, 'I'd have a look at it and probably eat it if I had stored it in the fridge. Dairy products, like yoghurt, if it's a month out of date and smelled and looked OK, I'd eat it. On the one hand,' he went on, 'manufacturers assume everyone is an idiot; and on the other, the public are very stupid to take these dates so seriously.' What Haskins proposes instead of the confusing jargon is a traffic light system, by which certain products are coded red, amber or green according to the level of risk.

In the United States there are no Federal laws requiring date labelling on food, except for some baby foods. The government prefers instead to focus on educating the public about how to treat food safely.[14] However, dating of some foods is legally required by more than twenty states; grocers and manufacturers across the land use a variety of sell-by dates, 'pull dates', 'expiration dates' and 'quality dates'. There are areas of the country where much of the food supply has some type of date and other areas where almost no food is dated. The result is a cacophony, almost as confusing for the consumer as the European system.[15]

However, Americans are wary of following Europe's example and enforcing food labelling – and the reasons for this are partly because they see how much unnecessary waste it can cause. In 2001 the United States Department of Agriculture floated the idea of requiring use-by dates on some foods. However, the Food Marketing Institute – an industry lobby group representing retailers and wholesalers – joined a number of expert opponents by pointing out that consumers were already confused by date markings so the government would have to complete an extensive education programme before any law could be rolled out. They also noted that there are too many variables – such as how the food has been stored and handled – that affect the speed of deterioration. Finally, they argued that:

If the Department applies worst-case standards for infectious dose and handling of the product in order to protect the most highly suscept-ible populations, food that may be safely consumed by the majority of the population will be destroyed. Furthermore, a use-by date based on worst-case assumptions may lead to a permissible shelf-life that is too short to allow reasonable distribution, sale and consumption of the product.

In 2000, Senator Tom Hayden of Los Angeles put forward a date-labelling bill for the State of California following a number of hysterical media stories about food safety malpractices. But it was dropped for similar reasons following opposition from an industrial lobby who argued, among other things, that 'Beef that darkens making it unappealing to the consumer can still serve as wholesome ground beef, while many meat products can be safely frozen well beyond traditional sell-by dates.'[16] This may be a symptom of corporate power in the US, but it may also be the reign of common sense over officious legislation.

As European legislators have been quick to recognize, public health must of course be protected. But whenever a policy is devised to avoid risks, the costs of that policy also need to be considered. The European law on food labelling, and its application by manufacturers and retailers, currently causes enor-mous quantities of waste. It may be that in their current form date labels do more harm than good, or at least more harm than is necessary. It is difficult to find people who are willing to take responsibility for balancing the interests of public health and waste reduction. In the UK, for example, the Food Standards Agency said in an interview in 2008 that it did not regard waste as part of its remit. Thankfully, this may now have changed.[17] There are at last significant moves afoot in the industry and in government to instigate a review of date labelling at the Euro-pean level. Date labels may in the future make it clear that food is still good *up to the end of* the date mentioned. In Germany the government decided that sell-by and display-until dates confuse customers, which is forbidden under the existing EC Directive and it has therefore banned them, and this interpretation of the law could be extended across the EU. However, for the most part, each body concerned with determining the shelf-life of products dances intricately around the others' agenda, and in many cases the decision is determined as much by fear as by science. When over-strict standards are applied in ways that unnecessarily increase food waste, it undermines the credibility of the whole system and risks turning it into one designed to protect corporations from litigation rather than one that protects people and the planet they live on.

5. Watching Your Wasteline

Jack Sprat could eat no fat,
His wife could eat no lean,
And so between them both, you see,
They licked the platter clean.[1]

Proverbial, seventeenth century

I open the fridge door of a friend. George is typical of his generation: living alone, eating alone and seldom cooking from scratch. Food has become a matter of convenience in his life. George considers himself a good environmental citizen. He has had insulation and double-glazing installed; he recycles his bottles and paper, and switches off lights when he leaves the room. But people who live alone waste 45 per cent more food per person than average households, so I have decided to put George to the food waste test.[2]

As George sits at his kitchen table pretending to read the newspaper, I prop open his fridge and pick my way through its contents. In the bottom compartment, I find a packet of baby spinach that has started wilting into a brown mass; on the shelf above is a half-eaten plate of pasta and a vegetarian lasagne ready-meal which has passed its use-by date; and in the door, a desiccated half-lemon which looks distinctly like it will never be squeezed again. In the kitchen bin I find quarter of a loaf of now-mouldy bread, the remains of a pot of yoghurt and a slimy slice of Cumberland sausage and smoky bacon quiche. George is sure that he wastes less than the average citizen – but studies in the US and UK suggest that most people are under the same illusion and fail to register how much food they really throw away. The study on consumer waste conducted by the waste organization, WRAP, found that more or less everyone in the UK wastes phenomenal amounts of food without realizing it. What is it that makes us do this?

Supermarkets, as we have seen, are very efficient at cleansing their public image by pushing waste up the food chain to manufacturers and others. Unsurprisingly, they also squeeze it downstream by offloading surplus onto consumers. One of the many ways they do this is through 'Buy One Get One Free', or 'BOGOF', deals. These can offer great value for money. But it is only good value if you need what you are actually buying. Instead, what many of us do is take home the 100 per cent extra free and then fail to eat it because we did not need it. The supermarkets have thereby offloaded a product of which they had a surplus, convinced us we are getting a bargain, but in fact just generated a lot of waste which has to be disposed of. And we walk straight into the trap. The same phenomenon applies to large value packs of produce and over-sized ready-meals. We get a lot for our money, but we are often being induced to buy more than we are going to use. Thankfully, the supermarkets have begun to listen to customer dissatisfaction on this issue, particularly from disenfranchised single-occupancy households. In 2008 in the UK there was a trend showing a considerable decline in BOGOF promotions in favour of price reduction deals which can be a more constructive way of dealing with surplus and reducing waste.

Attributing the ultimate causes of wastefulness to either the supermarkets or customers is a fraught chicken-and-egg scenario. When we consumers waste food, is it entirely our responsibility – or does some blame lie with the supermarkets' sophisticated and subtle marketing systems? When the British Prime Minister, Gordon Brown, gave his speech to the G8 summit in Japan in 2008, he drew attention to the supermarkets' role in making us buy more than we need and linked the resulting food waste to the global food crisis: 'If we are to get food prices down,' he said, 'we must also do more to deal with unnecessary demand such as by all of us doing more to cut our food waste.'[3] The Liberal Democrat environment spokesman, Steve Webb, put it more bluntly: 'Supermarkets make it harder for householders to avoid food waste, while throwing away large quantities of edible food through poor stock management. They refuse to stock small portions, which are essential for the growing number of one-person households, and offer too many buy-one-get-one-free deals on perishable goods.'[4]

Dr Tim Lang, professor of food policy at City University in London, sees waste as embedded in the structure of modern food supplies: 'Food pours out of the supermarket machine, and it ends up being dumped on consumers,' he says, 'it has to get down people's throats and consumers are colluding with that: the abundance model of food has been built into consumer culture. Supply is dictating demand; the tail is wagging the dog.'[5]

The WRAP study on the food we waste has given a greater insight into the reasons behind people's behaviour than has ever been available before, so the question can now be answered in a much more informed way. But the extent of the problem shows that the blame can hardly be laid entirely with the supermarkets. To begin with, WRAP found absolutely staggering levels of waste in the home. Based on protracted examination of the bins of more than 2,000 households, it discovered that the British throw away 6.7 million tonnes of food every year. That is less than the combined wastage in the food industry, but it is still the single biggest contributing sector.

Some of these 6.7 million tonnes of waste should not really be called 'food' at all – as they have been in both the media and WRAP's own publications – because nearly a fifth consists of things like orange peels and tea bags which no-one eats. But the British still manage to throw away 4.1 million tonnes of food which definitely could have been eaten if managed properly. That includes food left over after cooking too much and food left on people's plates, but also a lot of unopened and untouched packets of fruit, vegetables, cheese, meat and ready-meals which people have dumped in the bin, sometimes even before their best-before or use-by date: 484 million unopened pots of yoghurt, 1.6 billion untouched apples (27 apples per person), bananas worth £370m (imagine buying your annual share of £6 of bananas and dumping them in a bin all in one go) and 2.6 billion slices of bread are among the most startling instances of consumer indifference to wasting food.

A further 1.3 million tonnes was 'possibly avoidable food waste' such as bread crusts, potato peels and so forth which people clearly make a deliberate choice to discard but which could physically have been eaten. Technically, these items are edible and nutritious and arguably ought to be included in total estimations of food waste. In a food emergency – for example in a war or in the event of environmental calamity – it would be reasonable to encourage people to mash potatoes with the peels on, and to eat bread crusts (since they represent around ten per cent of the entire loaf). Saving them would reduce demand for valuable foodstuffs – and in the First World War, for example, the British government did issue a propaganda poster: 'Don't Waste Bread! Save two thick slices every day'.[6]

The total 5.4 million tonnes of avoidable and possibly avoid-able food waste which householders produce comprises a whop-ping 25 per cent of all the food people buy to eat at home.[7] This does not include the further millions of tonnes wasted at work or by leaving food on plates in canteens, fast-food outlets and restaurants. Studies have found that in canteens about one fifth of all food is wasted, while others find that 24–35 per cent of school lunches end up in the bin.[8]

Colleagues in other European countries simply cannot believe these sums. I corresponded with one member of the German Environment Agency who guessed, on the basis of the estimated contents of separate bins for biodegradable waste provided to half the German public, that food waste from households in

Germany amounted to just 15 kg per person − less than one seventh of the 112 kg of food waste per capita found by the WRAP study in the UK.[9] An Italian waste collection scheme found that on average households separated the equivalent of 73 kg of food waste per person per year into their biodegradable waste bins − though there is no indication of how much other food waste was created or how much of it was edible.[10] Audits of household garbage bins in Australia suggest that householders throw away around AU\$7.8bn of food each year, or around 13.1 per cent of total purchases.[11] A study in Sweden found that if supplies were examined from the plough right to the plate, 50 per cent of all food disappeared.[12]

One study in the US found that households waste 14 per cent of their food, though others estimated the figure to be nearer to one quarter.[13] The British − accustomed to wagging their fingers at their more profligate American cousins − throw away at least as big a proportion. However, while the percentage of food wasted in US homes could be lower than in the UK, US homes actually buy more so the quantity discarded is also higher. Americans throw away 96 kg of edible food each year, as compared to the British, who throw away 70 kg (avoidable only).[14] Americans simply buy, eat and throw away much more food than Europeans.

Such discrepancies between different developed nations is hardly credible, and one problem is that the various studies have used different criteria and methods. In the WRAP study, when a whole item of food was thrown away − for example a whole orange or chicken− the entire mass of the item was counted as 'avoidable' food waste. In the US studies by contrast, the inedible portions − the peel of the orange or the bones of the chicken − were deducted and only the 'edible portion' counted as wasted food.[15] Still more importantly, whereas the American study that found domestic losses of 14 per cent measured the weight and the *calorific* value of the food wasted, the one that estimated waste at 26 per cent measured only the *weight* of the food, while the WRAP study measured the weight and converted that into a proportion of the *monetary value* spent on food. Since fruit and vegetables, for example, have a high monetary value, a low calorific value and a relatively large mass, these different approaches inevitably produce different results. Until more accurate, extensive and comparable studies are conducted in different countries, we simply will not know whether Britain really is the food waste capital of the world, or whether it is representative of the Western norm.

Before we try to imagine possible solutions to the problem of food waste, it is vital to understand the reasons behind people's behaviour. WRAP has done an enormous amount of research into the reasons why people buy more food than they eat and this has proved an invaluable tool in their campaign to reduce waste.

One important phenomenon is what food psychologists often describe as 'Good Mother Syndrome' – always wanting to make sure that there is enough choice and plenty of food to satisfy the family. A well-stocked larder has, in many cultures since antiquity, been a principal signifier of status and affluence. Anybody who is entertaining guests or even just cooking for the family would rather cook more than enough than run out. In that way, we operate a bit like a supermarket: projecting an image of unlimited abundance. Anything less and we might appear stingy, or too poor to offer enough food. The cost of buying and cooking more than is likely to be needed is, for many people, insignificant in comparison to the shame of failing to provide adequately for guests, spouses or children. The expectation of hospitality in many different cultures includes the notion that everyone should have more than enough of everything.[16] This problem could be alleviated by enhancing people's understanding of portion sizes, which could be achieved by improving home economics education in schools as well as within families, and food retailers can also help by selling food in appropriate quantities and indicating portion sizes on the label. The second approach would be to demonstrate that providing ample food does not mean we have to throw away everything that is left over after a meal. A renaissance in bubble-and-squeak (typically, refried potato and cabbage, but really any scraps left lying around) is a nice idea and a number of new cookbooks and internet sites now aim to re-educate people in such frugal housekeeping methods.[17] Encouraging people to take home surplus after a wedding party, or even their own leftovers in restaurants is another measure which could benefit everyone.

The way we feed children is also crucial in determining how much food we waste.[18] It has become an almost indelible image of child-rearing in the West – a toddler sitting in a high seat with food smeared from ear to ear, having scattered most of their dinner on the floor. Many of us will remember being told to finish what was on our plates as children. One environmentally conscientious family I know heat their house on scrap wood, and for years lived on food reclaimed from supermarket bins. But their two-year-old, my godson, is a talented waster of food – after a meal the kitchen is splattered with carrots he has taken for missiles or bowls of cereal he has chosen to turn upside down. This is great for the pet dog and the family's flock of chickens. But for most parents it is just a mess that needs to be tidied into the bin. For years I regarded this as an unavoidable phase in all of our lives – until I lived in India and saw how children are fed there. Instead of giving a child their own plate of food, many Indian parents hand-feed them gobbets of the food they are themselves eating. At first this struck me as time-consuming and indulgent, until an Indian friend pointed out that this cuts down on food waste and saves mess. Youngsters learn in these early stages of their life how to treat food – disposable or valuable – and the way they are fed is surely influential. My godson's mother, on the other hand, says that tipping food overboard is a passing phase – and that as much for parents' sake as for their own, children have to learn to feed themselves, however much mess this creates.

Child-rearing may be problematic but there are other areas of our lives over which we have more control and many of the decisions that lead to food being wasted happen before the food even gets into the kitchen – while we are still in the shops. Hoarding surplus food has been a useful reflex in the past, and it probably developed as an evolutionarily advantageous instinct millions of years ago, as a response to unstable and seasonal food supplies. As I shall discuss later, it was the drive to store surplus that originally made agriculture – and consequently settled civilizations – possible. Today, buying more than you need no longer has these adaptive advantages, because there is always food in the shops. (Even if one day societal collapse or war interrupts supply, having bought an extra bag of oranges that week will not help.) Something so deep-rooted in our cultures, and even in our natures, can hardly be blamed either on individuals or on food corporations. But supermarkets know how to tap into that primeval hoarding instinct of ours and induce us to buy more than we need. Their marketing teams have spent millions finding out what works best at triggering our purchasing impulse; it may take an equal effort and investment to encourage people to overcome that impulse and to buy only the food they are actually likely to eat.

Spontaneity has become one of the most cherished features of many people's lives in the Western world. Shopping lists have fallen out of fashion so people end up buying things they already have at home. People now buy food in bulk, often at weekends, without any clear idea of what they will be doing later in the week. Carrots and beans bought fresh on Saturday with the intention of making a healthy meal during the week end up never being cooked and are discovered at the back of the fridge on Friday, while the minced beef that has been sitting in a plastic box all week has gained an unappetizing tinge of grey. At the time it just seemed easier to get a take-away or microwaveable ready-meal rather than cook the food that was already in the fridge. The cheese opened for a sandwich has dried out, cracked or gone mouldy, and while it may still be edible, buying a fresh one is more appealing. It may generate a temporary twinge of shame or regret as these items are dropped into the garbage can, but it doesn't alter these patterns of behaviour during the next big shop.

In fact, in people's long-term memories, it hardly even registers. Even though almost every single person in the Western world regularly wastes food, many are unconscious of it. They still perceive waste as something exceptional, rather than a routine resulting from habitual ways of organizing their meals.

Environmental and social organizations, and even some governments around the world, have realized that convincing people to waste less food is one effective way of reducing our impact on the environment and liberating resources for other uses. But the critical question which remains is how best to convince people that it is worth changing their behaviour? What arguments, strictures or initiatives will actually induce people to stop throwing food away? Here, I shall raise a few possible answers to this question. WRAP has focused on showing people the financial value of the food they throw away. Beyond this, one can try and convey the value of food by focusing on its nutritional content, which illustrates what that food could mean to hungry people if it weren't being sequestered in rich countries' landfill sites. There is also the environmental damage – such as the release of greenhouse gases – caused by growing the food. Arguably the most fundamental way of understanding the true cost is to look at the land and the resources used in producing the unnecessarily wasted food, and imagine how these resources could be used if they weren't growing food that is wasted – which is the focus of chapter 6 and a subject I shall revisit in chapter 12.

WRAP's principal method of trying to change people's behaviour by appealing to their wallets has apparently already started to yield some results. In UK households, food thrown away is worth £10.2bn a year, amounting to an average of £167 per person or £420 per average-sized household. In the US, studies found that the waste from consumers amounted to $54bn of food each year.[19] None of these figures include the cost of disposing of food waste, which is paid for by taxes.[20]

Accumulating the total cost of all this waste into annual lump sums may well trigger some people to concentrate on saving cash by avoiding waste. But, for many, the financial decision to buy more than is needed has become ingrained in their purchasing habits. People are familiar with the experience, week after week, of occasionally throwing out unopened or partially eaten foods and know, consciously or unconsciously, when they fill their baskets in the supermarket that they are paying for something they have only a 75 per cent chance of eating. They still decide it is worth parting with the money. A household that wastes £420 on food could use the money for something else – but many would still prefer to buy the food and throw it away.[21] Perhaps if we had a keener idea of what that money could buy, it might help. For example, if UK consumers halved their waste and sent the saved money to, say, Pakistan, it could buy nearly enough flour for the entire population.[22]

But the fact remains that the abundant availability of relatively cheap food has made us negligent about wasting it. In developed nations people buy more food now than ever before. In the US, between 1983 and 2004, the amount of food available in shops and restaurants rose by 18 per cent from 3,300 to 3,900 kcal per day for each person in the country.[23] Europe is not far behind, with EU food supplies on average reaching over 3,500 kcal per person per day in 2003.[24] Some of this increase goes towards expanding waistlines; most of the rest is used to fill waste-bins.

Even though the quantity of food bought has rocketed, the proportion of household income spent on food in the home has declined as people get wealthier and food has (until very recently) become steadily cheaper, falling by 47 per cent between 1961 and 1992.[25] In 1984, British households devoted an average of 16 per cent of their spending to food; today the proportion has fallen to an average of just 9 per cent, though the poorest UK households spend nearly 15 per cent.[26] In the US families spend 10 per cent of their disposable personal income on food in the home.[27] Among the world's poorest people, such as those in Pakistan, spending on food can be around 75 per cent of income.[28]

Wasting money on excess food may actually be something that comforts consumers. It is the very fact that we can afford to buy food even if we may not want to eat it which subliminally gives us a sense of affluence – a cosy buffer between us and hunger. Having surplus, even in excess of what is ever likely to be needed, can be reassuring. It is a normal phenomenon, but one which can become pathological: at that point it is called greed. This goes some way to explaining obesity caused by over-eating, and also why people buy more food than they eat. But this is putting the best possible gloss on our behaviour: in truth we have simply become lazy and negligent about food, and we are blind to the true costs of wasting it.

For as long as spending on food remains such a small proportion of people's overall expenses, being informed of the financial costs may not in itself be enough to convince people to change their behaviour. Perhaps a recognition of the other non-financial costs associated with wasting food would help. Affluent consumers in the Western world have become disconnected from what food really is, where it comes from and what its production entails. We have come to regard wasting it as simply a factor of what we can afford, rather than what can be sustained by the planet and the other people on it. Many do not realize the environmental impact; even fewer are aware that wasting food causes hunger elsewhere in the world.[29] Gordon Brown intimated this in his speech to the G8 summit in Japan, but in order to understand how waste causes hunger it is necessary to set forth some of the basic ways in which the global food supply system operates.

So in terms of taking food from the mouths of the hungry,

how significant is the food wasted in rich countries? One way of looking at this is to calculate the nutritional value of the food being wasted. This allows us to quantify the number of people that could theoretically have been fed on that food. It is difficult to imagine a million tonnes of food, but converting that measurement into the number of people that could have been fed on it makes it more comprehensible, and the value of that food more vivid. It can even help to provide a clearer idea of how many people the world would really be able to feed if people cut down on waste. It is also a more absolute measure than financial cost, since the price of food varies a great deal between countries whereas nutritional needs remain relatively constant.

The detailed studies conducted in Britain and America allow us to calculate the nutritional content of wasted food with some precision. For this reason, my great environmentalist friend Laura Yates and I produced a database of figures listing every type of food wasted by UK householders – from tomatoes and rice to poultry, pork, bread slices and cakes. We gave each item its established calorific value – that is, the amount of calories contained in each kind of food – 138 kcal per 100 g of rice, 744 kcal for butter, 47 kcal for apples, 75 kcal for potatoes, and so on for about 200 different kinds of food. We then multiplied the calorific value by the amount wasted. I did the same for American food losses according to data from the United States Department of Agriculture.

Instead of conjuring up imaginary extra people to feed, it seems more relevant to think about the actual hungry people in the world and consider what all the food we waste might mean to them. According to the United Nations, in 2007 there were 923 million undernourished people in the world, almost all of whom (907 million) live in developing nations. Ricardo Sibrián, Senior Statistician at the Food and Agriculture Organization of the United Nations (FAO), has calculated how much extra food these people would need to satisfy their hunger and lift them

out of malnourishment. The average calorific deficit for malnourished people in the developing world came to 250 kcal per
person per day – a level of undernourishment that is called
the 'depth of hunger'.[30] Supplying an average undernourished
person with an extra 250 kcal a day above what they are getting
already would allow them to attain a minimum acceptable body
weight, and perform light activity. Malnourishment causes children to be stunted and it retards brain development; it damages
the immune system and sometimes leads to death by starvation:
250 kcal extra a day on average would be enough to prevent
all this.

Cynics will argue that there is no connection between food
being wasted in rich countries and the lack of food on the other
side of the world. Their argument may have been stronger in
the past, when famines were sometimes more to do with local
conditions – such as war or natural disasters – than global shortages. But there has long been a connection, and the food crisis
of 2007–8, largely caused by global shortages of cereals, has
made this even more evident. It is now abundantly clear that
fluctuations in consumption in rich countries affect the availability of food globally and this impacts directly on poor people's
ability to buy enough food to survive.

In attempting to explain the food crisis of 2007–8, academics,
policy analysts and journalists concentrated on the 'new pressures' which triggered the soaring prices – the diversion of food
grains to make biofuels; the increase in developing-world meat
consumption (which requires crops such as wheat, oilseeds and
maize to produce); the drought-related harvests in some grain-
exporting countries such as Australia; oil price increases; and the
highly debated influence of financial speculation.[31]

Although the increased demand from biofuel and meat production had its most direct impact on the price of cereals, with
the cost of major staples such as grains and oilseeds doubling in
the two years to July 2008,[32] this also had an indirect impact on
other foods as well. When the price of one kind of food – wheat,

for example – goes up, poor people turn to other kinds of food – such as rice – which in turn stretches the supply of those foods and causes their prices to rise too. When all food prices increase, the poor in particular resort to the kinds of food that yield the highest number of calories for the amount of money spent: in other words they turn from fresh vegetables and meat to cereals such as wheat, maize and rice, and this increases pressure on supplies, which in turn increases prices further.

The squeeze on global supplies caused average food prices around the world to rise by 23 per cent in 2007 and by 54 per cent in 2008,[33] and this contributed to pushing an estimated 44–100 million additional people into chronic hunger, and could increase child mortality rates in some countries by as much as 5–25 per cent.[34] If this was caused partly by shortages ensuing from increased demand, it follows that alleviating shortages lifts people out of hunger. I am not, of course, suggesting that tomatoes gone mouldy in people's fridges could be shipped out to hungry people in Africa. This spurious connection assumes that the food in rich people's homes or overstocked supermarkets had no other potential destiny than ending up in rich countries in the first place.

The material connections linking consumption and demand in all parts of the world are quite different. The case for this is most easily demonstrated by cereals. Cereals – principally wheat, rice and maize – have global prices which affect the cost of food in the markets of Africa and Asia just as they do on the shopping aisles of the United States and Europe.[35] Wheat, for example, is a staple across the world. It is an internationally traded commodity, governed by global prices which are in turn determined by the balance of global supply and demand. Europeans buy grain from North and South America, south Asian countries buy it from Australia, others buy it from Kazakhstan, and so on. If the crop of a large grain-producing country fails – as happened in Australia in 2006 and 2007 – less grain will be available on the world market, and this can push up prices globally. If America uses more grains for fuel, or the Chinese use millions more tonnes to feed livestock, this can also stretch supplies. By the same principle, if Western countries divert millions of tonnes of cereals into their rubbish bins, there will be less available on the world market. If they stopped doing so, there would be more and it would be more likely to be affordable. The focus on the new surges in demand from biofuels and meat production was to the exclusion of underlying core problems in the global food system, such as waste.

The UK both imports wheat and exports it: about 6 per cent of supplies come from the high-protein varieties preferred for bread-making grown in France, America or Canada; but British farmers export even more wheat to other countries. The US is one of the largest grain exporters in the world. What the US and the UK import and export depends on how much is used within those countries – and how much is thrown away. Since food supply has become a global phenomenon, and particularly when demand outstrips supply, putting food in the bin really is equivalent to taking it off the world market and out of the mouths of the starving.

In the UK alone, householders throw into their domestic bins enough grain – mainly in the form of bread – to alleviate the hunger of more than 30 million people – that is, it could have supplied them with that extra 250 kcal a day they need to avoid malnourishment.[36] American retailers, food services and householders throw away around a third of all the grain-based foods they buy, which would have provided enough nourishment to alleviate the hunger of another 194 million people (and that still does not include the industrial waste in manufacturing and processing).[37] If you include arable crops such as wheat, maize and soy used to produce the meat and dairy products that are thrown away by British consumers and American retailers, food services and householders, then this comes to enough food to have alleviated the hunger of 1.5 billion people – more than all the malnourished people in the world. That grain – if we had

not outbid the poor for it – could have stayed on the world
market; people could have bought it, and eaten it.

Adding up all the different kinds of food wasted in UK
households, including the indirect waste of grains used to produce meat and dairy products that were thrown away, the total
number of people that could have been lifted out of hunger
comes to 113 million.[38] In other words, each of us in Britain
wastes enough food in our homes every year to supplement the
deficient diets of nearly two of the world's malnourished people.
In the US, the food thrown out by householders, supermarkets,
restaurants and convenience stores (including grains used to
produce discarded animal food), would have been enough to
satisfy every single one of the world's malnourished people two
times over.[39] If food wasted by consumers and the food industries
of the UK and Europe are estimated and added to that total
there would be enough food to satisfy the needs of all the world's
hungry between *three and seven times over*.[40]

In the seventeenth century, John Locke argued that if someone took more food than they needed and let it spoil, 'he took
more than his share, and robbed others.' This is what we in the
West are doing on a global scale; we sequester the land and other
common resources of the world to grow food that we end up
wasting. According to Locke, this annuls our right to possess
both the land and the food grown on it.

Imagine living in a closed room with five other people. One
of the people in the room is malnourished, and one of them is
much richer and more powerful than everyone else. The rich
person eats more than everyone else, and keeps aside enough
surplus food to fatten his pigs and cattle. He is also very negligent
with his food, hoarding it in his corner, and sometimes forgetting
to eat it before it goes mouldy. He throws away more food than
the hungry person needs to regain his health and strength.

We do live in a closed room, the Earth, on which we can
grow in any one year a finite (though variable) amount of food
– and currently the rich outbid the poor for it, sometimes merely

to waste it. It is understandable that we have not yet learnt to appreciate how our everyday actions affect people on the other side of the planet. Sympathy is partly a visceral emotion, and it is weakened by the distances between our actions and those who are affected by them. This kind of global consciousness is relatively new, and societies always take time to absorb big ideas, particularly when they are uncomfortable ones. It is also hard to overcome the tendency we have inherited – in social traditions and probably in our genes – to appropriate territory and resources that ensure the comfort of our own group at the expense of others.

It is too easy to resort to condemnation and outrage. We are all human and most of us have a sense of our own decency and fairness. We have to look at our conduct as a feature of human behaviour, which has to be understood – in its origins and the social, economic and military means by which it is perpetuated – for us to see why we are not only complicit in these actions, but actively engage in them, to the detriment of other humans and the natural world.

Being held responsible for the unintended consequences of our deeds is problematic at the best of times and the water is even more muddied when we either know beforehand what these unintended consequences will be, or would know if we informed ourselves. Our sense of personal responsibility is further dissipated when the ultimate causes are due to collective behaviour. What difference do we as individuals really make?

A survey conducted in Australia in 2005 found that 60 per cent of people felt guilty about buying and then wasting items such as food; only 14 per cent of respondents said they were not much bothered or not bothered at all.[41] But rather than feeling guilty, we should feel empowered by the sense of responsibility. It is a relief, in many ways, that we can enhance the lives of the world's hungry by doing something so easy as buying only the food we are going to eat, and eating whatever we buy.

It may be undesirable, not to mention unfeasible, to expect

everyone to consume only the minimum they need to survive. But the gratuitous sacrificing of others' most essential requirements to fill rubbish bins with food should surely become, in time, socially unacceptable. In that all of us in the affluent world deprive others of food necessary for survival, often for the most whimsical motives such as filling our fridges with food we never eat, it is difficult to escape from the conclusion that in this remote, indirect and ignorance-muffled way, I, you, and everyone you know behave murderously towards fellow humans.

6. Losing Ground: Some Environmental Impacts of Waste

this proud and troublesome Thing, called Man, that fills the Earth with Blood, and the Air with mutherous Minerals and Sulphur.[1]

Thomas Tryon, 1700

Calculating the calorific value of the food being wasted in rich countries illustrates the enormity of the problem. But for many types of food this does not represent the full extent of the impacts of waste. If a tomato in Britain is wasted, it is not the nutritional content of the tomato that matters so much as the resources used to produce it. What could be grown instead if those resources were not being expended on producing food that no-one eats?[2]

Fresh produce wasted in Europe and America is connected to global pressures on agricultural resources because demand for land in one part of the world has a knock-on effect on land use elsewhere. When demand for agricultural commodities soared in 2007–8 this raised the value of agricultural land in Brazil, which increased the incentive for people to cut down the Amazon rainforest to create more agricultural land.[3] In Kenya, when demand for sugar cane increased, partly owing to bio-ethanol production, this led to the decision in 2008 to drain and plough 21,000 hectares of the Tana River Delta which is an ecologically invaluable wetland inhabited by lions, hippos, primates including the critically endangered Tana red colobus, sharks and 345 species of bird, and depended upon by thousands of farmers and fishermen. This encroachment will cause what local groups have called 'an ecological and social disaster'.[4] When Europeans and Americans buy around 25 per cent more food than they eat, they are taking up land which could be used to satisfy demand for other agricultural products.

The impact of global consumption patterns has been demonstrated even more starkly by the trend of richer countries buying or leasing land in the developing world. For example, in 2008 the South Korean firm Daewoo Logistics sought to increase the security of the nation's food supply by buying a ninety-nine-year lease on 1 million hectares of land in Madagascar to grow food which will be exported to South Korea, where phenomenal amounts of it will be wasted by consumers almost as profligate as Americans and Europeans. This agreement will inevitably result in deforestation and the dispossession of small farmers in Madagascar. A company part-owned by a Japanese firm has bought 100,000 hectares of farmland in Brazil and numerous other such deals have gone through in recent years covering at least 7.6 million hectares of agricultural land worldwide. The head of the FAO, Jacques Diouf, has called this kind of deal, where poor nations produce food for the rich at the expense of their own hungry people, a new kind of 'neo-colonialism'. It is true that developing countries could benefit from investment in agriculture, but sequestering land solely for the satisfaction of foreign consumers is not the way to achieve this.[5] The use and waste of any agricultural produce requires the use of agricultural land, and almost all cultivable areas in the world are in some way connected to the global network of supply and demand.

Some will resist the idea that demand for food in rich countries can cause the exploitation of land and people elsewhere, and certainly there are those in the developing world, such as suppliers of food, who may profit from our profligacy. Where agricultural intensification can be supported, increased demand need not cause deprivation or environmental degradation and can contribute to rural economies. But currently there are few measures in place to ensure that this is how increased demand is satisfied. Growers that benefit from unsustainable demand are in the minority and their gain may be short-lived; the vast majority in the developing world suffer from the stretch on food supplies that is exacerbated by rich countries' wastefulness.[6] Besides, developing economies are better supported by offering suppliers a fair price for their produce and ensuring this is fairly distributed to labourers in good working conditions, rather than paying for exploited labourers to over-exploit their land to produce unnecessary surplus for rich countries. Affluent customers do not immediately suffer when over-worked soil dries up and becomes infertile or water tables fall beyond the reach of irrigation pumps – as has occurred in large stretches of deforested South America and is happening rapidly in Asia. Buyers can move their custom elsewhere, but farmers cannot generally move their farms.

If we take the amount of energy and land used to grow food that ends up being wasted, and propose instead that those resources be used to grow something useful, we would have a deeper sense of what we are depriving the world of by wasting so much. For the moment I shall restrict this just to the area of land used, categorized into different types of land, such as good horticultural land, arable land or land usable only for grazing or biofuel production; and the fuel used on farms pulling tractors around, manufacturing nitrogen fertilizer, heating and other farm inputs, which can be measured in joules or calories of 'primary energy input'.

Much of the food wasted in rich countries has what can be termed a low 'resource-to-calorie efficiency ratio'. Producing crops like tomatoes, dairy products and meat uses a lot of resources on average – land, fuel and water – when compared to the amount of calories yielded. If those resources were used instead to grow cereals, there would be a much higher yield of calories. For example, it takes an average of around 31 million kcal of primary energy input to grow a tonne of tomatoes with a calorific content of just 170,000 kcal. By contrast, it takes just 600,000 kcal of primary energy input to grow a tonne of bread-wheat which contains 3–3.5 million kcal, an energy input:output ratio 918 times higher.[7] In other words, the energy that goes into growing the 61,300 tonnes of perfectly good tomatoes that people throw into their household rubbish bins in the UK, is equal to the amount it takes to grow enough wheat to relieve the hunger of 105 million people.[8] This is not in itself a serious proposition for substituting one with the other; it is merely illustrative of how valuable the resources we waste are. Energy alone maketh not food – you also need land and water among other things, and tomatoes, of course, are not grown for their calorific content, but because of their gastronomic importance and their valuable micronutrients, all of which are essential for our survival, perhaps, and our happiness. But tomatoes that end up in the bin are not contributing to any of these things, and reducing waste would therefore liberate these resources for more useful purposes.

It is also fair to say that few tomato growers – with their top-quality horticultural soil, greenhouses and specialized equipment – will realistically consider growing wheat instead (though they could decide to grow other kinds of fruit and vegetable that might otherwise be produced elsewhere). This is, however, the kind of decision that a potato grower might make if, say, demand for potatoes dropped because everyone stopped wasting so many of them. Much of the land used to grow potatoes in the UK and elsewhere in the world is eminently suitable for growing cereals like wheat and these two crops are often grown in rotation on the same soil. In any given year, some farmers will decide whether to grow wheat or potatoes, and they will base this decision on their prediction of demand and profit. Indeed, in 2008, some farmers in the UK and elsewhere saw that the global demand for wheat was very strong, and decided to maximize their production and divert energy and attention to growing that crop. This wheat contributes to global supplies and will in 2009 probably help bring wheat prices down, making it more affordable for the impoverished of the world. I am not proposing that it is better in general to grow wheat rather than potatoes; neither am I speaking of growing cereals to donate as food aid; nor do I mean to suggest that growing extra wheat is the only other potential use for the land and resources. The land might alternatively be used to grow crops to improve self-sufficiency in other foods that are currently imported, or it might be used to grow biofuels, or potentially agriculture could be de-intensified to reduce dependence on fossil fuels and improve natural habitats. Growing wheat is just one of many ways in which land could be put to better use than growing food that no-one eats.

If household potato wastage not including waste in the industry was halved in the UK from the current level of 358,500 tonnes (only including entirely avoidable potato waste, not peels and so forth), this would liberate 5,400 hectares of good-quality arable land (calculated using parameters published by the British government). Were this to be diverted to wheat production, the potential yield would be 36,000 tonnes of wheat, or enough food to save 1.2 million of the world's malnourished people from hunger. Because wheat uses less energy and water to grow than potatoes, in addition to the extra food produced, there would also be a saving of 7.5 million tonnes of water, and reduced greenhouse gas emissions equivalent to 14,340 tonnes of carbon dioxide.[9] It is worth bearing this in mind next time you consider leaving potatoes on your plate or discarding half a bag of chips.

In addition to representing parcels of land, food can be seen in terms of the water needed to grow it. Water scarcity is one of the world's most pressing problems, and household water-saving measures are being encouraged in many countries. But wasting food represents a waste of water far greater than what we use in baths, toilets and washing machines. If 25 per cent of the world's food supplies are being unnecessarily wasted, this represents a loss of water withdrawn by farmers from rivers, lakes and wells (irrigation water only, not rain) amounting to approximately

675 trillion litres, or easily enough for the household needs of 9 billion people using 200 litres a day.[10]

This is not the end of the story, for the figures on farm inputs do not take into account energy and water used further down the supply chain, in manufacturing, refrigeration, transport, storage and cooking, which can amount to another half of all energy used in the production of food. According to some calculations, one calorie of food saved can result in a sevenfold reduction in the energy use across its lifecycle.[11]

According to WRAP, the carbon footprint of all the food wasted by households in the UK amounts to the equivalent of 18 million tonnes of carbon dioxide. This 'top-down' calculation is based on the premise that 20 per cent of all greenhouse gas emissions come from the production and preparation of food.[12] This being the case, food and drink in the UK would be responsible for about 130 million tonnes of carbon dioxide emissions, of which the proportion consumed in the home would account for 90 million tonnes.[13] If one third of food purchased is thrown away and 61 per cent of this could have been eaten, then food wasted unnecessarily in homes would account for the equivalent of 17 million tonnes of carbon dioxide emissions. In addition to the emissions from food production and preparation there are also those that come from food waste as it decomposes in landfill sites. Most of this decomposition happens in the absence of oxygen, and therefore occurs anaerobically, releasing methane which the British government estimates to be the equivalent of 2 million tonnes of carbon dioxide.[14] Again, only 61 per cent of this can be attributed to food unnecessarily wasted, so it is safe to say that the total emissions from food wasted is 18 million tonnes, equivalent to the emissions from 5.8 million cars or 21 per cent of cars on UK roads.

However, as Mark Barthel at WRAP admits, this figure is extremely cautious. Firstly, it does not include the category of 'possibly avoidable' waste like bread crusts. Secondly, there are good reasons to believe that food production and preparation are in fact responsible for a good deal more than the stipulated 20 per cent of all emissions. Studies calculating the total emissions associated with each kind of food are in their infancy. But an alternative method of calculating the carbon footprint of food has been employed by the authoritative Environmental Impact of Products (EIPRO) project, conducted for the European Union's twenty-five member states (before the most recent expansion to twenty-seven states). Figures from both the EIPRO project and WRAP's assessment included emissions from food produced overseas and imported, which is an important consideration since the UK imports (in net terms, by value) around 51 per cent of its food.[15] But the EU study also accounted for less-industrialized European countries where food production comprises a higher proportion of emissions overall because there are fewer other polluting industries. The study concludes that food production and preparation in the home contribute to 30 per cent of all emissions. If food eaten in restaurants and other food service outlets is included, eating is responsible for 41 per cent of emissions.[16] If wastage levels by consumers of around a quarter of all food is representative – as found by WRAP's study in the UK and by the Department of Agriculture in the US – this would suggest that 10 per cent of all greenhouse gas emissions in these countries come from producing, transporting, storing and preparing food that is never eaten.[17]

Even this figure may still be an underestimate because the EIPRO study did not account for emissions due to land use changes associated with food production, such as the destruction of the Amazon rainforest. This category of emissions is difficult to calculate and has been left out of most studies. One team at Aberdeen University did recently try to take these considerations into account and found that agriculture alone could be responsible for between 17 and 32 per cent of global anthropogenic emissions. To this would be added all the other emissions associated with food production and consumption, such as manufacturing, refrigeration and cooking – which in Britain at least are responsible for around another 10 per cent of emissions.[18] By current estimates, therefore, food consumption could be responsible for between 26 and 50 per cent of all manmade emissions.

Another landmark study included the impact of land use changes, and the results were astounding. In 2006 the UN FAO published its investigation into the environmental impacts of meat and dairy production, *Livestock's Long Shadow*. This showed that the global livestock sector alone was responsible for 18 per cent of all the world's anthropogenic emissions.[19] If animal agriculture is responsible for 18 per cent of emissions, then the total for food must necessarily be much higher. Meat and dairy foods produce a disproportionately high level of emissions considering their relatively small contribution to total calorie intake (less than 20 per cent globally[20]), but as a proportion of all emissions from food production animal-based food is responsible for lower emissions overall than is plant and vegetable food which is eaten in much larger quantities.[21]

The 'changes in land use' that the FAO's estimate took into account include cutting down rainforests in order to provide more land for cattle grazing and soy production, or, as often happens, for soy producers to appropriate existing farmland, pushing peasants further into the forest to slash and burn new areas. From the 1960s to 1997, about 200 million hectares of the world's tropical forest were lost, mainly through conversion to cropland and ranches.[22] In 2006 the World Bank estimated that deforestation alone may be responsible for 20 per cent of greenhouse gas emissions, and the Intergovernmental Panel on Climate Change suggests that land use changes could have caused up to a third of global warming during the 1990s.[23] Deforestation accounts for 75 per cent of emissions in Brazil, the fourth-largest emitter of greenhouse gases in the world – and most of this is for meat and soy production; the third largest emitter is Indonesia and this too is due to deforestation which is driven largely by demand for vegetable oil from palms.[24] Cutting trees in the

Amazon disrupts the local hydrological cycle and may now be dangerously close to drying out that entire tropical region to such an extent that the forest will die off and turn into semi-arid grassland, which would potentially release 55 billion tonnes of carbon dioxide into the atmosphere.[25] At a climate conference in March 2009 in advance of the Copenhagen UN summit, representatives from the Met Office's Hadley Centre revealed that a 4 °C rise in temperatures would kill off 85 per cent of the forest, and that even under the most optimistic climate change scenarios the die-back of large parts of the Amazon is already unavoidable and 'irreversible'. This will in turn create a positive feedback loop as forest loss will contribute still further to climate change.[26] Such eventualities are what make scientists worry that deforestation and climate change could reduce the food yields of many regions or even cause global crop failure.[27] In the long term, therefore, extending the agricultural frontier to grow more food may actually decrease the total amount of food we can grow on planet Earth.

In addition to carbon dioxide released into the atmosphere when agriculture encroaches into forests, there are emissions from soils when ancient savannah grasslands are ploughed over, when soil erodes or when wetlands are drained to grow more crops. Food consumption in America and Europe is directly linked to these processes. In just the seven years to 2004, the amount of meat Brazil exported increased by nearly 700 per cent. Between 1990 and 2001 the percentage of Europe's processed-meat imports that came from Brazil rose from 40 to 74 per cent.[28] Meanwhile, Europe's farm animals annually eat their way through 36 million tonnes of soybeans and soymeal imported almost entirely from South America, where increasing demand for soy and meat is the principal cause of deforestation.[29]

In 2006, Greenpeace achieved a landmark victory by getting some of the largest meat companies operating in Europe, including McDonald's, to sign a two-year moratorium on selling meat fattened on soymeal grown on recently deforested land.

In 2008 the moratorium was extended for a further year, and may be continued after that. However, deforestation (both legal and illegal) continues. Following increases in the price of soy, in the twelve months to July 2008, 1.2 million hectares of Amazon forest were destroyed, an increase of 4 per cent on the previous year (though less than the 2.7 million hectares in 2003–4, three quarters of which was illegal). The Brazilian government pledged to halve the illegal deforestation over the coming decade. However, the environment minister, Marina Silva, resigned in 2008 after claiming that she was no longer able to implement federal laws on forest protection, and there are continuing incidents of armed insurrection by loggers and their allies against enforcement officers.[30]

Most assessments of the environmental impacts of meat and soy consumption do not include these land use changes.[31] If these emissions were calculated, the carbon footprint of food consumption would go up substantially. In Britain, farm emissions constitute about half of all emissions associated with food consumption.[32] One study of dairy consumption in the Netherlands found that when the impact of land use change was included, on-farm emissions doubled.[33] But for products like soymeal, taking deforestation into account can multiply emissions by between fifteen and 300 times.[34]

Land use change is by far the single biggest impact humans have had on nature, and the Amazon is just the most biodiverse of the many frontiers where agriculture is being extended into valuable ecosystems.[35] It has been estimated that we have already reduced the world's 'natural capital' by two thirds. In 2003 land use changes were estimated to have been responsible for more than 70 per cent of this, in contrast to climate change at just 12 per cent (though this is set to rise in the future).[36] Over 80 per cent of all endangered birds and mammals are threatened by unsustainable land use and agricultural expansion.[37]

Even if individual consumers or whole nations decided to stop buying meat fattened on soy from recently deforested land, should they continue to consume as much meat as they do, their consumption would contribute indirectly to deforestation – because while they shifted demand to 'sustainable' sources, other consumers or nations without these scruples would buy the unsustainable produce.[38] Forests have to be protected either by internationally enforced conservation measures or by making it more profitable for people to keep their forests intact. In the meantime, it is essential to reduce the demand for resource-intensive foods such as meat, and one of the ways of achieving this is to stop wasting so much. It takes 8.3 million hectares of agricultural land to produce just the meat and dairy products *wasted* in UK households and by consumers, retailers and food services in the US.[39] That is seven times the amount of land deforested in Brazil in the past year. Demand for food contributes to the financial incentive to extend agriculture into forests. If we stopped wasting so much it is more likely that this incentive would shrink. It is true that other incentives for deforestation and other demands for food, which I address in later chapters, can and do play a part, but they all contribute to a net effect and curtailing any of them reduces pressure overall. By wasting food, we are funding the extension of agriculture into forests, wetlands and natural grasslands. By reducing waste we have the power to relieve this. We would not willingly tear endangered species of fish, butterflies, birds, trees, flowers, primates – or even un-known microbes, spiders and insects – out of their natural habitat and dump them uselessly in the bin or vaporize them into greenhouse gases. And yet that is exactly what we are doing.

PART II
Squandered Harvests

7. Farming: Potatoes Have Eyes

Neither shalt thou gather the gleanings of thy harvest; . . .
thou shalt leave them for the poor.[1]

Leviticus 19:9–10

As summer turns to autumn, potato farmers across Europe –
from Poland to France, Germany to Scotland – harvest the
fruits of their labour. The work began months earlier when they
drew their tractors through this land, ploughing the soil, soaking
it with fertilizer, and sowing potatoes deep into the ground.
Once the plants began to grow, they returned to their fields,
scattering pest-killing poison and anti-blight spray. Finally,
several months later, the crop is ready to eat. But most potatoes
grown in Europe and America are destined for conglomerates
that supply supermarkets and exporters, and thus, before they
reach consumers, they must pass a ruthless shape and size test.

Despite the energy and money they have put into growing
it, potato farmers are forced to discard a portion of their crop.
Harvesters leave behind them heaps of reject potatoes: outsized,
double-lobed, with eyes that seem to wink acknowledgement
of the variety of nature that even modern agriculture has failed
to quell. Some of these potatoes will be fed to pigs and other
livestock, but this – as all farmers know – is an inefficient way
of using food originally grown for humans.

There used to be a tradition of gleaning in many parts of the
world. It was sanctioned by several biblical instructions, such as
that in Leviticus: 'And when ye reap the harvest of your land,
thou shalt not . . . gather the gleanings of thy harvest; thou shalt

leave them for the poor and stranger.' This held for fruit left on trees and for grain that was spilled on the ground; farmers were obliged to let the poor take what was left in the fields after harvest. Classical rabbinical commentaries specified that the regulations should only apply to cornfields, orchards and vineyards, and not to vegetable gardens, and they argued that it only referred to poor Jews in Canaan, though in practice it continued elsewhere and non-Jews were permitted to join in. Today, in Israel, there are some Jewish communities still living out the ideal of *tzedakah* – charity, or justice – instituted in the injunction on landowners to allow the poor to glean, by recovering and redistributing surplus sandwiches for poor children.[2]

In Europe, it was part of the agricultural cycle that the poor, and then livestock, would be allowed to collect what was left in fields after the harvest. In revolutionary France, the ancient practice was given explicit protection and the Constituent Assembly hailed gleaning as 'the patrimony of the poor'.[3] At the beginning of the nineteenth century one legal commentator called these primitive rights 'a possession of many centuries', and Napoleon Bonaparte's Penal Code implicitly upheld them by specifying that it was a punishable offence to 'glean, forage or gather* in fields not yet entirely harvested and emptied of their crops' or to glean at night time.[4] If it was illegal to do so in these circumstances, there was a presumption, at least, that after the harvest was complete gleaners could forage during the day. The romantic appeal of this rural thrift was celebrated in paintings such as Jean-François Millet's *Des glaneuses* (1857) and Jules Breton's *Le Rappel des glaneuses* (1859). It would appear that the restrictive clause on gleaning remained in the French Penal Code right up until the most recent revisions at the end of the twentieth century – though its excision does not necessarily

---

* *Glaner*, *râteler* or *grappiller* – the three terms referred to different types of gleaning, from combing up fallen grains to plucking remaining fruit from trees and vines.

suggest that gleaning is no longer legal.[5] The same clause – originating in French law – still appears in the Penal Code of Belgium, the Ivory Coast, in Luxembourg (where the fine for breaking it is €25–250) and in Algeria, where the penalty is 30–100 dinar or three days' or more imprisonment.[6]

Such restrictions were part of the rural traditions in Britain also. Technically gleaners could be prosecuted for trespass on enclosed land, but cases were rarely brought to the court and it was so common for gleaners to be given tacit or overt permission that they generally regarded it as their right. Magistrates' records show that gleaning without permission was treated as a minor misdemeanour, and fines for stealing crops were rarely issued. As in France, gleaners were not allowed to enter fields until the harvest was completed, and some farmers left prominent sheaves of wheat as 'guard sheaves' to mark out the fields that were not yet open for gleaning.[7] In the nineteenth century, many parishes instituted a gleaners' bell which rang in the morning and evening to mark the permitted beginning and end of the gleaning day, thus ensuring that all gleaners had a fare share of the pickings. Nevertheless, as more and more of the land was enclosed under Enclosure Acts, farmers began to keep gleaners out of their fields, and by the beginning of the twentieth century the gleaners' bell had fallen silent in all but the most remote villages.[8] In the Soviet Union, the 'Law of Spikelets' apparently criminalized gleaning, under penalty of death or twenty years of forced labour in exceptional circumstances.[9]

Gleaning survives in some parts of the world, as the director Agnes Varda demonstrated in her acclaimed documentary about people who scavenge for food in France, *Les Glaneurs et la glaneuse* (2000). In 1999, when I helped harvest hay on a remote French cattle farm, the grandparents and children from the village followed the tractors raking up any stray grass the bailing machines had missed. For years afterwards, when I worked on farms back in Britain, I found it difficult to ignore the great wads of grass that are left to rot when tractors work alone. Most people in the industrialized West are now disconnected from the land, and few even among the poor glean in the fields. Those that do glean have often found that the bins of supermarkets yield quicker results. And yet, millions of tonnes of good-quality fresh produce are discarded annually in fields all over the Western world.

One way or another, supermarket standards in the West force some farmers to lose up to a third of their harvest every year. The UK's Soil Association once estimated that between 25 and 40 per cent of most British-grown fruit and vegetable crops are rejected by the supermarkets. Another study by the waste company Biffa estimated that a third to a half of British fruit and vegetables grown for supermarkets are rejected, largely because of the tight specifications regarding size, blemishes and appearance.[10] Forests have been burnt, hedges flattened, wetlands drained and ponds filled in – sometimes only to produce crops that end up being ploughed back into the earth. Modern farming has improved the total yield per acre. But the same industry has introduced standards that exacerbate waste, standards which have nothing to do with taste or nutrition. They are based on a subjective aesthetic – a hybrid monster generated in part by supermarkets, reinforced by fastidious (or impressionable) consumers, and now stamped with the authority of international trading standards.

This preposterous situation could be avoided without turning back the clock. Some farmers in western Europe and America have found markets for 'out-graded' crops and now sell much of them to food processors or catering trades. In the US, charitable organizations like Feeding America specialize in collecting crops that have become uneconomical for the farmer to harvest either because they do not meet the supermarket cosmetic standards or because they are surplus to requirement (usually these are overlapping categories, as I explain below). Waitrose, the British supermarket chain, recently launched a new line in 'knobbly' fruit, and farmers in the US are learning techniques to use the market to reduce their losses from abandoned crops. Farming is the most direct footprint humans leave on the land; if we are to tread more softly, we must begin here.

The exact quantity of waste from farms is the biggest unknown of all waste statistics. Researchers have had a stab at calculating total wastage in manufacturing, retail and catering sectors and at the domestic level. But agricultural waste at the first step in the food chain has been difficult to quantify. One reason for this is that waste separated on the farm is often not legally classified as 'waste' at all: because farmers can plough it back into their fields it does not have to be processed and treated in the same way as industrial waste. Nevertheless, when farmers are obliged to destroy crops they have grown, this is still a huge squandering of food, land, water, agri-chemicals and fuel.

Although I knew that the problem was acute, it was not until I started visiting farms that I realized how destructive it was. On a crisp December day in 2007, I traipsed through the mud and into the packing yard of M. H. Poskitt Carrots, a major supplier to the supermarket chain Asda.[11] The affable and down-to-earth Guy Poskitt came out to meet me, muttering that he felt 'honoured' that I had taken the train all the way to Yorkshire just to see him and his carrot farm. There in the yard, a conveyor belt was spitting out bright orange, washed and topped carrots into a large wooden container. They gleamed in the winter sunlight – gems extracted from the brown earth that surrounded us. We walked over and I selected one from the heap, crunching my teeth into it so that its pleasant and mild juice filled my mouth. 'These must be your high-end supermarket specimens?' I asked him. He looked me in the eye. 'No,' he said, 'those are the out-grades: they're going for animal feed.'

I tried to look unfazed, and picking up a handful of carrots, asked why they had been cast out of the human food chain. I could see no blemishes, they were a good standard size and they tasted wonderful. 'They have a slight bend in them: they're not perfectly straight,' came Guy's dead-pan reply. I held the carrots up to the light and squinted to see if I could detect their kinks. The carrots were straighter than any I had ever grown, and at least as good. I checked to see if he was joking and turned to a man nearby who was raking the pile down flat. 'Asda insist that all carrots should be straight, so customers can peel the full length in one easy stroke,' he explained. Any knobbliness or bend can get in the way of the peeler.

Guy took me into his packing house, where gargantuan machinery whisked columns of carrots up and around two storeys of the building. The first port of call for carrots, once they have been washed, is to pass through a pair of £400,000 photographic sensor machines which work with almost satanic speed and efficiency. The carrots fly in at one end on a high-speed conveyor belt and are hurled out the other, over a precipice and onto another belt. But in between, a camera searches for defects: to see if the carrot is bright orange enough, or if it has a bend or blemish, or is broken. Any specimen the camera spots which fails to match its pre-programmed ideal of carrotness is marked down as condemned, a jet of air is fired at it with infernal precision, and the misfit is blasted down into a chasm below, where it is swept off into the livestock feed container. Once the successful carrots have passed through the machine their ordeal is not over. They still have to travel along conveyor belts past serried rows of people grading the carrots into different streams, picking off whatever does not belong, and finally packing them into plastic bags to go straight off to the supermarket. A representative from Asda, dressed in white, stood at the final gate, selecting bags at random and scrutinizing them for compliance with their quality codes. She eyed me suspiciously as I made the rounds with Guy Poskitt by my side.

In total, 25–30 per cent of all the carrots Guy deals with are out-graded. He hands me a specification sheet for a typical load they just despatched: 29 per cent of the carrots were picked off. About half of these were rejected for physical or aesthetic 'defects' such as being the wrong shape or size, being broken or having a cleft – nothing to do with the eating quality of the carrot itself. The other half were out-graded for having blemishes due to cavity spot or carrot root fly, which is a more justifiable reason for out-grading than size or shape, though even most of these could easily be eaten, as indeed they are by many growers and consumers all over the world, after cutting out the small affected part.[12]

Guy Poskitt is disarmingly open about the waste he generates as a farmer, and I soon learn why. It is extremely frustrating, he tells me, to grow all these carrots, to go to the expense of pulling them out of the ground, topping and washing them, only to see a third of them rejected. He sells many of the over-sized carrots to food processors – but only the over-sized ones because these require less labour to handle. Secondary sales absorb about a third of all out-grades. Those which do not go to processors end up in the livestock bin. Sending carrots for livestock feed is better than sending them to landfill, but it is little more than a cost-effective disposal method. As far as Guy's bank account is concerned, giving them to livestock only recuperates 10 per cent of their value. The current system institutionalizes regular surfeits of quality washed carrots which could be put to better use.

Guy is adamant that a lot of the carrots he is currently forced to discard are perfectly good to eat and believes that there are plenty of consumers who would love to be able to buy them at a cheaper price. He even claims that Asda want to be able to provide customers with these cheaper carrots. But Asda have told him they are prevented from doing so by the quality controls imposed by the government's agricultural ministry, Defra. 'If it weren't for these stupid rules,' he says, 'we wouldn't have to waste so much.' He is referring to the British government's application of what are essentially European uniformity rules on fruit and vegetables.

These are the infamous standards bureaucrats in Brussels have devised to ensure that farm produce looks the same across the whole European Union, and standards like these also are imposed on imported produce by Japan and the US.[13] One of the worst aspects of the emphasis on cosmetic appearance is that crop technologists and growers are encouraged to concentrate on breeding uniform produce – rather than the best-tasting or most nutritious. It also means that farmers have to use more pesticides and fungicides than they would otherwise to prevent superficial blemishes from insects and other pests.[14]

In Europe, the strictest rules aim to classify produce into different classes – Extra, Class I and Class II – which retailers are obliged to adopt. But there are bottom lines below all classes, beyond which the EU for the last twenty years has not allowed produce to be sold or even given away to charities, except for direct sales on farms. For example, Defra stipulates that carrots of less than one centimetre diameter cannot be sold; bunches of carrots should all be of a uniform shape and size; and that it is illegal to sell a carrot that is forked or has a secondary branch – a naturally occurring feature.[15] Similar strictures cover numerous fruits and vegetables in the EU – from bananas and cucumbers to potatoes. Since the introduction of the laws, apples under 50 mm in diameter or 70 g in weight have been banned from the shops.[16] In the summer of 2008, one British wholesaler was forced to throw away 5,000 kiwi fruits for being 4 g lighter than the required EU standard 62 g cut-off, the equivalent of being 1 mm too thin. Tim Down, the kiwis' owner, could have been fined several thousand pounds if he even gave away his fruit. 'They are perfectly fit to eat,' he insisted. 'These regulations come at a time when rising food prices are highlighted and we're being forced to throw away perfectly good food.'[17]

Under this sort of pressure, in 2008, the European Commission finally announced a U-turn on cosmetic standards and from July 2009 these stipulations will be relaxed. These 'unnecessary rules', said the European Commission spokesman Michael Mann, have 'been rather silly . . . People [are] saying that prices are too high, [so] it makes no sense to be chucking food away.'[18]

However, the reforms may turn out to be less sweeping than has been imagined. The repeal only applies to twenty-six types of fruit and vegetables,[19] whereas the standards will continue to be imposed on ten major crop categories which account for 75 per cent of the value of EU trade. Apples, all citrus fruit, kiwi fruit, lettuces, peaches and nectarines, pears, strawberries, sweet peppers, table grapes and tomatoes – all these will continue to be subject to EC cosmetic standards. The glimmer of hope is that EU member states will now be permitted to exempt any of these products from the standards as long as they are sold in the shops with an appropriate label such as 'product intended for processing' or equivalent wording.[20] But member states still have to be convinced to exert this right, which may not happen given that the majority of states and some industry bodies opposed the EC's decision to relax the rules.[21] Then, even more trickily, supermarkets have to be convinced to stock the cheaper, previously out-graded produce. Even now supermarkets currently stock very little Class II fruit, so they are even less likely to stock significant quantities of out-graded produce. There is still a case for going further, and abolishing both national and international rules governing cosmetic standards altogether. If it makes sense to do this for a quarter of fruits and vegetables, why does it not make sense to do it for the remaining three quarters? As one potato grower who supplies Tesco put it to me, 'If there is a willing buyer and a willing seller, why would governments want to prevent the sale of safe and edible produce?'[22]

Although Guy Poskitt blames these official specifications for much of his farm waste, when I spoke to Chris Brown, Asda's head of ethical and sustainable sourcing, he categorically denied that these rules had any impact on their choice of carrots because Asda's quality specifications are still more demanding. This is despite the fact that Asda caters for some of the 'lower and medium demographies' and thus has more relaxed (and thus relatively less wasteful) specifications than other supermarkets. Each supermarket has its own criteria, and upmarket chains like

Waitrose and Marks & Spencer in Britain have even stricter cosmetic standards than most. For nearly all products, super-market requirements are actually more stringent than those of the EC. One EC insider who did not want to be named pointed out that Sainsbury's recent attempt to blame the EU for the waste of food through cosmetic criteria has been 'just a cheap publicity stunt. Blaming the EU is a very popular sport.'[23] As Michael Mann, the EC spokesman explained in an email, 'Pri-vate standards are stricter than public standards and are the major source of out-grading.'[24]

So why do Asda and the other supermarkets not create a cheaper range that includes some of these perfectly good carrots that Guy Poskitt currently discards? Chris Brown says that Asda aims to provide high-quality food affordable for everyone, but that the fundamental problem with misshapen produce is that 'customers won't buy them'.[25] This is the line parroted by super-markets all over the world to justify their arbitrary aesthetic standards. We are back to the chicken-and-egg conundrum of whether these superficial attitudes originate with consumers or with the retailers. There is some evidence, which supermarkets bear daily witness to, that if customers see a bag of wonky produce next to a similarly priced bag of glistening, uniform produce, they will tend to choose the latter. But what if you mix the two bags together, or sell the wonky fruit and vegetables at a cheaper price? One of the probable reasons why super-markets refuse to stock wonky, cheaper produce that would otherwise have been out-graded is because they make greater profits by selling more expensive 'high-class' ranges. If they stocked more of the cheaper produce, people would buy less high-end stuff. In other words, it is as much to do with super-markets satisfying their directors and shareholders as their customers.

But cosmetic standards are not the only way the supermarkets encourage farmers to waste large proportions of their crops. The first stage in waste happens even before the grading house. Guy

Poskitt tells me that his biggest problem is 'the tremendous pressure to meet contracts'. He is keen to emphasize how glad he is that working for the major supermarkets has transformed the carrot area from a 35-acre plot in his father's day to one that now handles 70,000 tonnes of carrots a year. He does not mind if I criticize the government, but does not want me to 'kick' Asda. As a major grower and packer, he has definitely been one of the winners from the rise of the supermarket cash cow, but it is, as he puts it, 'a tough game'.

Each year Guy will agree with Asda how many carrots he can supply. Most he will grow on his own farm; the rest he will buy in from all over Britain, particularly in the summer when his harvest is dwindling. But each growing season is unpredictable: there may be poor weather, or a particularly severe outbreak of cavity spot disease. On no account must he fail to meet his contract. The supermarkets are uncompromising in this respect. There are innumerable other growers they can turn to if he starts to falter, so he has to maintain supply no matter what. He has never failed to achieve his target, but, as he told me, he has 'come fucking close'. At one 'horrific' period he had to buy a thousand trays of parsnips at £18 each and sell them on to the supermarket for a loss, at £12.50. In other words, he took a hit of £5,500 to avoid disappointing his customer. If he failed to deliver the tonnage he had guaranteed, he would have been penalized and could have lost his contract altogether – and with it his livelihood.

To avert this disaster, Guy now plants far more carrots than he is likely to need. Indeed, he aims to grow 25 per cent more than he is contracted to supply just in case there is a bad growing year and he needs that extra buffer. Over-producing to avoid under-supplying supermarkets is absolutely standard practice in the agricultural sector for crops like carrots, sugar beet, brassicas, lettuces and so forth; as one NFU official told me, planting 140 per cent of actual demand is 'not an unstandard example of the industry being inefficient to avoid shortfall'.[26] In a good growing year, when there are no unusual rates of disease or adverse weather conditions, Guy Poskitt will harvest some of the surplus and sell to lower-end purchasers like wholesalers or processors. He could harvest the rest for other customers but – at £10 per tonne for animal feed or £35 per tonne for freezing and processing as opposed to the £250 or so per tonne he gets from retailers – it is not financially worth his while. The land, fuel, water and chemical sprays have already been used to grow the crop, but because of the rigidity of the retail market, it is not worth pulling out of the ground. So the carrots just get turned straight back into earth. In any year, there is no predicting how much that will amount to, but a safe estimate which Guy agreed with would be an average of 10–12.5 per cent of his crop each year.

In combination with the 25–30 per cent out-grades, that means that, minus the oversized ones that get back into the food chain through food processors, on average, 28 per cent of Guy's carrots fail to make it into the human food chain and are either ploughed back into the land or sent for animal feed. Of those that do make it, a further proportion are wasted by manufacturers, retailers and consumers. To give a very safe estimate, let us say another 5 per cent are wasted between the farm and the consumer (though processors of some vegetables have been known to waste up to 50 per cent of the produce they receive).[27] If consumer waste is added to this, it seems that of all the mass of carrots grown in the UK, from farm to fork, 58 per cent ends up as waste, leaving only 42 per cent to be consumed by people. For every carrot you eat, you have paid for at least one more to be thrown away.

Many other crops are wasted at similar rates. Fresh salad is even worse: packers, processors and retailers reportedly throw away about 40 per cent of salad in the UK, so once householders have discarded another 45 per cent of what they buy, for every serving of salad eaten two more have been thrown away.[28] A survey of British strawberry growers in 2006 found that 5–30 per cent of the crop, or an average of 16 per cent, were Class II and nearly all of these were left in the field to rot despite being good to eat.[29] A survey of apple growers in the UK by Friends of the Earth in 2002 found that 11 per cent of growers said that half or more of their crop was rejected, 3 per cent said that their entire crop was rejected, while just one third of growers met the supermarket standards with 80 per cent or more of their crop. Nearly two thirds said that some fruit was not even harvested and was simply left or dumped. Reasons for apples being rejected included minor skin blemishes such as natural russeting, apples being either too red or not red enough, being the wrong shape, too big or too small.[30] Remarkably little research has been done to assess exactly how much of each different crop farmers are forced to waste because of these cosmetic criteria. The task of finding out is made more difficult because the farmers fear the consequences if they speak out. As one farmer said, if their comments were 'leaked to supermarkets I would be delisted and so forced out of business'.[31] Without more research, calculating the waste of resources resulting from this phenomenon across the world is impossible, but suffice it to say, the system has plenty of slack.

Vegetable growing does not need to be so inefficient. In the United States, there are large carrot supply conglomerates that find markets for nearly everything they grow. Anything super-markets do not want goes to other less fancy shops; wonky carrots are chiselled down to make 'baby carrots'; anything that is unsuitable for that is turned into juice or condensed to make natural sweetener to be added to other foods. Once all these options have been exhausted, the tiny percentage of remaining carrots are fed to dairy cows, which produce orange-coloured milk. This in turn is used to make cheese or milk solids for the food industry – or is marketed straight to children at Halloween.[32]

In the UK, growers' associations have pointed out that greater communication between farmers across the country could sig-nificantly reduce the need to over-produce in order to meet contracts.[33] Furthermore, increasing the proportion of a farmer's crop that gets into the supermarket by just 5 per cent can increase the farmer's profit margins by up to 60 per cent.[34] Still more dramatic savings can be achieved by bypassing the supermarkets altogether. One study showed that organic fruit and vegetables supplied to supermarkets are out-graded at a rate of 30–50 per cent of the total harvest. Selling the produce directly to customers, through box schemes, farmers' markets and farm shops, cuts the out-grade levels down to just 5–10 per cent.[35] Not only does buying direct from farmers eliminate the unnecessary waste instigated by the supermarkets, but it can also reduce food miles.

Even with the supermarket-led system, an interesting example of what could potentially become standard practice occurred in the freak summer of 2007 with the British potato harvest. Heavy rain caused floods which washed away the soil, leaving potatoes to go green in the sun, while others rotted in the damp. Some crops split or cracked or had high moisture content, which does not make them inedible but unsaleable nevertheless.[36] Around 40 per cent of the nation's potato harvest was destroyed, and in some areas entire crops were wiped out.[37]

To make up for these catastrophic losses, one would expect retailers to have been forced to import potatoes from overseas. And imports did go up by a few tens of thousands of tonnes, but nothing like enough to make up for the millions of tonnes that were lost. Yet retailers were still able to source almost all their potatoes from Britain. Where did all these extra potatoes come from?

I was finally given the answer by Robert Baird, the procurement manager of Greenvale AP, one of the largest potato packers in the country. The supermarkets, he told me, simply relaxed their stringent cosmetic standards. Growers negotiated for supermarkets to accept 'under-sized' potatoes and the personnel responsible for receiving deliveries unofficially relaxed their tests. They let in potatoes that would normally have been rejected for being too big, too wonky, having eyes, or not being perfectly smooth and rounded: in other words, potatoes that looked like potatoes rather than billiard balls.[38]

But surely, if the supermarket mantra were true – that customers refuse to buy misshapen produce – there must have been revolt in the aisles. Weren't customer help desks inundated with complaints? I rang round the supermarkets and their suppliers to find out if this had been the case. None of them reported a spike in customer dissatisfaction.[39] It turns out that if you sell customers produce that is not homogenous, they don't mind. If it tastes good, *they hardly even notice*.[40]

As others in the farming sector have explained to me, the supermarkets often use their quality criteria merely as a pretext for rejecting deliveries when they are not managing to sell as much as they expected. If there is strong demand, they will take a delivery; if they are not selling so much or so many, they will reject it. By claiming that it is for quality control reasons, the supermarket avoids officially reneging on their purchasing agreement. It is just another way in which they maintain complete flexibility in what they choose to buy, leaving growers at their mercy.

To be fair, the potato industries of western Europe and America have become quite good at finding a market for at least some of the potatoes that the supermarkets reject. A lot still get wasted, but a proportion are sold on to processors or caterers who are happy to turn them into chips or mash regardless of what they look like. But in other countries, especially Poland and elsewhere in eastern Europe, this is still not the case, and there heaps of rotten potatoes are a common sight.

Even in the best-run industries, however, growers cannot always find enough people to buy their rejects. Without taking into account crops left in the field unharvested, British farms and processors produce just under 1 million tonnes of potato waste each year – or one sixth of the national supply of finished spuds.[41] The cost to farmers of producing that many potatoes is £54.5–63.6m each year.[42] Meanwhile, annual wastage in households comes to 358,500 tonnes of finished potatoes, plus 463,000 tonnes of 'possibly avoidable' potato waste like peelings, at a cost to consumers of £660m a year. The waste of potatoes in Britain (not including retail or catering waste) comes to 1.7 million tonnes, or the equivalent of approximately 30 per cent of all potatoes sold.[43]

Some of the potatoes wasted on the farm are sent to landfill or are left in the field; tiny but increasing amounts are used for anaerobic digestion to make biogas; while others are fed to livestock. But at present there is no record of how much ends up in these different channels.[44] I visited one producer in Kent who supplies Tesco. Not including any spuds that were out-graded in the field, they pick off approximately 30 per cent of the harvest for the supermarket but sell most of these 'pick-offs' to caterers in London. However, at least 8 per cent of the harvest is still consigned to livestock feed or landfill. They let me have a look at their pig potatoes (I took a photo); they had been in the sun for a while and a few of them had gone a little green, but most of them seemed to need no more than a scoop with the end of a peeler to remove holes or eyes (that is why peelers used to have pointed ends – to gouge out the eyes).

If these quality controls sound severe, they are an easy ride compared to those of the high-end supermarkets like Marks & Spencer – which, for example, will not usually sell potatoes smaller than 40–45 mm across. Phil Britton of Manor Farms, who supplies them, said that at the first stage farmers would expect to pick off 21 per cent of a potato crop in the field, often just to plough them back into the ground. A further 40 per cent would then be rejected by the supermarket, most of which would be sold to processing firms, and the rest (around 5 per cent) would go for stock feed. 'I don't know,' said Phil Britton when I asked him where this superficial strictness originated from, 'it's just there, passed down from father to son as it were.'[45] So when you see potatoes on sale in a posh shop like Marks &

Spencer, remember that for every one on the shelves, another has been rejected.[46]

Out-grades also cause growers anxiety because of the fear that the buyers might be market traders masquerading as livestock farmers; instead of buying the reject spuds cheap for animal feed, maybe they will sell them on to the numerous humans happy to eat them – evidence that there are plenty of people who are very happy to buy cheap potatoes and do not mind in the least having to pick out the odd knob or crevice.[47] In Germany, many potatoes that in Britain would have been dumped or sold on as animal feed are bought at a discount by shoppers.[48] But the supermarket-led system in Britain and many other countries has been unwilling to allow this, because customers would then buy fewer top-grade potatoes, which is where the highest profits are made. Farmers are thereby made to discard their valuable crops, consumers on a budget are denied access to cheaper food, and the environment is needlessly exploited.

Another fear of those in the potato trade is that dropping standards could damage demand. Because pasta and bread (and to some extent rice) – the other main Western sources of starch – are processed foods, they can be made to look uniform and perfect and easy to use in a way that raw spuds cannot. Potato people worry that if visual uniformity slips, customers will buy those other starches instead. But supermarkets wield huge marketing power; it would not be beyond them to encourage customers to buy all types of potatoes, including the cheaper ranges with the 'knobbly ones left in'.

The inflexible demands of supermarkets are out of sync with the unpredictable nature of food production. Natural variation affects the growth of the plants we eat. In order to become sustainable, food supply has to adapt to these ecological realities. It is the retailers' reluctance to bend which results in the madness of calculated over-production and aesthetic out-grades. If it really were the case that resources such as land, fuel and water were infinitely available, then forcing farmers to over-produce and discard part of their crops would not be a problem. This situation is going to change: indeed, some of the big retailers have already started making moves in the right direction. It is up to us – their customers – to encourage them on this path; to give them the confidence to abolish absurd cosmetic standards that have nothing to do with taste or nutrition. We are not going to use our carrots as rulers or our potatoes to play snooker, so a slight bend or knob is of no consequence.

It does not take a radical environmentalist to find current levels of waste deplorable. Nick Twell, recently retired member of the British Potato Council committee and an important figure in a potato company with an annual turnover of £260m, voices a view increasingly held in the industry: 'The input into growing the crop is phenomenal, everyone is looking over his shoulder about carbon footprints and reduced food miles, so why throw out a large proportion of what you've grown, which is perfectly edible, because someone doesn't think it looks right?' Twell is unambiguous that it was 'the supermarkets' fault' because they, in his Orwellian phrase, 'educated the public to expect perfect potatoes'.[49] Or, as Philip Hudson, chief horticultural advisor at the NFU put it, supermarkets claim they are just responding to customer demand, but 'there's a degree of manipulation by the retailers.'[50]

In a frank admission of the industry's culpability, Twell insists that eating out-graded potatoes not only makes economic sense, but has also become a moral obligation. 'Populations are still rising, and we have got to feed people. But you worry what the future generation has in store. We can't keep using resources as we are, and controlling waste is one of the ways to deal with that.' He emphasizes that there are global dimensions to the problem: 'I think if you saw some of the potatoes going out you'd think it was absolutely horrendous, and if you showed them to people from poor African countries they would wonder what we were up to, throwing away food of that quality.'[51]

Not all crops are judged primarily on their appearance. Birds Eye, for example, tests peas to check they are sweet enough. Richard Hirst, a farmer and Chairman of the NFU Horticulture Board, is one of its suppliers. To maintain sweetness, he says, a pea harvest has to be delivered for freezing within 150 minutes of the first pea being picked. If they fail the sweetness test – for example, if the weather has been too dry – Birds Eye can reject the entire crop. But the Birds Eye contract stipulates that suppliers cannot sell those peas to anyone else, even if there are willing buyers for lower-grade peas. Doing so, as Hirst points out, would be bad for Birds Eye's business because it would be under-selling the brand; and this in turn would be bad for him because Birds Eye is his main customer. Instead, Hirst and other farmers like him use the rejected peas for seed or animal feed and sometimes, if weather has been adverse, they have to resort to ploughing their crops back into the ground. Although this does involve excluding peas from the market that could be eaten by people, in Hirst's view it is unavoidable and nothing like the waste caused by fickle cosmetic standards imposed on fresh produce.

In 2008, Hirst left an entire crop of fresh spinach, representing an investment of £25,000–30,000, to wither in the field just because a sample contained a few harmless blades of grass which had grown up amongst the spinach; and this was mainly because he has stopped using some herbicides that have been banned in the EU (though they are still used on crops imported for consumption in the EU). A generation ago, Hirst reckons, people would not have been troubled by a little grass in a bag of spinach. Now, one of the biggest causes of waste, he thinks, is 'people's perception of what's edible'. Hirst also supplies lettuces to Tesco, which will only sell them for three to five days after they have been picked, even though they are likely to be fresh for much longer. Despite having secondary markets for Tesco's rejects, Hirst wastes 15–20 per cent of his lettuces because they are either the wrong colour, too small, not perfectly round or slightly damaged, or because they have a bit of earth on them. If just 10 per cent of a pallet-load of around 700 lettuces fail to meet the standard, the whole lot is rejected.

Wheat, meanwhile, is tested neither on taste nor on appearance primarily, but on gluten content and a measure of protein content called the Hagberg levels, which ensure that the grain is ideal for bread-making, or alternatively, depending on the variety grown, for biscuits. Anything that drops below these stringent tests by flour mills ends up being either exported to less fussy nations who make unleavened bread, or, more usually, turned into livestock feed.[52] The inefficiency of taking cereals that could be eaten by people and diverting them instead to fatten intensively raised animals is such a drain on resources, and so widely practised to satisfy the inordinate Western demand for meat, that to deal with the issue would require the space of another whole book. Certainly for now it would be worth considering the benefits of keeping a greater proportion of cereal production in the human food chain and reducing the amount we feed to animals. As far as consumers go, that means eating less meat and dairy, and ensuring that the meat we do buy has been fed on genuine by-products and pastures that cannot be put to better use – rather than wastefully stuffed with cereals and pulses that require huge inputs to produce.

It might be assumed that fruit and vegetables would only be out-graded and wasted in affluent countries where food and other resources are abundant. But the power of the super-markets' uniformity requirements extends across the globe to growers of exotic fruit and vegetables in developing countries. In Latin America, the West Indies and Africa, heaps of bananas are left to ferment at the harbour or are tipped into rivers, where they cause eutrophication and death to aquatic life. The estimate of the proportion of bananas wasted because they are too straight or bent, too small or too large, varies from 20 to 40 per cent of total crops, or around 20 million tonnes. This is despite the fact that many of them would be perfectly good to eat. At the very least, some of the out-graded bananas could be processed to form starches as an ingredient for other foods.[53] When I first moved to Delhi, I was delighted and amazed by the bananas on sale there – small and sweet, with several different types unknown in the markets of Europe. It is bizarre that the size, shape and taste of food should have been dictated to us by what is most profitable for supermarkets.

Even odder, in that respect, is the position of Western consumers with regard to produce grown in their own native place. Supermarkets, with their homogenizing notions of what the perfect apple or pear should look like, are excellent at bamboozling their customers into paying for insipid unripe versions of the rich variety of local fruits that every year rot on the tree from neglect. Every autumn shoppers in England walk past parks and gardens where apples are dropping off the trees, and into supermarkets to buy apples packed in plastic bags (some of which are imported from as far away as New Zealand). Our forebears spent a good deal of time nurturing strains and planting fruit trees, which centuries later still yield an abundance of fresh fruit. The sight of windfalls rotting on the ground has become so common that we do not even notice it today. In other European countries, there are companies that tour villages and towns, offering a fruit-juicing service; not everyone is quite so out of touch with autumn harvests as the British.

We are so used to the countryside being privately owned, it is easy to forget that, until three or four hundred years ago, much of the agricultural land in Europe was held in common. Then the rich and powerful began enclosing it, arguing that they would improve agricultural yields, which in turn would benefit everyone. But, as Locke argued in the seventeenth century, if they let their harvests go to waste, they should lose those rights: 'if either the grass of his enclosure rotted on the ground, or the fruit of his planting perished without gathering, and laying up,' wrote Locke, 'this part of the earth, notwithstanding his enclosure, was still to be looked on as waste, and might be the possession of any other.'[54] Locke chalked out a manifesto for freeganism: if it is going to waste, you have a right to take it.

When I moved back to Sussex, where I grew up, I could see apples and pears being wasted in numbers almost as enormous as those of the ready-meals and baguettes I had foraged from the supermarket skips of Euston and Southwark. I fantasized about the amounts of delicious chutney, jam, pickle, tarts and Kilner jars of bottled fruit that this vegetable wealth could make. Soon after joining a Sussex group of the internet sharing site Freecycle, I sent out a message: 'WANTED: APPLES'. Huge numbers of people replied. Many of my neighbours, it seemed, had orchards groaning with more apples than they could ever use. One couple emailed to ask if I could come over as soon as possible: their garden was carpeted with so many rotting apples they were gathering them into sacks and driving them down to the local dump.

My old friends Charlie, now a tree surgeon, and Frank borrowed an old-fashioned oak and cast-iron mangle and press from a colleague. We set them up on the village green, piled up the apples we had gleaned from nearby orchards, and squeezed 550 pints of apple juice. Kids from the neighbourhood queued up to spin the mangle arm round, squishing the apples into pulp, while others twisted the vice down to press out the juice. That night, everyone took home as much fresh apple juice as they could drink; the remaining 500 pints were left to ferment. A few months later we had our reward: strong, pure and un-adulterated local cider, fuel for parties throughout the year. This cider-pressing party has now become an annual autumn tradition.

Fruit trees – especially the older quirky varieties that yield a deliciously assorted fruit which never sees the inside of a super-market – are a glorious part of our inheritance. A grassroots response to this disused treasure would make a small impression on the amount of fruit we currently import and the amount of industrially produced booze we buy. As anyone who has scrumped in the old orchards of Somerset, or from roadsides in London or East Anglia will know, trees like this are also one of the secrets to happiness.

Besides gleaning at autumn time, people have substantial power to instigate change across the industry. Growing your own fruit and vegetables or buying from farmers' markets would help cut out a degree of industrial grading. But even when filling up a trolley with supermarket food, you can have an effect. If one out of every hundred customers demanded knobbly fruit or vegetables every time they went shopping, supermarkets would start to rethink their absurd standards. If the customer complaint desks – there is one in every store – regularly heard consumers bemoaning the fact that all the potatoes, parsnips, lettuces and carrots looked the same, week in week out, and that the only apples available were Cox and Granny Smith, that it was actually pleasant to be able to buy bananas of different shape, colour and curvature, they might start to market produce that reflects what actually grows in the ground and on the trees in England, Kenya and Guatemala. Supermarkets do actually listen to their clientele. If they knew that they might lose custom to a chain that held less ludicrous cosmetic criteria, they would change the way they viewed every different crop.

The other major cause of waste in the farming sector – which has ramifications for the entire food supply chain – arises from the way agricultural subsidies have been organized. The modern subsidy system can be traced back to the food shortages in Europe and America during the Second World War which motivated governments to invest heavily in agriculture. Prior to the war, national food supplies depended on international trade – both transatlantic and from Europe's empires. Arguably, wealthy nations' agricultural systems were designed only to cope with 'average' years. By 1945 it was clear that this was complacent: it was time to readapt. Governments realized that food should be produced indigenously to guarantee supplies for the whole population, regardless of international prices.

Europe's Common Agricultural Policy (CAP) assured farmers that anything they grew that could not be sold on the market would be bought up at a high guaranteed price, well above the world market price. So effective was subsidized production that from the 1960s farmers were growing more than the market demanded, and by the 1980s the CAP had become a victim of its own success. Farmers were churning out lakes of wine and mountains of grain and butter that Europe's population could never consume. Surplus was often exported, and the EU paid export refunds to bridge the gap between the EU price and the world price. This subsidized export was blamed for undermining the local markets of the developing countries who received it, and disincentivizing production in those countries. It swiftly became a source of controversy. Why should taxpayers and consumers pay farmers to produce food that nobody wanted? To tackle the problem of over-production, the European Economic Community came up with the idea of set-aside, whereby farmers were paid to leave at least 10 per cent of their land uncultivated, which also produced substantial benefits for wildlife. This policy remained in place right up until 2008 after the spike in prices for agricultural commodities, when farmers were allowed to return to their full productive capacity, though in 2009 there has been debate over the reintroduction of set-aside once again.

As well as set-aside, other reforms began in 1992 (the so-called MacSharry reforms), continued in 1999 (Agenda 2000) and were advanced further in 2003. Under these reforms, the EU cut guaranteed prices and even got rid of them for a number of products, thus reducing the incentive to over-produce. Subsidies for exporting surplus have also been falling consistently and in the WTO negotiations the EU has offered to phase them out entirely from 2013 as long as other big exporters discipline their own export support programmes. Guaranteed prices for some products and export subsidies for dairy products, eggs, beef, poultry and pork are still there as a safety net, and this means that farmers can still be slightly more relaxed about the financial

The contents of the bins outside a local village shop, Sussex, England, 2008.

The contents of the bins of a local organic fruit and vegetable shop, Sussex, England, October 2006.

The contents of the bins of a village Co-op store on 8 September 2008, including 86 bread rolls, 12 loaves, 248 cocktail sausages, 42 sandwiches, 30 peaches, 36 oranges, 117 tomatoes, 1 melon, 1 cauliflower, 9 lemons, 6 sausage rolls, 6 pork pies, 11 cheese twists, 10 croissants, 45 tubes fromage frais, 4 pots yoghurt, 4 pints milk, 18 scotch pancakes, 1 chocolate cake, 1 toffee cake, 2 packs crisps, 7 sauces and dips, 1 smoothie, 1 pack chocolates.

Contents of the bins of a Marks & Spencer supermarket in central London, April 2009, sprayed with blue dye.

A selection of fresh produce from the bins of a local Waitrose, July 2008.

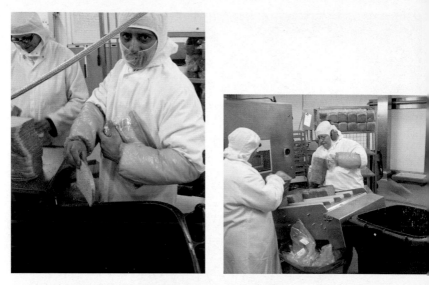

Marks & Spencer requires their sandwich supplier, Hain Celestial Group, to discard 4 slices from each loaf, the crust and the first slice at either end, amounting to 13,000 slices of fresh bread every day from a single factory.

Pastry trimmings in the Ginsters cornish pasty factory, Cornwall, UK.

Richard Hirst with his crop of spinach, rejected by the retailers and left to rot in the field, Norfolk, England, summer 2008.

Potatoes rejected for cosmetic reasons at a potato farm that supplies Tesco's. Kent, England, 2008.

Surplus tomatoes dumped on farmland in Tenerife.

Surplus oranges in California, USA.

Thousands of imperfect bananas lie in a drainage ditch, dumped by United Fruit. Rio Estrella, Costa Rica.

risks of over-producing beyond demand than they might otherwise have to be.[55] For example, in 2009 the EU expects to have to buy up some grain, butter and milk powder, and in January 2009 it temporarily reintroduced export subsidies for dairy products. In the ten years to 2008, the EU spent €38bn on export subsidies, though this has declined from nearly €5bn in 1998 to under €1bn in 2008.[56] By and large EU farmers now produce roughly at world market levels. Instead of getting subsidies according to how much they produce, farmers are now paid per acreage and are expected in return to follow environmental, animal welfare and food quality standards. This often means farmers are being paid to de-intensify their operations to provide benefits to natural ecosystems and valuable landscapes, which improves the amenity value of the countryside, as well as encouraging slack in the agricultural system.

The US, by contrast, has not reformed its agricultural subsidies in the same way, and farmers are still paid to produce more than a free market might otherwise demand. If farmers are being paid subsidies to pump out more food than can readily fit down the gullets of the American people, it is no surprise that large proportions of the excessive food supply leak out at the seams and bottlenecks in the system. Some of the surplus food is exported as foreign aid, increasing amounts are being diverted into biofuels, vast quantities are wastefully fed to livestock, some of it is converted into fat by a population plagued with obesity, and much of the rest simply ends up in the bin. There is nowhere else for it to go.

8. Fish: The Scale of Waste

But still another inquiry remains . . . whether Leviathan can long
endure so wide a chase, and so remorseless a havoc; whether he
must not at last be exterminated from the waters . . .

Hermann Melville, *Moby Dick* (1851)

I am standing in the waters of the English Channel, where the
sea laps at Brighton beach, wearing only cotton shorts and a pair
of boots, waiting patiently for a fish to tug at my line. As my
wife lies on the pebbles nearby, dozing in the sun, children pass
with ice creams and tinny music drifts towards me from the pier.
We hardly ever buy fish but Alice loves to eat them. I want to
take enough home today to freeze or smoke so we can enjoy
them throughout the winter.

But today nothing bites. Another two hours, my feet are
turning numb and I still haven't caught a single fish. Alice has
caught the sun, it is almost dusk and she is now ready to go
home. I am just drawing in my line in defeat when another
fishing trawler moves across the salty horizon in front of me. It
is followed by a huge flock of gulls, and I watch as they swoop
down on the discarded by-catch that are being thrown back into
the sea. There go the dead fish that I have failed to catch.

Just as in farming, waste is a chronic problem in the other
domain of primary food production – fishing. The most acute
type of waste in the marine fishing industry is the 'discards'
whereby unwanted fish that are too small or the wrong species
are thrown back into the sea.[1] In many fisheries, an estimated
70–80 per cent of these fish perish in the process.[2] It may come

as little surprise that Europe wastes more than almost anywhere else in the world:[3] the EC estimates that between 40 and 60 per cent of all fish caught are thrown back into the sea. Greenpeace suggests that of the 186 million fish caught by European fleets, 117 million are thrown back.[4] In the North Atlantic, an estimated 1.3 million tonnes a year are thrown overboard;[5] in the North Sea, annual discards have been estimated at just under 1 million tonnes.[6] Up to half of all haddock caught in the North Sea are chucked away while 50,000 tonnes of whiting are rejected each year. Beam trawlers targeting flatfish in the North Sea end up throwing away 70 per cent of their total catch, while for shrimp trawlers the rate is as high as 83 per cent. The market value of the cod, haddock and whiting thrown away by UK trawlers alone comes to €75m – equivalent to 42 per cent of what fishing fleets actually bring ashore.[7] For every kilo of Norwegian lobster or scampi sold, 5 kg of by-catch are dumped back into the sea.[8]

Many of these discards consist of species that are very good to eat, but which the fishing industry has yet to market successfully. Species like saithe, conger eel, gurnard, whiting, pouting and dogfish are thrown away in thousands of tonnes. I have caught all of these species off the coasts of Britain and can vouch that they are magnificent eating. It is insane that we throw their carcasses away.

Still sadder is the routine practice of chucking away some of the most valuable species of all. Sardines are not appreciated in some European countries and are routinely discarded by some fishing fleets, even while others go out to catch them.[9] Despite the fact that North Sea cod is classified as 'threatened and/or declining' by the OSPAR Convention for the Protection of the Marine Environment of the North-East Atlantic, Britain's fishing vessels still throw distressing quantities of it away.[10] This is largely due to the ill-governed and badly designed European common fisheries policy, which allocates to member states a quota of each kind of commercial fish they are allowed to bring ashore, which is then divided and traded between individual fishing vessels. If a vessel accidentally hauls in more than its quota of cod, apart from illegally landing its catch, there is little its crew can do but hurl these valuable but generally dead or dying fish back into the sea. Most pernicious of all, because skippers know they are limited in how much they can bring ashore, they often keep only the highest-grade specimens and throw back anything that will make less money, even if they are legally allowed to sell them.[11] Norway forbids discarding any of the most commercially important species of fish in its waters, but this has meant that non-Norwegian trawlers have just returned to EU waters and dumped the less profitable specimens there. One British trawler, the *Prolific*, demonstrated this flagrantly when it was filmed by Norwegian coastguards in 2008 tipping 5 tonnes, or 80 per cent of its haul of cod and other white fish, into the sea after leaving Norwegian waters.[12]

When fish stocks are exploited faster than they can regenerate, the average size of the fish caught declines as shoals are comprised of a high percentage of juveniles. One result of this is that a higher proportion of juveniles are caught and thrown back into the sea, decreasing the ability of stocks to recover and increasing the problem of discards still further. In the waters west of Ireland and Scotland in 1999, fishing fleets from Ireland, France, Spain and the UK rejected (by number rather than weight) 80 per cent of the whiting they caught. Spanish trawlers, meanwhile, throw away huge quantities of horse mackerel, mackerel and blue whiting because of weak demand, size and quota restrictions. The FAO itself has observed that in some European fisheries the more strictly the quota system is enforced, the more discard waste is produced. To make matters worse, EC legislation does not require that its fishing fleets keep a record of discards of all species, so assessing the full extent of the damage is very difficult and most EC estimates are based partially on data provided by observers who are only boarded on a sample of vessels.[13] If the aim were to enforce wastefulness, it would be hard to design anything more effective than the European quota system.

Under such ineffective management policies the world has succeeded in bringing three quarters of all marine fish species to the brink of, or actually below, sustainable population levels. According to the authors of an article published in *Nature* in 2003, the oceans have already lost more than 90 per cent of large predatory fishes such as cod, salmon and tuna, and a group of fourteen academics writing in *Science* magazine concluded that, on current trends, by the year 2048 all species currently fished will be extinct.[14] Just 4 per cent of the world's oceans remain in a pristine state – the rest have been affected by a lethal combination of over-fishing, climate change and pollution.[15]

It is possible to eliminate much of the damage caused by inefficiencies and waste. Firstly, the fishing industry and the retailers could inform consumers about the severity of the crisis in fish stocks, and encourage them to try some of the many other kinds of fish which are more abundant and are currently treated as by-catch. If consumers continue to see cod and skate on the market, without any warnings about the unsustainability of current fishing practices, they will continue to buy them. Organizations such as the Marine Stewardship Council (MSC) and the Marine Conservation Society (MCS) provide information on which fisheries are being managed sustainably, and list fish that can be purchased and eaten without contributing to the chronic demise of the oceans.[16] It turns out that sustainably sourced fish is often cheaper than conventional fish – the opposite trend to most 'ethical' food labels. For example, one of the substantial outlets of fish certified as sustainable by the MSC is Tesco's 'Value' frozen white fish, which is comprised of Alaskan pollock.[17] The popular British frozen food brand, Birds Eye, has also launched a sustainable fish finger made from MSC-certified Alaskan pollock instead of cod.[18] Worryingly, surveys of Alaskan pollock in 2008 showed a dramatic decline in stocks, which could be taken to mean that as soon as people turn to a sustainable

fishery, it quicky becomes overfished; but the MSC claims that
the decline is caused by natural fluctuations and catches have
been limited accordingly. According to the MSC, Alaskan pollock fisheries avoid any waste of fish, achieving by-catch rates
as low as 0.5 per cent and that all edible portions of the fish
caught are fully utilized in seafood for human consumption,
while inedible portions are used for other purposes.[19] Since the
financial crisis and food price rises of 2008, traditionally shunned
fish like whiting, gurnard and saithe have begun to gain in
popularity. For the first time in fifteen years, Waitrose in the
UK has started to stock whiting, which costs a third less than its
relative the cod, and sales of other more economical fish have
increased.[20] As long as the stocks of these fish are well-managed
(though in Europe they often are not, and whiting are now
particularly badly overfished), encouraging consumers to shift
from over-exploited species like cod to more abundant and
currently wasted species would be an effective way of improving
the efficiency of the fishing industry.

The second measure would be to undertake a radical overhaul
of the quota system. This has proved to be a political minefield.
But, on a scientific basis, it has been demonstrated that there
are better ways of conserving fish stocks while ensuring the
profitability of fishing fleets. Currently less than 1 per cent of
the world's oceans are designated no-fish zones. And yet, in the
few places where marine reserves have been introduced, such as
the Philippines, critically threatened fish stocks continue to
recover exponentially even two decades after the establishment
of the reserve, and fishing industries have been saved from fishing
themselves into extinction.[21] Tracts of the ocean in which no
fishing is allowed provide habitats which act as spawning grounds
and nurseries for stocks, thus increasing the number of fish
available for fishers to catch beyond the borders of the reserve.
On one reserve off the coast of New Zealand's North Island,
commercial fishing fleets catch more lobsters in the five kilometres adjacent to the reserve than in one hundred kilometres

of unprotected coastline. Snapper are fourteen times more numerous and reach sizes eight times larger in the reserve than outside it. Dr Callum Roberts, an expert on marine conservation in the UK, believes that 10–20 per cent of oceans – starting with migration routes and feeding grounds – should be no-take zones.[22] Marine reserves are also more enforceable than quota systems. In Europe, illegally landing an over-quota catch is common practice. It is impossible to keep a tally on every vessel at all times in the vast expanse of sea, so skippers daily engage in the tragedy of the commons, over-exploiting common resources to the long-term detriment of all, including themselves. If, on the other hand, portions of the ocean were marked out as no-fish zones, it would be relatively easy to use existing satellite systems to detect any fishing vessel that strayed into protected waters.[23] It is for this reason that environmental organizations, such as Greenpeace, have drawn up a marine reserve plan that would give total protection to 40 per cent of the North and Baltic Seas, based on scientific assessments of spawning areas and other vulnerable stretches of the sea. Despite a manifesto pledge and majority public support, in 2007 the British Labour government fell short of launching a full marine bill under which marine reserves would be created, and instead offered only a draft bill.[24]

Everyone – from European Members of Parliament to the skippers themselves – is frustrated by the problem of discards. The UK Minister for Marine, Landscape and Rural Affairs, Jonathan Shaw, believes that 'The crisis of throwing fish back is immoral.' The European Fisheries Commissioner, Joe Borg, agrees that 'Discarding means a waste of great quantities of valuable fish. It is a major environmental scandal that we must strongly tackle.'[25] Richard Lochhead, Scottish Cabinet Secretary for Rural Affairs and the Environment, says, 'I am appalled and frustrated at the scandalous level of waste and the economic and environmental madness discards represent. In what other industry would it be acceptable to throw away so much of what is produced?'[26] But there is still chronic disagreement about the best way of tackling the problem.

A recent EC legislative initiative to 'eliminate discards in European fisheries' was, according to Karoline Schacht of the World Wildlife Fund (WWF) in Germany, 'torpedoed by the powerful fishery lobby of some member countries . . . That is an outright scandal.'[27] At the end of 2008, the EC actually increased the quota for North Sea cod by 30 per cent on the basis that stocks have just begun to recover slightly, even though they are still recognized in the scientific community as being severely depleted. Thanks to fierce diplomatic pressure from the Norwegians, the EC simultaneously undertook to try and reduce discard levels in the North Sea by making it mandatory for fishing vessels to use selective fishing gear, including eliminator trawl nets which allow cod to escape. In addition, there will be a ban on 'high-grading' (when fish of a legal landing size are thrown away to make room for larger more valuable specimens) and parts of the sea will be closed during the spawning season. The EC has also now committed to work towards a complete ban on discarding fish when the Common Fisheries Policy is reformed in 2012. This is extremely promising, though in 2007, when the EC raised cod quotas by 11 per cent on the condition that fishing fleets reduce cod discards to 10 per cent of catch through voluntary measures, discards actually increased to 40 per cent of total catch rather than decreasing – so there is no guarantee that the new measures will be successfully introduced or enforced.[28] The EC has been talking about solving this problem since 2002: the urgency of the situation does not allow room for any more prevarication.[29]

Estimating the total quantity of fish discarded around the world is a controversial subject. In 1994, the FAO estimated that 27 million tonnes of fish were being discarded each year, around a third of the amount of fish consumed by humans.[30] An update in 2005 indicated that there had been improvements. The methods and definitions used in these two surveys were different and therefore the figures are not truly comparable, but the update concluded that between 1992 and 2001 the global marine fishing industry discarded 7.3 million tonnes of fish each year. The FAO emphasizes, however, that a lack of recorded data makes such estimates tentative; for example, North and South Korea and Russia provide no records, and illegal fishing discards are necessarily unverifiable.[31] Furthermore, the 2005 report did not include fish wasted onshore (for example, when roe fisheries sort through herring and discard the males); nor did it include fish killed in the sea by fishing gear (such as the substantial mortalities caused by scallop dredging and the innumerable fish killed by abandoned nets and traps). Nor did it include fish that die in trawl nets and fall out of the net before reaching the deck, even though studies have shown that these 'dropouts' may approach 50 per cent of an entire catch in some fisheries.[32] Nor did it include the substantial mass of jellyfish, sponges, coral and exotic rarities like sea-snakes which are hauled in by trawl nets and never recorded. It did, however, include the discard of sharks after their fins have been cut off for sale at a high profit; the rest of the body is thrown back into the sea.[33]

In March 2009 the FAO published its latest crude estimate, suggesting that the total 'could be more than 20 million tonnes globally (equivalent to 23 per cent of marine landings) and growing'.[34] Other organizations, such as the United Nations Environment Programme in a report published in February 2009, claim that total discards are around 30 million tonnes a year, and that humans eat barely more than half of all fish caught.[35] Charles Clover, in his book *The End of the Line* (2004), concluded that once waste from discards, spoilage, fishmeal and inedible portions are taken into account, the amount of fish-based protein actually consumed in the world comes to only around 10 per cent of the marine animals actually destroyed annually in the oceans.[36]

Nevertheless, it is apparent that in recent years the phenomenon of throwing by-catch overboard has declined in some areas, and this at least demonstrates that improvements can be made. Some of this reduction has been due to real advances in the management of fish stocks: some fish that used to be discarded are now landed and sold for people to eat, partly because conventional species have been so over-exploited they are no longer available and thus other species have taken their place on the market, and partly because of the popularity of processed fish products such as the Asian fish paste *surimi*. Some fishing fleets have adopted more selective methods, ensuring that fewer undersize and unwanted fish are caught in trawl nets. Bold and sensible laws have been passed in some countries (Norway, Iceland and Namibia) prohibiting discards altogether, while other fisheries have set by-catch quotas or established 'no-trawl' zones. Some species with high by-catch rates have been fished less intensively, and there have been efforts to improve the survival rate of by-catch returned to the sea.[37] All of this has demonstrated that concerted global action can achieve a more sustainable management of the oceans.

Australia and the United States have achieved discard reductions owing to high public awareness of the problem, and effective legislation and market co-operation to encourage responsible fish stock management. In the seas of Alaska, for example, access to some areas is permitted on the condition of using by-catch avoidance technology. Placing bans on discards entirely, or targeted bans on specific species or areas and at certain times, have also been effective.

Iceland has saved its fish stocks from the worst effects of over-exploitation by granting each fishing vessel an individual transferable share of the national allowable catch for each species. But unlike the EU quota system, which encourages or mandates the discarding of economically less valuable fish, in Iceland it is illegal to dump any by-catch at all. Instead, the whole catch must be landed and recorded as part of that boat's quota. If a boat exceeds its quota, it must either buy a share of another vessel's quota, or 'borrow' part of its quota for the following year. The good sense of this policy, in contrast to the topsy-turvy logic of the common fisheries policy, is a principal reason why Iceland and Norway have refused to join the EU. Iceland's recent economic crisis, which has brought it to the brink of joining the EU at the time of writing, may spell an end to this haven of sensible fisheries management. WWF has called for a complete ban on discards throughout the EU, arguing that in some fisheries 90 per cent of discards could be avoided through use of better technology, such as nets that allow undersized fish to escape, and that any discards which do arise should be landed and counted against a fleet's quota, as in Iceland.[38]

In addition to banning discards, some economists argue that a key to the success of fisheries' conservation is enforcing property rights over marine resources. Under international law, coastal nations only have exclusive rights over waters up to twelve miles from their shore. The high seas and their contents are still common property and there is thus little incentive for any individual to take a unilateral decision to limit their exploitation. Enfranchising nations or communities with ownership rights could mean that they have more reason to look after the long-term interests of the fish stocks.[39] One negative example of what happens when a nation ceases to exert control over its fishing rights is Somalia: there the absence of an effective government meant that its waters were exploited without regulation, very often by European tuna-fishing vessels. The result has been a collapse in fish stocks, which has been one contributing factor in driving Somali fishermen into piracy.[40]

The method employed for catching various species of fish likewise has a critical impact on the amount of non-target fish that get killed. Incredibly destructive bottom trawls catch sea creatures indiscriminately as the nets are dragged along the ocean floor with crushing weights that demolish millennia-old coral reefs, sponges, fish habitats and anything else in their path. Studies show that obtaining a kilo of sole in this way involves killing another 16 kg of other marine animals.[41] Imagine if the standard method for killing cattle were to use bulldozers to haul a net through the countryside, uprooting trees and hedges, smashing ancient monuments and exterminating along with the cows everything from badgers and stoats to lapwings and barn owls, all of which perished slowly by suffocation. This is what happens every day under the cover of the oceans, despite more sensitive alternative ways of catching fish being available. Bottom set gillnets and longlines which remain stationary on the sea floor can provide both practical and economical alternatives for some species. A very small number of commercial fisheries have already developed and implemented simple modifications to existing gear, such as making the 'trawl doors' (the rolling weights that hold the nets down) lighter or designed to reduce contact with the sea bed. But the vast majority of bottom trawlers continue to use the most outdated and harmful equipment.[42]

Fishing for tuna on a longline (which can be up to 125 kilometres long with thousands of hooks) gives rise to a discard rate more than seventy-one times greater than fishing for them on an ordinary pole and line.[43] Exchanging traditional J-shaped hooks on longlines with new 'circle' hooks has been shown to reduce the number of accidentally killed turtles by 90 per cent without reducing the catch of target species, and setting the lines at a depth of more than 100 metres has avoided catching sharks as well as turtles.[44] For each year in the 1990s, 40,000 albatrosses were killed while trying to eat bait on these hooks; and trawling for yellowfin tuna was killing 400,000 dolphins a year. However, some new 'dolphin-friendly' methods of catching tuna, which can involve surrounding and scooping up entire marine habitats, can actually kill even more kinds of other fish, turtles and sharks.[45] For 15,721 tons of tuna caught using these dolphin-friendly methods in the eastern Pacific, fishing fleets killed 15,737 tons of sharks, rays and other fish – a by-catch rate of over 50 per cent. By-catch in tuna fisheries regularly includes critically endangered and vulnerable species such as loggerhead, leatherback, hawksbill, gulf ridley and green sea turtles, minke and humpback whales, great white sharks, sting rays, spotted eagle rays, shortfin mako, and great hammerhead and numerous other types of shark.[46]

The fishing of freshwater paddlefish is banned in most states of the US, but, where it is permitted, by-catch rates reach 92 per cent of total catch – and even those mature females that are targeted are marketed mainly just for their roe, which can get $143 per kilo. In Tennessee, shortening the catching season by eight days to prevent fish dying in warmer waters helped to reduce by-catch mortality, but skippers are still allowed to use monofilament nets that cause greater losses than alternative more obtrusive multifilament nets.[47]

In contrast to European flatfish trawl fisheries, American plaice and witch flounder fisheries have comparatively low discard rates of 8.7 and 18.8 per cent respectively. However, there are still some US fishing fleets throwing away huge quantities of fish. For example, in the Gulf of Maine and the north-eastern United States, silver hake trawlers shed 41.7 per cent of their catch by weight. Despite efforts to alleviate the problem, the Gulf of Mexico shrimp trawl fishery has about the worst discard rate of any individual fishery in the world, throwing back around 480,000 tonnes of snappers, emperors and turtles, among other species.[48]

Shrimp fisheries all over the world suffer by far the biggest by-catch rates. In the Asian tropical shrimp trawl fisheries, huge quantities of juvenile and small fish are caught in the nets despite the availability of proven by-catch reduction devices that use metal grids and net meshes to allow these fish to escape. Trawlers catching shrimps in tropical waters account for 27 per cent of the world's discards. The global average shrimp-trawling discard rate is around 62 per cent of total catch, but in the worst areas the rate averages 96 per cent; in other words, for every kilo of shrimps, 24 kg of other fish are thrown away.[49]

In addition to discards, an estimated 10–12 million tonnes of fish per year are reportedly wasted through spoilage.[50] Many fishing fleets and ports, particularly in poorer parts of the world, lack the facilities and infrastructure to keep fish fresh before it gets to market. In Karachi, Pakistan's economic capital and major port, heaps of fish can be seen on the docks, rotting in the sun. In some African fisheries, when discards, accidental losses and spoilage are added together, waste may be around 40 per cent of landings.[51] In the developing world, assistance in investment in equipment for handling, preserving and transporting fish could be the most effective way of maximizing efficiency. Traditional methods of smoking and salting, or refrigeration and canning, can reduce losses; keeping fish cool with ice can even be made more effective by insulating containers with local materials such as coconut fibre, sawdust or rice husks.

Despite the apparent good news that the practice of discarding by-catch is in decline, one of the main reasons for this reduction is almost as worrying as the original problem: it is being mushed up into fishmeal and fed to farmed fish and livestock. Although the production of fishmeal has declined by 30 per cent since 1994 (largely because of a drop in anchoveta catches), the UN estimates that 33.3 million tonnes of fish – 36 per cent of the total amount of fish caught in the world – are destined for non-food uses, mainly fishmeal and fish oil for feed.[52] Pigs, poultry and other livestock eat a little less than half of the world's fishmeal (partly because in the EU livestock are no longer allowed to be given food waste containing meat by-products, which I discuss in later chapters). The rest is consumed by the booming fish-farming industry, which produced 51.7 million tonnes of fish in 2006.[53] Most fishmeal is made of ground-up anchovies, menhaden or sardines; but, as these run out, fleets are increasingly using small specimens of commercially valuable fish. The plummeting populations of seabirds such as kittiwake, guillemots and puffins in British coastal waters can be at least partly attributed to the disappearance of sand eels, which have been dredged for salmon feed. Even krill – the tiny shrimp-like zooplankton that fill the oceans – are now filtered out of the sea using fine-mesh nets and turned into fishmeal, particularly prized by salmon farmers because it gives salmon a desired pink appearance. Krill are near the bottom of the marine food chain, and although industrial krill fishing is still in its infancy and currently sustainable, stepping up this exploitation could cause a collapse of whale, seal and penguin populations which depend upon them for food.[54]

It may be better to use fish as animal feed than to discard them altogether. But trawling fish out of the ocean to pulp and feed them back to other fish or livestock can be a substantial waste of resources, and since it means that fishers can find a market for undersized juvenile fish that they might otherwise have more incentive to avoid catching, it can be very harmful. It takes on average 3 kg of fishmeal to produce a kilo of farmed salmon. This level of inefficiency looks unwise when one considers that many of the fish turned into fishmeal are already delicious fish that could be eaten as they are. Even sand eels – as I discovered after a day of catching them off north Wales – make superb ready-made fish-fingers. Aquaculture can make a good deal of sense when the captive fish forage on wild resources, as is the case for around half the world's inland and marine fish farms, and relatively high levels of efficiency can be attained by farming herbivorous and omnivorous fish with a minimum of fishmeal, for example, tilapia, catfish, milkfish and some though by no means all carp;* but it very rarely makes sense to feed edible fish to carnivorous species, such as tuna, salmon, shrimps, prawns and freshwater eels. Moreover, warm-blooded livestock convert fishmeal into meat even less efficiently, wasting up to 80 per cent of it. This happens most extensively in regions such as Latin America, where populations have a cultural preference

---

* Although it is possible to raise some of these species with minimal or no fish-based feed, carp consume around 42 per cent of all the world's fishmeal and tilapia consume 10 per cent (FAO 2009, p.144).

for meat over fish, and in Europe and North America, where they can afford to buy imported fish of a type they prefer. Fishmeal is an acceptable use for some by-catch species and it is certainly a good use of unwanted fish bones and entrails, but as far as most by-catch species are concerned it should be a secondary option, after finding a human market.

Better still, of course, would be to leave over-exploited fish in the ocean.

9. Meat: Offal isn't Awful

'You gonna waste that crab gut? You're a pussy, besides, Jimmy.'[1]
Bunk Moreland in *The Wire*, HBO (2004)

In addition to rich countries throwing away nearly half of their
food supplies, around 40 per cent of the world's cereals, including wheat, rice and maize, are fed to farm animals.[2] That comprises around 700 million tonnes of cereals, in addition to which
livestock consume more than 500 million tonnes of roots, tubers,
fishmeal, brans, pulses, oilseeds and – most importantly – oilcakes
such as soymeal.[3] In all, around one third of the world's arable
land is given over to growing livestock fodder, and the proportion is ever-increasing.[4] In modern intensive-farming systems
– which do vary massively in methods and efficiency – it takes
in the region of 10 kg of cereals to produce one kilo of beef and
5 kg to produce a kilo of pork. The most efficient land animals
at converting feed into flesh are modern breeds of chicken,
which put on a kilo of weight for every 2 kg of grain.[5] Modern
egg production can also have a conversion ratio of just over 2:1.[6]
Globally, we give more than three times more food to livestock
than they give us back in the form of milk, eggs and meat,
meaning that on average livestock lose over 70 per cent of the
calories in the harvests fed to them.[7] This is not 'waste' in the
same sense as throwing food into bins; but in that those cereals
could be used to feed far more people directly than can the meat
produced, it is an inefficient use of resources. Food, and the land
used to grow it, is expended in order to satisfy people's preference for succulent meat, which the world's rich countries eat
in quantities that exceed what is either good for their health or

that of the planet. The result is a squeeze on world food supplies, which exacerbates food price rises and consequently hunger for the poorest in the world. As discussed in chapter 6, it is also driving the rapid extension of the agricultural frontier into natural habitats, and it contributes significantly to global warming.

Meat production has multiplied by more than two and a half times since 1970, and now the combined weight of cattle on earth exceeds that of humans.[8] Global production rose from 27 kg of meat per person per year in 1974–76 to 37.4 kg per person in 2000. By 2050 it is projected to rise to over 52 kg per person.[9] By then, cereal consumption by livestock may rise to well over a billion tonnes, which the UN optimistically estimates would be equivalent to diverting enough food for around 3 billion people.[10] If these crops were not being fed to animals, all of them could be used to feed people directly. Cereals supply around half of the calories consumed by humans all over the world,[11] and if we ate less meat and fewer dairy products, there would be much more food to go round.[12] The soaring global demand for meat has certainly given many people in the developing world better diets and it has provided the stimulus to increase arable cultivation around the world. But global consumption levels are now driving deforestation, global warming and food shortages.

Meat also requires much more water in its production than do vegetables. It takes 500–4,000 litres of water to grow a kilo of wheat. But a kilo of meat takes 5,000–100,000 litres.[13] To produce 1,000 kcal of vegetable food takes on average approximately 0.5 m³ of water; it takes about eight times that to produce the same amount of animal-based food.[14] With water scarcity presenting one of the most serious threats to human survival in many parts of the world, including south Asia and sub-Saharan Africa, continued high levels of meat consumption will almost certainly lead to a loss of human lives.

As developing-world countries get richer, their populations are inexorably adopting the meatier, milkier diet of affluent countries. Between 1980 and 2002 in developing Asian countries, per capita consumption of protein from livestock increased by no less than 140 per cent. And yet, for these nations to reach the level of consumption now enjoyed in industrialized countries, they would have to eat another three and a half times more than they do now. In 2003, Americans were eating twenty-four times more meat per person than Indians. At 123 kg of meat per person per year, Americans exceed the British appetite for 83 kg per person per year, the Dutch for 67 kg, the Chinese 55 kg, the Japanese 43 kg, Ugandans 10 kg and Indians just 5 kg. So it is up to rich countries to address their consumption first, rather than pointing the finger of blame at developing nations' increasing meat consumption – as innumerable media articles and political leaders (such as George W. Bush) did during the food crisis in 2008.[15]

On first examination, it may appear that the increase in livestock farming in tandem with rising incomes implies that the desire to eat more meat is an unalterable fact of human nature: people eat more meat whenever they can afford it. However, a closer examination reveals significant differences between countries. For example, it is estimated that by 2020 Indians will consume 8 kg of meat each per year, whereas by current trends the Chinese are likely to consume 73 kg. Indians, by contrast, will consume the equivalent of 105 kg of milk each per year, while the Chinese will imbibe little more than 31 kg.[16] Since dairy cattle can be five times more efficient at turning feed into edible calories than beef cattle are at turning it into meat,[17] this suggests that India will increase its consumption of animal food much more efficiently than China.

Important also are the methods of rearing animals. Where traditional livestock farming persists, animal agriculture can still be a net contributor to food supplies. In sub-Saharan Africa and south Asia, for example, livestock are typically raised on crop residues and extensive non-arable grazing land, and only 10 per cent of the grain supply is used for feed. In some developed countries, by contrast, two thirds of average grain production is devoted to livestock.[18] India's livestock, at the same time as performing most of the labour on farms and supplying vital fertilizer and fuel, also yield back in meat and dairy products 70 per cent of the food they consume. By contrast, in the US, livestock on intensive farms produce unmanageable quantities of slurry, which becomes an environmental hazard, they typically sit in barns all day doing nothing, and then they give back only 20 per cent of the food they consume.[19] Around a third of that is then wasted by shops and consumers, so only around 13 per cent of the calories in the original arable harvest is actually consumed by people.[20] In the US, even bearing in mind the rich sugary diets of the American people, the nation's livestock eat roughly twice as much food as the Americans themselves.[21]

Some meat is still produced in the industrialized world without this inordinate reliance on feed concentrates – British hill-sheep farming still relies basically on extensive grazing and can even be good for local habitats. On the other hand, there is a trend, also in developing countries, towards cereal-based meat production which, in countries such as Pakistan and India, is contributing to local food shortages.[22] In that sense, the increasingly affluent middle classes in developing-world countries are doing to their poor what consumers in the affluent West are doing to the world as a whole.

The impact of these trends concerning meat consumption is so vast and pervasive that I have addressed it in various chapters throughout this book. In the appendix there are tables and charts indicating just how much food we deprive the world of by intensively rearing so many farm animals. It is clear that the single easiest way we could reduce our individual impact on the environment, and liberate the greatest amount of food, land and water for other uses, would be to eat less meat and fewer dairy products from grain-fed animals.

However, this argument for efficiency is categorically different from the main thrust of this book, which focuses on efficiencies that can be made without anyone having to sacrifice those foods they enjoy. To achieve the benefits outlined in this book, all we have to do is stop throwing so much food away. But here too, the resource intensity of meat production is a critical issue.

The EU-wide study Environmental Improvement of Products (IMPRO), found that meat and dairy products are responsible for nearly a quarter of all environmental damage done by consumption in the EU, including global warming, ozone depletion, acidification, toxicity to both human and natural environments, and eutrophication (and this did not even include deforestation and other land use changes). The conservative estimate for how much meat and dairy consumers waste suggested that a full 4 per cent of all kinds of environmental damage caused by the EU can be attributed to just the portion of meat and dairy products consumers throw away.[23] In the current global system of production, we fund the destruction of the Amazon rainforest and drench the American plains with agrichemicals to grow soybeans and cereals for animal feed, ship them around the world to fatten our livestock, truck the animals to a slaughterhouse and kill them, chop them into pieces, wrap them in plastic and keep them in a fridge for several days. Then we throw a huge proportion of them in the bin – as if their lives, and all the resources expended in raising them, were of no significance.

Wasting meat and dairy products is often an underappreciated problem, because in studies showing how much food is wasted vegetables and fruit represent a large proportion of the total mass. For example, the US government points out that fruit and vegetables constitute 28 per cent of food losses from retailers, food services and consumers.[24] Similarly, in the UK, WRAP found that 40 per cent of the food waste thrown out by householders was fruit and vegetable matter. WRAP has consequently focused much of its attention on encouraging consumers to keep fruit in the fridge and other measures designed to extend shelf-life.[25]

This is very useful advice, and dispensing it demonstrably reduces profligacy. But in terms of tackling resource waste and environmental impact, reducing the amount of meat and dairy products that are discarded could yield even better results. One academic study published in 2004 found that in Swedish canteens an average of 20 per cent of the food was wasted, or enough for 287 million portions every year. The authors calculated that the arable land needed to produce all the food that was later discarded in canteens would be equivalent to 40,000 hectares or 1.5 per cent of land under cultivation. Although meat made up only 20 per cent of all the food losses by weight, it accounted for no less than 91 per cent of the 'wasted' land, because so much land is taken up growing fodder for livestock. They estimated that if the same levels were true for the fifteen countries which then made up the EU, the wasted arable land would be in the order of 1.5 million hectares, an area roughly equivalent to the total land under cultivation in Belgium – and that is from just one small sector in the food industry.[26]

We all have the power to reduce the waste of meat and dairy products in our homes. But here I shall concentrate on one particular inefficiency in the meat industry: the waste of offal – the animals' livers, kidneys, lungs, hearts, tongue, brains, blood, cheeks, tails, feet, ears, testicles, stomachs and intestines. For centuries people across the world have invented ways of cooking these into an array of ingenious forms, from paté and blood-pudding to tail-soup and tripe-saucisson. The trend over the past few decades, particularly in Britain and America, has been for consumers to shun them all and the best fate most of these delicacies can expect in the Western world is to end up as dog food. Homer Simpson once observed that 'Animals are crapping in our houses! And we're picking it up! Did we lose a war?' He might have said the same thing about the finest morsels we serve our pets for dinner.

In place of this variety, consumers in the past few decades have grown accustomed to eating only the muscular tissue –

the meat on an animal – from which supermarkets have learnt they can make the highest profits. In this respect, Britain is far worse than its Continental neighbours, and the US is still more wasteful. As a proportion of meat consumption, offal eating in Britain and the US has roughly halved in thirty years. The French eat on average nearly 100 kg of meat each a year, but alongside that they enjoy 9 kg of offal; for every kilo of meat, the British eat two thirds less offal than the French and the Americans 90 per cent less. The French thereby get over 8 per cent more food out of an animal than the Americans. The Chinese eat less than half as much meat as Americans, but they eat more than three times the amount of offal per capita. The Chinese and the South Koreans are using that resource around eight times more efficiently than Americans.[27]

In France, tripe, tête-de-veau, tongue terrine, goose gizzard and even cow's udder are still available in many restaurants and butchers; in Spain, crispy pigs' ears or golf-ball-like bundles of wrapped-up intestines are on display in tapas bars for punters to nibble with their beers. In the remote hilly region of the Cévennes in France, where I worked in my teens on a cattle farm, the peasants whom I lived with thrived on a daily diet of livers and cows' cheeks served with home-made mayonnaise. In Kazakhstan I have enjoyed the national dish of horse meat stuffed into horse intestines, and their simple but delicious roadside favourite, broiled sheep's head. Outside Beijing I have gnawed freshly fried chicken feet on a railway platform for breakfast; in Japan I have had crab-gut sushi. In the western Chinese city of Kashgar, I was astonished by a wonderful dish, daintily called 'sheep's organs': street vendors display their wares in a pyramidi-cal structure, with a huge mass of yellow broiled lungs (which have the consistency of creamy-polenta) festooned with garlands of stuffed intestines, stewed stomach and grilled livers. Even in London the offal connoisseur can find satisfaction at restaurants such as Gourmet San, an unassuming Chinese restaurant in Bethnal Green which graces its menu with spicy pig's intestines and fish head, barbecued rabbit's kidneys, pork lungs in chilli sauce, whelks and sliced pig's ear, chicken gizzards and braised pig's feet. My favourite dish of all is *takatak*, served on the streets of Pakistan, so called for the noise the knife makes as it chops through a delicious assortment of brains, testicles and other offal which are fried up on a big steel pan.

The waste involved in shunning these body parts was highlighted in 2008 when the Chinese pushed for and won a new trade agreement with the UK: to export unwanted pigs' trotters halfway across the world for the Chinese to chew on.[28] In South Korea, I once made the mistake of asking an industry figure what the country did with all its slaughterhouse waste: 'We don't have any slaughterhouse waste,' came the straight-faced reply. There, the bits we throw away in the West are sold at premium prices.

When I slaughtered my own pigs earlier this year, I was as keen not to forgo any of their body parts as I had been to fatten them on waste. Having completed the slaughter, quickly and painlessly at home, a team of us set about re-creating the traditional tit-bits. We collected the blood in a large basin, turned the guts inside out to make sausage casings, roasted the brains and crispy ears in a wood-fired oven, cured a year's supply of bacon and ham, chopped the stomach into a classic Cantonese stir-fry, and used the fat and trimmings to supplement a variety of other dishes such as head brawn (an old favourite I used to make for my father). Under the enthusiastic orchestration of my friend the author Martin Ellory, we made exquisite delicacies including 'rolled spleen' (a recipe to be found in Fergus Henderson's *Nose to Tail Eating*) and crimson *morcilla* sausage containing blood, cream, rice and nutmeg (based on one by Hugh Fearnley-Whittingstall).

That these delights do not appear more regularly on our tables is not just a tragic loss of fine gastronomic traditions; it is also a substantial waste of resources. We now waste between a third and a half of each animal we kill, much of which could have been eaten if we were not so squeamish.[29] Anything that is not fed back to animals has to be sent for specialist rendering, incineration or other processing at a punitive cost. Raising animals is a resource-intensive process in itself; we should at least be maximizing our use of their carcasses. In the UK alone, 100,000 tonnes of blood, representing 20,000 tonnes of protein are available each year, and only a fraction of it is eaten by people: our livestock shed it, the least we could do is eat it.[30]

The perception that offal is awful is a recent, local phenomenon brought about by conditioning under a culture which has been so disconnected from the origins of food that we flinch at the remembrance that meat comes from a dead animal. Organs and ears remind us that what we are eating once had eyes and a face. Most people find it gross, but it is surely far more grotesque to show disrespect to the animals we kill by discarding some of their most edible parts. And as I found while teaching courses on how to prepare squirrels for consumption, even vegetarians who had previously not eaten meat for years can see that it made no sense to discard the nutritious and delicious hearts, livers and kidneys. If each of us replaced some of the prime-cut meat we are used to with offal, it would be one delectable way of maximizing the efficient use of the animals we raise.

10. Moth and Mould: Waste in a Land of Hunger

The ox hath therefore stretched his yoke in vain,
The ploughman lost his sweat, and the green corn
Hath rotted . . .

William Shakespeare, *A Midsummer Night's Dream*, II.i

'It is the coming of *Qayamat* (Doomsday),' said Sana, a middle-aged Pashtun woman with ten children living in a mud-brick settlement on the outskirts of Islamabad, Pakistan's affluent capital. It was May 2008, and Sana's complaint – 'We no longer have enough to eat' – could be heard in slums across Pakistan. Her children clustered around her, a young daughter tugging at her long chador, while her teenage son stood guard, adding his voice to the lamentations of his mother. Soon, a crowd had gathered – relatives and neighbours, men as well as women, who had stepped out from under the flimsy curtains that shielded their dwellings from view. They each added their own tales of woe to Sana's: wages not paid, labourers laid off, flour prices rising beyond their reach and children – especially little girls – not being given enough to eat.

The city that most visitors to Islamabad see is one of grand villas, a sequestered diplomatic enclave and high-rises purpose-built for foreign staff. Though the planners did their best to keep the poor out of view, still they come into the city – to work as maids, drivers and, in the case of Sana's adult sons, as security guards. 'But now we cannot afford to buy food,' Sana cried, her arms stretched towards the sky.[1]

The most immediate cause of Sana's hunger was the strain on

global food supplies that emerged after the harvest of 2007. In the year to March 2008 an additional 17 million Pakistanis sank into food insecurity, swelling the numbers of the hungry to 77 million, nearly half the country's population.[2] More than a third of children in Pakistan are stunted by malnutrition while 420,000 under the age of five die every year, a number set to rise further if more food is not made available soon.[3] And yet these Pakistanis live in a world where more food is being produced than ever before,[4] where millions of tonnes are thrown away every year, and where relatively painless adjustments could take the pressure off the markets in which women like Sana have to buy bread for their children.

Pakistan usually grows just about enough grain to feed its own population and it has a surplus of rice and cotton for export. It also has an abundant supply of fruits and vegetables which thrive all year round in the balmy climate and fertile soils – irrigated from the mighty river Indus and its tributaries, which fan out over the country. In Baluchistan, the most arid and hilly province, extensive orchards nevertheless yield healthy crops of apples and dates; the province of Sindh on the Arabian sea produces mangoes from as early as May; the Punjab has square miles of citrus orchards to complement its vast grain production; while the North Western Frontier Province (NWFP) and the Northern Areas produce cherries, almonds and mulberries from spring through to autumn. Goats, sheep and cattle roam the uncultivable hilly or arid parts of the country, converting scrub and grassland, straw and chaff, into high-quality meat. In all, Pakistani farmers produce an enviable range of about forty different kinds of vegetables and twenty kinds of fruit. Unlike in Europe, a patch of ground in Pakistan can yield up to three harvests in one year. And yet, partly because of inadvertent wastage, millions of its population go hungry.

Islamabad is situated in the Punjab, the nation's breadbasket and the backbone of its economy. Not far from Sana's home, just beyond the concrete and mud patchwork of Islamabad's suburbs, I came across Umar Hayat, a farmer squatting in a field, stroking his grizzled beard and smoking a hookah while his cousin sipped tea beside him. They looked out across the sway-ing ears of golden wheat, contemplating the area they had cut, and the wall of wheat that still remained standing. Their sickles lay on the ground, sweat clinging to the wooden handles, the blades moist at the cutting edge from the straw's residual sap. As Hayat rested, three fellow workers continued their labour in the sun, rhythmically hacking and throwing fistfuls of wheat to the ground. Weeds were disentangled from the ears and each precious bundle laid in the sun to dry. In the neighbouring plot, another team of men were threshing their harvest with an automatic blower which fired grains into a heap, while the chaff floated slowly down.

With so many eyes watching the process, it is difficult to see how any grains of cereal could go astray. But in fact the initial processes of harvesting, transporting and storing in developing countries lose somewhere between 10 and 40 per cent of har-vests. Millions of tonnes of grains, fruit and vegetables perish – largely through a lack of basic agricultural infrastructure and training. If Western countries are anxious to increase global food supplies, this is one of the first places they could look.

Often it is very simple things that are required. Hayat and his workers say that whereas a combine-harvester would put them all out of a job, 'All I really need is one of those barrels that stand twelve feet from the ground – big and round like this,' signified Hayat with his hands. 'A grain silo?' I asked my transla-tor. 'That's right,' came the reply. 'Then I could safely store my share of the harvest, and sell any surplus when the market is right.'

Hayat is on the second rung of the social ladder of the Pakistani peasanthood. One level above a day labourer, he is a crop-sharer, who cultivates the fields of a feudal landowner and in return keeps half the harvest. But since Hayat has no capital, he cannot afford to buy farm equipment. The most he has ever been able to invest in grain storage was eight years ago when he spent Rs 2,500 (about $50) on metal bins. This meant that his wife no longer had to build the mud-and-straw dome *bharolas* in which south Asian farmers have traditionally stored their grain, and his bins are certainly safer containers than the jute bags used by many of his compatriots. Although this puts him at a significant advantage, it is also true that metal drums need to be carefully designed for grain storage, particularly if the grain has not been dried properly or if there is insufficient ventilation, because moisture condenses on the inside of the barrel causing mould. But Hayat says he has solved this problem, as well as that of rodents and insects, because he always drops six poisonous pills into a matchbox and seals them in with his wheat. This, he assures me, will keep it fresh for up to ten years. These are aluminium phosphide tablets, an effective fumigation method widely used in the West, but nevertheless a dangerous substance that has caused scores of deaths from mishandling.[5]

Despite being relatively well-equipped, Hayat's drums only store enough grain for his own consumption and that of his twenty-strong extended family. He has nowhere to keep any surplus, and because he invariably has debts and bills to pay off, he has to cash in his wheat as soon as it is harvested. Inevitably, since all other share-cropping farmers are doing the same thing at exactly the same time, this means he is always forced to sell his wheat when the price is at its very lowest. Only those with plenty of money, Hayat explains, like his landlord, can afford to store large quantities of grain for longer periods. There is a paucity of recent research on the current situation in Pakistan, but many of the problems are still the same as when, in the 1980s, 80 per cent of Pakistani farmers were found to be losing up to 7 per cent of their grains in storage because they were damaged by insects and other pests. Much of this 'spoiled' grain is fed to livestock and is thus not entirely wasted, but, even as animal feed, infested grain is a severely degraded resource.[6]

Although Hayat's long-term grain storage problem could be

solved with a small loan or credit facility and some expert advice, he and millions like him are constrained in their effort to boost their incomes while also contributing to storing their country's food supply.[7] Instead, all his surplus ends up in large storage units owned by grain mills or government agencies. But the tragedy is that even these are insufficient. Extraordinary though it seems in an overwhelmingly agricultural country, a large proportion of the nation's grain stays outside under tarpaulin sheets, often without permanent plinths to hold it off the ground, or is stored in old and leaky barns, where it goes mouldy and is eaten by weevils, grain-boring beetles, moths, birds and rats. In a report published in 1994 which surveyed earlier studies, grain stored for two years in state-run facilities was being destroyed by insects at an average rate of 9 per cent, and in some cases up to 15 per cent.[8]

Wheat is by far the most popular staple food in Pakistan. Around 20 million tonnes of it are grown every year. Most Pakistanis eat it every day in the form of the numerous unleavened flat breads south Asia is famous for – chapatti, paratha and naan. But despite the prevalence of malnutrition in the country and increasing shortages, there are still debilitating gaps in the efforts to increase the national food supply. Pakistanis take advantage of many modern agricultural technologies – from new crop hybrids to nitrogen fertilizers, fungicides and pesticides. So why has there not been matching investment in relatively simple aspects of post-harvest technology even though investing in grain storage has frequently been a cost-effective way of preventing losses?

I arrived in Pakistan overland from China in the late spring of 2008, and traversed the country from its northernmost tip in the Karakoram mountains down to the Punjabi plains. From the border crossing, the road passes through a barren rocky land inhabited by marmots and ibex, until at last the scree gives way to the shimmering green tongue of cultivated land which is the Hunza valley. At this high altitude the wheat was standing stiff and green, and the potato fields were bare, waiting for the new shoots to appear. The villages that cling to bends in the river are oases, surrounded by terraced fields, and beyond the fields bright green poplar woods supply timber and fuel. Men and women labour in the fields together in this comparatively liberal Ismaeli region. Drystone walls line the fields whose furrows and ridges are hand-moulded so that every square yard is under production. In contrast to the litter-strewn shabbiness of much of the rest of Pakistan, these peaceful settlements appear at first sight to be mini-paradises. But this tidiness, the minute care devoted to each clod of available soil, is also a sign of desperation. The communities in this marginal rocky place only just manage to cling on to the world, between snowy peaks and the thundering river below. Wasting a patch of soil is an unthinkable luxury.

And yet the Northern Areas suffer from post-harvest losses as badly as anywhere else in Pakistan, largely because this remote land has received little development funding. In one wide survey of post-harvest losses in Pakistan, research was carried out in the other four provinces – Punjab, Sindh, NWFP and Baluchistan – but none in the Northern Areas at all.[9] Rainfall in these mountains is less predictable, making cut sheaves vulnerable to spoilage. Most farmers have only very small holdings and lack the funds to purchase modern storage facilities or grain-drying equipment. By consequence, at least 10 per cent of the wheat harvest, and sometimes much more, goes to waste each year.

NWFP, a relatively impoverished state, has many of the same problems. Zahir Shah, who works for a development NGO near Besham in Kohistan District, is troubled by the crops that are being destroyed as a result of the two-year-long militant uprising in the neighbouring valley of Swat, but he is also concerned about what happens after the harvest because of long-term infrastructural problems. 'Grain is stored in covered places in sacks piled up on the ground. People do not even put them on pallets to keep the moisture away,' he explains, 'One reason for this is lack of awareness, but even when traders are aware, they are worried that investing to solve the problem will not maximize profits. But clearly this would help alleviate our food insecurity problem.'[10]

This is a classic case where investment may not appear to be a priority, or even viable, for an individual trader; but economically for the nation as a whole, it would pass any reasonable cost-benefit analysis because of the urgent need for larger food stocks.[11] Access to markets is the other obstacle to efficiency in the region: the surplus of small farmers is such a tiny proportion of the market share that they can never get a good price for it, and this discourages production. If small farmers joined up in co-operatives, says Shah, they would have more power. The agriculture minister for NWFP agreed, saying that despite local people's diets lacking fruit and vegetables, 'due to non-availability of proper preservation, most of what we do grow is wasted.'[12] These areas exhibit the difference between waste incurred in rich countries through apathetic neglect and that generated in poor countries simply through lack of funds or easily available knowledge.

As I travelled south through the mountains, the season changed with the altitude, and instead of green shoots and newly ploughed soil, there were soon hillsides full of people harvesting wheat by hand. Foothills eventually gave way to the enormous plains of central Pakistan, where miles of cereal crops stretch out in a seemingly endless expanse of food cultivation. Here in the comparatively wealthy Punjab, some farmers have abandoned hand-scythes in favour of small and battered combine-harvesters. It remains controversial whether these industrial beasts reduce post-harvest losses or not. On the one hand, they increase the capacity for mechanized bulk handling, which can be a way of ensuring grains get to stores and markets more efficiently. On the other, the design of some of these old machines can mean that straw, which is a valuable source of animal fodder in the sub-continent, is lost. Indeed, at Rs 100 ($2) for a 40 kg bale, straw represents a significant chunk of the total value of the crop.[13] Thus in Pakistan there are both proponents and opponents of increasing mechanization, and it is often the clash between the two which results in bottlenecks and waste in the food supply system. For example, some Western-style grain stores have been built to receive loose grain harvested and transported industrially, but these have proved incompatible with the manually transported jute sacks still widely used in Pakistan.

Faisalabad is a former British colonial town (laid out in the shape of a Union Jack) in the centre of the Punjab, and it is here that the Agricultural University of the Punjab's National Institute of Food Science and Technology is based. When I visited the chairman, Dr Faqir Mohammad Anjum, he was keen to talk about post-harvest losses. 'If we need more food,' he pronounced, 'we should at least save what we are currently producing: this should be the main thrust of the government.'[14] The director of the university, Professor Iqrar A. Khan, likewise argues that preventing post-harvest losses would be enough to make Pakistan entirely self-sufficient in wheat. 'The government should enhance its storage capacity,' he wrote in the national broadsheet *Dawn*; adding that credit should be offered to farmers so that they can store their crops.[15]

The UN's World Food Programme calculated that 12.5 per cent of Pakistani wheat was lost from the field to milling, not including retail and consumer waste.[16] Sahib Haq, who manages the UN's food emergency programme in Pakistan, told me of some pitiful cases he had witnessed in which the harvest from entire districts was destroyed because peasants stored their grain in primitive underground pits covered in tarpaulin. Floods came and rotted the harvest.

In addition to losses in grains, Dr Anjum highlighted the waste incurred in the dairy and meat industries. Pakistan is the third-largest milk producer in the world but of its 38 billion litres per annum it wastes up to 15 per cent. The government's monolithic concentration on increasing production, he said, ignored the more pressing and simple problem of losses in dairies, slaughterhouses and butchers. The country should be focusing on increasing the capacity for cooling and processing, he said, and on training meat handlers.

Pakistan is also the fourth-largest date producer, but, again, the lack of processing facilities means that its share in the world market is negligible. Neighbouring Iran has successfully invested in waxing facilities to extend the shelf-life of dates for export, but Pakistan has seen much of its harvest spoiled by flies and high temperatures in marketplaces. Many older traditional traders still sit in open-air bazaars with their produce out on display, exposed to the sun and insects, rather than storing it in 'invisible markets' where stock is sold in bulk on the basis of samples.[17]

Nor has Pakistan invested in the lucrative processing of fruits into juices and other value-added products. Dr Anjum has his eye on by-products like citrus peels and mango stones which in Pakistan are left to rot, even though countries such as Denmark import them to extract useful materials such as pectin, and other developing countries such as China have a booming export trade.[18] Even waste among consumers is still a problem. One of Dr Anjum's colleagues told a story of visiting a mosque's madrassa education establishment where chapattis surplus to requirements were stored in a backroom and sold for cattle feed. 'Since seeing that,' he said, 'I have never donated money to mosques.'

Stories of unnecessary agricultural losses are replicated in many developing nations. In addition to all the food wasted in affluent countries, there are millions of tonnes of food that could be saved even in hungry parts of the world. The Green Revolution in the 1960s and '70s brought new crop strains, machinery, pesticides, fertilizers and other chemicals to world agriculture, and these boosted yields dramatically. Western corporations have made a fortune exporting their hi-tech agricultural solutions. What has been left out to a very large extent is the simpler stuff of grain stores, drying equipment, fruit crates, refrigeration and other essentials of post-harvest technology. These offer less in the way of corporate profits, but they could yield greater benefits for overall food availability. Expensive high-yielding strains have even sometimes been part of the problem: traditional varieties were adapted to the environments in which they were grown and stored, having a lower moisture content in ripe grain and thicker husks resistant to rodents, insects and moulds, which meant they could survive in storage until the following season's planting.[19]

This neglect of post-harvest losses is one of the mysteries of world agriculture. As the agronomist Professor Vaclav Smil puts it, there is 'an inexplicable lack of attention . . . paid to post-harvest losses'. While the world's largest agricultural electronic database, *Agricola*, lists tens of thousands of publications relating to increasing crop yields, a search of 'post-harvest losses' for the 1990s comes up with just twenty items.[20] In 1981 the FAO suggested that reducing post-harvest losses 'draws its importance not only from a moral obligation to avoid waste, but also because . . . it requires fewer resources and applies less pressure to the environment in maintaining the quantity and quality of food than through increased production'.[21] But despite the World Food Conference declaring it a development priority as long ago as 1974, and a UN resolution being passed the following year calling for a 50 per cent reduction in post-harvest losses over the following decade, it still remains a vastly under-funded dimension of the development process.[22] Foreign aid dedicated to improving developing-world agriculture has fallen globally from 20 per cent of Official Development Assistance in the early 1980s to 3 or 4 per cent by 2007,[23] and only 5 per cent of investment in research and promoting agricultural improvement is directed at reducing post-harvest losses.[24] As the FAO declared, 'It is distressing to note that so much time is being devoted to the culture of the plant, so much money spent on irrigation, fertilization and crop protection measures, only to be wasted about a week after harvest.'[25]

Rich countries have invested heavily in overcoming these

accidental losses in their own countries, particularly since the Second World War, when improving domestic food supplies was considered a paramount political necessity. The fact that they have merely re-introduced similar levels of waste by imposing unnecessarily stringent cosmetic standards, or because they can afford to over-produce, should not distract from the fact that the efforts to reduce accidental losses have been enormously beneficial. Ways of drying and storing grains, beating the weather, transporting fresh fruit and vegetables, preserving, cooling and processing products so that they last long enough to reach consumers have been a major success story in Western agriculture. Under optimal weather conditions in rich countries, staple grain crops such as wheat can be harvested with losses as low as 0.07 per cent.[26] As long ago as 1802, one philanthropic organization in England expressed delight over the nation's improved agricultural technology and the investment of capital 'to form and establish a storehouse of corn and dry food . . . as a guard and security, against the period of scarcity. By this has been obviated an inconvenience, under which every individual in this land did formerly suffer; that of the annual waste and profusion of food immediately after the corn was gathered.'[27]

But poor nations are still struggling to save their crops from the ravages of nature. Nearly every bug, fungus, bird and rodent on the planet wants to get its metaphorical hands on the produce of cultivation. It has been the constant effort of humans since the origins of long-term food storage more than ten millennia ago to fend them off. But the battle is still raging.

Published figures on the exact levels of waste have usually been based on out-dated estimates, and very few precise studies have actually been made – which is a symptom of the neglect this issue has received. Reliability of data is also often questionable because figures are sometimes manipulated – either to exaggerate losses to encourage aid donors to part with their money, or to minimize them so as to avoid political embarrassment. In 1993, China lost 15 per cent of its grain harvest; up to 11 per cent of the nation's rice was being destroyed because peasants stored it in poorly maintained buildings.[28] In Vietnam, similarly, 10–25 per cent of rice is normally lost, and in extreme conditions this can rise to 40–80 per cent.[29] Across Asia, post-harvest losses of rice average around 13 per cent, though in Brazil and Bangladesh losses are recorded at 22 and 20 per cent respectively.[30] Vaclav Smil estimated that if all low-income countries were losing grain at a rate of 15 per cent, their annual post-harvest losses would amount to 150 million tonnes of cereals.[31] That is six times the amount that the FAO says would be needed to satisfy all the hungry people in the developing world.[32] And yet experts suggest that it should be possible to bring developing-world post-harvest losses of cereals and tubers down to just 4 per cent.[33]

When grain is stored in bad conditions, even the salvaged portion, while edible, will be nutritionally degraded – levels of amino acids such as lysine can fall by up to 40 per cent in storage, as can thiamine and carotene.[34] Raw statistics on food losses therefore underestimate deficiencies in available nutrition.

In the year 2008, *Homo sapiens* became a majority urban species. Food now has to travel further from farms to mouths. A farmer who was once producing only for his local village may now be trucking his produce to cities hundreds or even thousands of kilometres away, requiring technology and know-how with which farmers and traders are not always familiar. Many economies have recently liberalized trade, partly because of pressure from the World Bank, and consequently grain storage, which was a state concern in countries such as Pakistan, is now being run by private traders who often lack expertise. Many of these problems can be solved merely through the dissemination of knowledge.

Supplying these newly populated, rapidly growing cities with plenty of quality food is essential if widespread social turmoil is to be avoided. In the past, food shortages in urban areas have helped to trigger numerous revolutions – Paris in 1789, several

European cities in 1848, Russia in 1917. The food riots of 2008 in Egypt, Cameroon, Côte d'Ivoire, Senegal, Burkina Faso, Indonesia, Madagascar and Haiti are no exception.

Cereal losses are particularly damaging because of their role in supplying the staple calories for the majority of the world's human population. But the level of waste in perishable foods is often far higher. Dairy produce is highly susceptible to waste owing to a lack of technology such as refrigeration and pasteurization on farms and in markets. In east Africa and the Near East alone, milk losses amounted to $90m in 2004; in Uganda they constituted 27 per cent of all milk produced. Provision of modest levels of training and some equipment has the potential to raise income for farmers and improve local diets, and would remove the need to import dairy products into the region: dairy imports in the developing world as a whole increased by 43 per cent between 1998 and 2001, which the FAO claims is 'unnecessary and could be reduced by the simple expediency of post-harvest loss reduction'.[35]

The lowest hanging fruit of all in terms of quick cost-effective savings is, appropriately, in the fruit and vegetable sectors. Waste of these valuable foodstuffs occurs even in places where people are not getting anything like enough of them to eat.[36] Fruit and vegetables not only supply vital micronutrients but make the predominantly herbivorous diet of many of the world's poor more palatable and enjoyable. The African staples cassava and yams have a short shelf-life and there is little tradition of transforming them into more stable products like flour, so they rot in the barns of the hungry. Similar examples exist all over the world – a tragic loss of income for farmers, and of healthy, tasty food for undernourished populations.

Sri Lanka reportedly loses fruit and vegetables at a rate of 30–40 per cent, or 270,000 tonnes, annually, with a value of approximately 9 billion Sri Lankan rupees (US$100m).[37] In the main produce market in the capital city of Colombo alone – a city where thousands cannot afford to buy enough fresh food for a proper diet – the Municipal Council discards an estimated 11 tonnes of fruit and vegetables every day. Three quarters of the nation's fruit and vegetable post-harvest losses, argues the Sri Lankan Institute of Post-Harvest Technology, could be eliminated through relatively simple measures. At present, a great deal of the country's abundant supplies of fruit are thrown into poly-sacks and trundled on bumpy roads in the tropical heat many miles to market, by which time much of the farmer's hard work has been reduced to a sweet and sticky mess. Introducing reusable wooden or plastic crates in which fruit and vegetables can be carefully stacked at the point of harvest – as is done in affluent nations – would solve this problem almost immediately. Likewise, much could be achieved through education; for example, teaching farmers the right moment at which to pick fruit to maximize shelf-life. The exact position on the stem at which to pluck particular fruit can also have a significant impact on its susceptibility to decay. In markets, cooling systems can be developed by using nothing more complicated than shade and water.[38] Adopting new methods such as these can make a huge difference. In recent years a variety of projects in Sri Lanka have cut waste levels from 30 down to 6 per cent, and increased farm incomes by up to Rs 23,000 ($256) per hectare. However, despite this good work, efforts in Sri Lanka are chronically underfunded; even the government, which is subsidizing reusable plastic crates for fruit farmers, can only afford a fraction of the number required.[39]

According to one Pakistani study, the prevalent method of harvesting mangoes in south Asia using a stick and a locally made bag can cause a lot of fruit to fall to the ground and could be improved by more careful harvesting with a blade or crook with which to cut fruit free. A bruise incurred at this stage may not be visible within the first day or so of harvesting, but it soon becomes a broken defence mechanism that lets in an army of insects, fungi and bacteria. In Pakistan alone, it is estimated that mangoes worth more than Rs 1bn are wrecked each year, and

half of this could be avoided by better harvesting techniques.[40] As
Daniyal Mueenuddin – a Pakistani novelist and mango grower in
the Punjab – told me, 'it's all about motivating and monitoring
the men involved in the operation.' To illustrate that this was
no new idea, he quoted Robert Frost's 1914 poem, 'After Apple
Picking':

> There were ten thousand thousand fruit to touch,
> Cherish in hand, lift down, and not let fall.
> For all
> That struck the earth,
> No matter if not bruised or spiked with stubble,
> Went surely to the cider-apple heap
> As of no worth.[41]

Pakistan and Sri Lanka's giant neighbour, India, faces many of
the same problems, but on a much larger scale. India is the
third-largest agricultural producer in the world, growing 41 per
cent of all the world's mangoes, 30 per cent of its cauliflowers,
23 per cent of its bananas and 36 per cent of its green peas, and
it is the third-largest cereal producer with 204 million tonnes of
food grain each year. With an annual output of 90 billion litres,
it extracts more milk from more cows than any other country.
And yet, it has only a 1–1.5 per cent share in global food trade
and only processes around 2 per cent of its produce, in contrast
to some developed countries which process 60–70 per cent.
Estimates suggest that 35–40 per cent of its fruit and vegetables go to waste.[42] According to one media story in 2008,
P. K. Mishra, the secretary in the ministry's Department of
Agriculture, claimed an even higher figure of 72 per cent losses.[43]

Such figures seem incredible, and it is true that there are many
food-processing companies who stand to profit from government development grants and may therefore make exaggerated
claims. But it is nevertheless clear that there are fundamental
problems. The Agricultural Minister Sharad Pawar blames it on

'huge post-harvest losses arising out of inadequate storage, cold chain and transport infrastructure'.[44] For example, a survey of fruit and vegetable wholesale markets in India revealed that 17 per cent had no covered shops and only 6 per cent had cold-storage facilities.[45] (The cost of stalls often varies according to which are in the full blaze of the sun and which in the shade.) In total, it has been calculated that Rs 518bn (about $12bn) of food is wasted in the Indian food industry every year, much of this owing to the lack of infrastructure.[46]

Some attempts to solve these problems have failed, but those which were well designed and executed have transformed rural societies.[47] The micro-credit facilities of Grameen Bank, pioneered by 2006 Nobel Peace Prize-winner Muhammed Yunus, have helped villagers to invest in rural enterprises, and similar low-interest loans have been directed at building infrastructure that reduces post-harvest losses. The 'Village Community Granaries' scheme in Madagascar helped 27,000 small farmers store 80,000 tonnes of paddy rice, increasing output by 50 per cent.[48] In Benin, beans were placed in hermetically sealed storage containers, which meant that insect larvae infestations were asphyxiated; yams were stored in houses on stilts to help control humidity.[49] In rural Nigeria, major losses in cassava occur during traditional methods of harvesting (14 per cent), handling (9 per cent) and processing (23 per cent). But in the 1990s the International Institute of Tropical Agriculture (IITA) invested in village processing centres in Nigeria that more than halved the losses in processing and cut labour hours by 70 per cent.[50] In the wake of the recent food crisis the government of the Philippines – one of the countries worst hit by rice price rises – announced heavy new investment in rice-drying machines to address the losses of 25–50 per cent (by value, taking into account losses in quantity and quality) suffered by south-east Asian rice growers.[51] In Timor, the UN has funded local blacksmiths to construct hundreds of small grain silos and given training to farmers and householders – all in an attempt to save food that is already being produced.[52] During the mid-1980s, the UN helped 9 per cent of Pakistani farmers in non-irrigated areas to invest in metal grain storage containers like Hayat's in order to replace jute bags and *bharolas*, and this cut those farmers' storage losses by up to 70 per cent. Simultaneous projects aimed at eliminating rat infestations boosted yields by 10–20 per cent.[53] But a vast proportion of farmers continue to use sub-optimal methods of grain storage and are still at the mercy of moths, rodents and mould.

So if the severe problem in Pakistan could be alleviated by such simple things, why is more not being done? I conducted a series of interviews with ministers and secretaries in the departments of agriculture and food in Islamabad, as well as policy advisors from the UK and US. Despite the fact that Pakistan was suffering acute food shortages at the time, the majority of the government and non-government personnel I met seemed to be nonchalant about the potential for reducing post-harvest loss. Some ministers were unwilling to talk about it at all, possibly to avoid accepting responsibility for the food crisis at a time of political turmoil. President Musharraf's government had just been voted out by a populace disenchanted with the cost of living and in particular by the high price of flour; in July 2008 one survey found that for 70 per cent of the population the most pressing issue was not democracy, corruption or terrorism, but inflation, with food prices up by one third. The new government had been voted in partly on the promise to bring change, but it was failing to deliver.

My quest to find someone who would talk frankly about the country's shortcomings took me one rainy evening to a house at the foot of the Margalla hills in the north of Islamabad, home to Muhammad Shafi Niaz, retired Founder-Chairman of the Agricultural Prices Commission. Retired people, as I found time and time again in my enquiries, are often the most independent critics of the status quo. 'Avoid speaking to government officials,' Niaz told me, 'they cannot tell you the truth because they are afraid. I also suffered sometimes because of talking freely, but I survived until now. I am lucky.'

As advisor to Musharraf's cabinet between 2000 and 2003, Niaz had warned then of the lack of agricultural storage in Pakistan. But despite the ministry employing several highly paid experts, the strategy documents the government produced in his view 'were not up to the mark'. Niaz agreed to inspect grain stored across the country and found that though much of it was infested with weevil, it was still being milled into flour and sold on the market. 'My driver,' he said, turning to the servant beside him, 'needs to buy flour in these shops; I have gone with him, and the flour there is black and so bitter as to be inedible. Many people are being made sick after eating this grain.'

Niaz assessed new measures to increase storage, but what he found was cause for more concern than if nothing was being done at all. There had, he said, been some efforts to remedy the shortfall by constructing four or five industrial-size grain silos. He visited one of them in Quetta which had been designed so badly that it was impossible to withdraw grain from the bottom section, so residual grain was left there, going mouldy. Another outside Karachi had been all but completed several years ago but, on a visit in 2001, he found that the lock was rusty: nothing had ever been stored there. He rang the army general responsible. The general refused to say why he had received the money to build the silo without ever putting it to use.[54] One head of a development NGO in Pakistan expanded on this story with a more menacing suggestion: 'For some mysterious reason,' he said, 'investment is non-existent. There are beneficiaries of the grain shortages in Pakistan – those who import and export.'[55]

Governments all over the world use grain storage as a way of rewarding certain powerful lobbies. In rich countries, governments have historically given generous subsidies to farmers even when this stimulates over-production, and have then used tax-payers' money to stock-pile the resultant grains.[56] In some

developing countries, civil servants and the political elite control food storage, and this can be used as a means of dispensing patronage to anyone involved in food procurement and
distribution.[57] Thus, while a grain silo for a farmer may seem
like nothing more than a twelve-foot-high metal drum, to
those in power it is an instrument of economic and political
independence which they are not always willing to encourage.

There is one more indirect cause of the food crisis in Pakistan
related to waste – the putative connection between hunger there
and gratuitous waste of food in far-off, wealthy countries. Its
weight came home to me one evening in Islamabad, at a high-
class function to which I had been taken by a group of ex-pat
friends. After paying Rs 2,000 ($40) at the door – enough to
buy a Pakistani five months' worth of flour – guests were ushered
onto a carefully trimmed lawn where tables had been laid for
dinner. After eating a more lavish meal than I had seen for
months, washed down with an array of banned alcoholic beverages, I started chatting to a Somali-born official from the World
Food Programme. 'The privileged of this world,' he spat, 'are a
bunch of selfish, wasteful motherfuckers. Someone should write
a book about this food waste issue. Get the facts out there.'[58]
We wandered into the food-hall, where a dozen vats of luxury
dishes sat getting cold. There were piles of seafood, beef stews,
chicken fricassees, salads, and enough rice and bread to feed
a hundred hungry people.

Outside the walls of the compound the next day, I spoke
to share-cropping peasants as they sweltered in the sun, and
slum-dwellers who could not afford enough flour for their
children's meagre sustenance. In Pakistan, demand for food
was outstripping the supply in the market, pushing up prices,
and people were going hungry. They were suffering from the
same spike in food prices as the rest of the world, and for some
of the same reasons. The proximity of this wasted feast of the
elite to the want of those on the other side of a brick wall was
grotesque. But is the food wasted in the West at even more

lavish and numerous occasions not just as responsible for hunger in Pakistan?

In theory, Pakistan should be relatively well insulated from the global strain on food supplies that affected the rest of the world so dramatically in 2008 because it is nearly self-sufficient in wheat and has trade barriers which aim to protect it from global market fluctuations. The government controls imports to protect its farmers from cheap grain grown overseas; it also controls exports to prevent farmers selling their grain before local demand has been fulfilled. If the government stopped doing this – as ultra-liberals in the West demand – the world's rich would be able to outbid the poor for food, leaving nothing for Pakistanis to live on. The government also guarantees the selling price of grain to protect farmers from fluctuations. Fixed pricing is also supposed to protect consumers from food price rises.

However, none of these measures works perfectly. In a country like Pakistan, imports and exports do occur, both with and without government permission. The greater the disparity between prices inside and outside the country, the greater the incentive for smugglers and corrupt officials to initiate cross-border trade. Even a small amount of exposure can send prices up or down and thus food prices in Pakistan demonstrably follow global trends. The food market in Pakistan has never been completely free from fluctuations in world prices; and 2007–8 was no exception: the country both exported and then later had to import millions of tonnes of grain, both under government-granted contracts and also through the activity of smugglers along the borders with India and Afghanistan. The same is true in nearly every corner of the globe – and thus virtually no-one is free from the effects of global food price fluctuations, and thus indirectly from the effects of Western consumption and profligacy.[59]

I asked the group that had gathered round Sana's slum what they thought about Western levels of waste. 'Why are you wasting?' demanded Sana's son Mohammed: 'You should think

of someone else's world. You could send that food to the poor
in other countries, or even to your own poor.' 'Every parent
tries to make sure their children get enough food,' said Sana's
brother, 'but if this continues, people will come out onto the
streets and start robbing and rioting because no-one can go
without food.'[60] His prediction rang true. Army troops have
been deployed in Pakistan to prevent mobs from seizing food in
fields and warehouses. There is not much Pakistanis can do
about global food shortages or Western wastefulness other than
riot – or draw international attention to their plight, probably
by starving to death in significant enough numbers to make
a news story or to force a politician's hand.

Appalling though it is, in some ways it is encouraging that
millions of tonnes of food currently go to waste unnecessarily
both through the indifference of the affluent and the accidental
post-harvest losses in developing countries: it means that much
more food could be made available comparatively easily. If the
world needs to bring more grain onto the world market, the
vast pit of spoiled grain in developing countries would be a
sensible place to begin foraging. The developing world would
benefit from investment in agricultural technologies to prevent accidental losses, while the industrialized world should rein
in its profligacy. These distinct measures to address two very
different kinds of waste could help to improve the lives of the
poor. It is more sustainable to increase available food by reducing
waste than by chopping down virgin forests to increase cultivable
land.[61] International aid agencies, governments, individual
donors as well as food corporations and consumers both in
affluent countries and the developing world can help to make
more food available without the need to chop down a single
extra tree.

11. The Evolutionary Origins of Surplus

And in the seven plenteous years the earth brought forth by
handfuls. And [Joseph] gathered up all the food of the seven years,
which were in the land of Egypt, and laid up the food in the cities
. . . And the seven years of dearth began to come, according as
Joseph had said: and the dearth was in all lands; but in all the land
of Egypt there was bread.[1]

Genesis, 40:47–54

Many people assume that society's blasé attitude to wasting food
is a recent phenomenon and that in the past people were more
frugal, and food was too valuable to discard. If this were true,
rectifying our current levels of waste would simply be a matter
of reverting to earlier customs. But the history of human wastefulness has deeper roots than late capitalism or consumer culture.
Waste is a product of food surplus, and surplus has been the
foundation for human success for over 10,000 years. Everything
we call civilization depends upon it.

When we talk about food waste, it is essential to differentiate
between inevitable inefficiencies and gratuitous wastage that
actually harms our long-term prospects. Some waste is adaptive
and desirable; some is maladaptive and destructive. If we are
currently indulging in the destructive kind, what are the social
and evolutionary forces that make us behave with such apparent
irrationality?

Archaeological records suggest that some early humans actually treated their food with a profligacy that matches that of
modern supermarkets. When people first walked southwards

across the American continent from Alaska down to Patagonia around 12,000 years ago, they encountered continent-sized herds of docile animals. In contrast to the animals of the African savannah, which had evolved alongside our ancestors for 2 million years, American species had no previous knowledge of human predators and hence a very limited capacity to evade them. Equally, having never faced such easy prey before, the human hunters probably had no idea how to regulate their hunting sustainably.

Giant sloths, woolly mammoths and bear-sized rodents fell to the collective human onslaught. Seventy-five per cent of America's large animal species were wiped out in barely more than a millennium, with climate change being a debated contributor to their demise. Archaeological remains of mammoths hunted by humans at this time reveal that only a small proportion of bones show signs of having been butchered. Early hunters could have cut the carcass into strips and dried it out to preserve the meat, but instead it seems that they left much of it to rot. It was apparently more convenient to move on and make a fresh kill than go to the effort of preventing dead meat from reaching its use-by date too rapidly.[2] The fact that this practice led to the extinction of their preferred prey is an alarming legacy – and it is one we, in our own way, are still pursuing by our over-fishing of the oceans. The half-finished carcasses of woolly mammoths strewn across the continent in the wake of America's first humans are the forebears of modern fish discards and supermarket garbage bins, packed full of butchered animals and stale groceries, all sacrificed on the altar of human rapacity.

After most American megafaunal species were extinct, people had to seek other sources of food; it was apparently in this context that the evolution of agriculture in the Americas occurred. Hunter-gatherers who had formerly collected the wild ancestors of maize and potatoes started to domesticate and cultivate them, replacing the nutrition that had previously come more exclusively from gathering wild plants and hunting large animals. A parallel scenario occurred in the Fertile Crescent – stretching from modern-day Jordan to Iran – where agriculture had emerged thousands of years earlier. There, people turned to cultivating grains after the enormous herds of gazelle that once roamed the region had been seriously depleted – again by either hunting or climate change, or more probably a combination of both. In Australasia, the first human inhabitants burnt entire forests to capture a few large beasts as they escaped – most other animals were left to go up in smoke, though by the time their rampage was over there were no more suitable species left to domesticate.[3]

Mass extinction of large animals testifies to two things: humans' efficacy as hunters and their disregard for the sustainable use of resources. In the past, when faced with abundance, humans have often reproduced exponentially and gorged themselves on all available resources. In this regard, we resemble other species – exploding rabbit populations, or the cyclical blooms and decays of marine plankton. However unsustainable this may seem, in the past these short-term bonanzas have provided enough food to boost human population growth. They caused the extinction of numerous wild animals and the permanent destruction of large ecosystems. But they also created the conditions under which human settlement developed, agriculture emerged, and the path to modern civilization was beaten.

Just as hunter-gatherers sometimes over-hunted their prey, so when humans turned to agriculture, they often over-exploited the environment until the land became barren. This has occasionally resulted in the collapse of entire civilizations as their resource base became depleted. In innumerable technical articles and the books *Collapse* and *The Third Chimpanzee*, Jared Diamond gives many examples, including the Maya of Central America, the inhabitants of Easter Island, the Anasazi of New Mexico's Chaco Canyon, and the Mediterranean civilization surrounding Petra in the Fertile Crescent itself.[4] But, conversely, the strain on resources sometimes drove people to new levels of ingenuity. As Esther Boesrup argued in the 1960s and '70s, population growth has often stimulated human innovation, producing new technologies and, over the long term, increasing agricultural productivity and standards of living. Constantly over-reaching the supply of food has been, according to Boesrup, an incentive for technological and social development.[5]

In a territorial species such as *Homo sapiens*, the size of a population is a crucial factor in determining its ability to defend or enlarge territory. A group that lives sustainably and keeps its population in check may merely discover that it is outnumbered and overpowered by a neighbouring clan which has grown large enough to overwhelm it. A group that over-exploits its territory may destroy the resource base it depends upon, but if this temporarily boosts its population, it may find itself able to conquer the territories of its neighbours. It is a risky business and unfortunately it rewards unsustainable rapacity, but this is one successful route that humans have taken in their monopolization of the earth.

Around 13,000 years ago, antecedents of the first farmers in the Fertile Crescent developed ways of storing wild grains in pits – and later in ventilated granaries – which (mostly) kept food dry and prevented seeds from germinating.[6] Surplus stored in this way could be used to provide food throughout the year, for trade, and for distribution at feasts which cemented alliances between different peoples. This continuity in the food supply also allowed people to raise more offspring and to live a more settled life, rather than shifting around in a persistent nomadic search for food. Semi-permanent and permanent settlements appear in the archaeological record at around this time, and these gave rise to the greatest revolution in human history – the domestication of plants for food and the development of arable cultivation. At first, plant domestication probably happened by accident: discarded seeds from wild gathered plants grew up where they fell on the ground near human settlements, and people gradually realized the benefit of deliberately scattering them for cultivation. Residing in one place meant that crops grown in this way could be guarded and collected when ripe, while the cultivators lived on grain stored from previous harvests. The revolutionary symbiosis between humans and grasses yielding edible seeds emerged.

The creation and storage of agricultural surplus meant that some members of social groups could specialize in occupations that were not directly related to food production. Artisans, soldiers, priests and chieftains could be fed on the spare food, and so social specializations and hierarchies developed in tandem with growing supplies. The more non-food producers a population could sustain, the likelier it was to be able to defend its territory and invent technologies that would further its aims. (Even in the modern world, European and American powers found that their success in the two world wars of the twentieth century depended as much on their ability to produce food as the sagacity of their generals – leading to propaganda slogans such as 'Food is a weapon: don't waste it! Buy wisely – cook carefully – eat it all'.)[7] Agriculture spread around the world partly by neighbouring groups observing and learning the new technology: but arguably more significant was the fact that people who practised agriculture and grew surplus reproduced faster and conquered anyone who did not.

Sustaining population growth, division of labour and military prowess are the first rationale for the production of food surpluses. Above and beyond these requirements, a population would be well-advised to grow even more food than its basic nutritional needs in case of extraordinary times of scarcity.[8] As the cultural anthropologist Marvin Harris argued, 'An established principle of ecological analysis states that communities of organisms are adapted not to average but to extreme conditions.'[9] The thinking behind this principle is that any population that is not adapted to extremes will die out every time there is a freak environmental event, such as a particularly cold winter or dry summer. In good years, surplus could be stored against scarce harvests in the future – as in the biblical story of Joseph, who, warned by Pharaoh's dream, kept aside 20 per cent of the harvest for seven years and thus averted famine in Egypt. Major grain-producing countries still lay up stocks – currently about 20 per cent of what is actually used – to ensure stable supplies.[10] Producing surplus every year may look like a horrendous waste of good food: but what if one year, or for several years, a catastrophe eliminated a sizeable chunk of our agricultural output? By constantly over-producing, all we would have to do in such a year is waste a little less to avoid inconvenience. This, as we have seen, is exactly what happened after the destruction of nearly half the British potato harvest in 2007.

But the surplus of the West today appears to exceed the population's nutritional needs to such an enormous degree that it is difficult to believe that the level of surplus is either necessary, healthy or safe. If we are to make the global food supply more efficient, we need to determine more carefully the margin between the safety net of essential surplus and unnecessary waste.

The first question to ask is: how much food do we actually need? Taking into account the different requirements of men and women, children and adults, on average in the West – where there is a large ageing population and prevalence of sedentary urban lifestyles – the FAO estimates the minimum energy requirements of western Europeans and Americans to be between 1,900 and 2,000 kcal per person per day.[11] Agronomists reckon that in order to guarantee food security, nations should aim to supply around 130 per cent of nutritional requirements. A supply of 2,600 to 2,700 kcal per person per day would therefore be sufficient for affluent countries.[12] The question of desirable surplus has not received anything like the attention it deserves, and it would require extensive historical surveys of human populations to test how successful this level of surplus has been in insuring against famines. But from the evidence and expert opinion available, it seems that 130 per cent of needs is a reasonable safeguard, though clearly all sectors of the population also need adequate access and entitlement to food, which is a chronic problem in many developing countries. As I shall discuss in more detail in the next chapter, the shops and restaurants of Europe and the US make available to their populations a smorgasbord of nourishment between 3,500 and 3,900 kcal per person per day, or up to 200 per cent of what they physically need. If the edible grains and pulses currently fed to livestock were included, the total available food supply in the US comes to over 400 per cent of the country's energy requirements, and just about every European country is well above 300 per cent.[13] What is the purpose of all that extra food?

Current levels of over-production in the West exceed anything that would be deemed desirable from an agricultural or public health perspective. There may be an argument for individual nations to reduce imports and even increase local production to enhance food security, but on the issue of actual food available to consumers the surplus in rich countries is clearly excessive. So the next key question to ask is *why* we do it?

Marvin Harris argued that 'there generally are good and sufficient practical reasons for why people do what they do.' Though he was critical of Western over-consumption,[14] it was with this pragmatic assumption that Harris attempted to explain some of the world's most counter-intuitive eating habits, arguing that they had evolved as a functional adaptation to environmental conditions, and that they invariably served human material interests. Others, such as Vaclav Smil, have judged that the current level of food waste 'is among the most offensive demonstrations of human irrationality'.[15] But, according to the logic of Harris's theory, modern Western culture ought to obey his laws of pragmatism no less than any other. So what happens if we apply his thinking to the conundrum of wasted food surplus in the modern world? Rather than appalling evidence of idiotic profligacy, are the mountains of rotting meat, croissants and cauliflowers actually evidence of Western capitalism's intricate wisdom? Could food waste actually serve society's interests in hidden ways?

Harris himself identified a number of societies where over-production and over-consumption appeared to have practical advantages. One was the ceremony known as 'potlatch' observed among native American peoples, such as the Kwakiutl, in the American north-west, Canada and Alaska. In the potlatch cere-mony, chiefs invited guests from neighbouring villages and gave away box-loads of fish and whale oil, dried fish, heaps of blankets, furs and ceremonial masks. Fish oil would be poured onto the fire or guzzled in competitive feasting events. Even entire houses were reportedly burnt down in what many believed was a megalomaniac urge to display wealth and power. A potlatch feast was judged a success only if the guests could 'eat until they were stupefied, stagger off into the bush, stick their fingers down their throats, vomit, and come back for more'. Rival chiefs were spurred into competition, hosting their own potlatch giveaways, and failure to match a rival led to loss of prestige. European onlookers assumed that this was a senseless waste of valuable goods, and the Canadian government outlawed the practice from 1885 to 1952.[16]

Despite the assumed 'irrationality' of potlatch, Harris argued that it in fact accrued sophisticated material benefits to society as a whole. Comparable institutions, he observed, could be found in other cultures. In Melanesia and New Guinea, the village 'big man' encourages his friends and relatives to extend their yam gardens, catch extra fish, gather more pigs, and then, in one big feast, he gives away all the surplus goods. Harris believed that the big man was benefiting society by squeezing people into producing more than they otherwise might. 'Under conditions where everyone has equal access to the means of subsistence,' he writes, 'competitive feasting serves the practical function of preventing the labor force from falling back to levels of productivity that offer no margin of safety in crises such as war and crop failures.' The giveaways also serve the function of redistribution between villages that have enjoyed different levels of production as a result of their varying microenvironments – good fishing years on the coast can compensate for bad growing years on land or hunting in upland territories.[17] One retort to Harris's theory could be that Melanesians and the Kwakiutl simply enjoy an occasional blow-out like the rest of us. But this does not in itself explain why so many humans have cultivated or evolved an enjoyment of producing and eating more than their bodies require. This demands an explanation, and Harris's theory applies as much to modern Western cultures as to the Kwakiutl and the Melanesians.

The modern global food system does resemble potlatch in many ways. In industrialized nations, a similar custom goes under the name of food aid: offloading surplus to countries that have a deficit. Food aid donations from Western countries such as the US have been a vital safety valve for domestic over-production, saving farmers from bankruptcy. In 1961 the Kennedy administration had to deal with the greatest food surplus in American history, and it was this that led to foreign aid policies under which, for example in 1966, one fifth of the US wheat harvest was sent to India. At first glance, it might look like donor nations are motivated by altruism. But, just as with potlatch, donors accrue prestige: one only has to look at the proud announcements by industrialized nations of the number of tonnes of food they give away to see that surplus is presented as generosity. Indeed, Western powers in the late nineteenth and early twentieth centuries really did present potlatch chiefs with unmatchable 'gifts' of flour and blankets in an early example of the political leverage of food aid.[18]

Further benefits accruing to donor nations can be summed up by the Eskimo proverb: 'Gifts make slaves just as whips make dogs.'[19] Food donations in the modern world often help stave off famine, but they can also create dependency. If the survival of a poor country's population is threatened by food shortages, then food aid will tend to make them dependent on donors.

They may have to pay for this with political complicity, or through what might be considered inequitable trade agreements. In the US, the Agricultural Trade Development and Assistance Act (1954) made this perfectly clear: sending food to Africa and Asia opened up new markets for American exports, and the threat of denial could be used to exert political and economic pressure.[20] To address developing nations' loss of independence, aid agencies such as Care International in 2008 called for an end to non-emergency food aid.[21]

Over-production and over-consumption in the modern world have reaped material benefits for individuals as well as whole nations. When a powerful person or nation gives away food or throws a lavish feast, they increase their prestige and their number of friends or followers. In *The Theory of the Leisure Class* (1899), the Norwegian-American critic of Western conspicuous consumption Thorstein Veblen wrote that 'Since the consumption of these more excellent goods is an evidence of wealth, it becomes honorific; and conversely, the failure to consume in due quantity and quality becomes a mark of inferiority and demerit.'[22] Individuals certainly behave in this way, but businesses like supermarkets similarly pile their shelves high with innumerable products, aiming to increase their customer base by demonstrating that they provide more abundance than their rivals.

As with the Kwakiutl, over-eating in the West has been another outlet for surplus – but this time on a far greater scale. Two thirds of Americans are overweight, half of those are obese, and nearly 8 per cent suffer from the related condition of type 2 diabetes; Europeans are well on the way to joining them. The steep rise in obesity in the United States since 1980 is strongly correlated with the increasing food supply.[23] Experiments done on rats suggest that eating sugar and fat triggers the release of a chemical in the brain that makes the consumer feel good.[24] For the entire history of mammalian evolution, this has probably been a helpful adaptation, encouraging us to eat when food is in abundance and lay up stores of fat against periods of dearth. (Our agricultural systems, in this sense, replicate what our bodies evolved to do millions of years ago.) However, for the inhabitants of affluent nations, there have been no major food shortages for decades, and so this evolved instinct to over-consume is constantly elicited. These problems are probably best managed by encouraging people to eat less and more healthily, and trying to make healthier food more affordable relative to unhealthy energy-dense food (fresh fish and fruit cost up to five times more per calorie than fast-food meals and soft drinks).[25] But it is certainly the case that supplying more food than we can possibly eat contributes to the problem of over-eating.

Beyond over-eating, we fatten in addition an unprecedented number of livestock, and still there is more food than we can use so we throw a sizeable proportion of it away. Again, this follows a long-established way of using up surplus, but on a far larger scale than ever before.

As early as 1798 the founder of modern demography, Thomas Robert Malthus (1766–1834), drew attention to the vital role that surplus and luxurious consumption played in maintaining slack in the food supply. Malthus looked at the agricultural systems of China and India, where he saw enormous populations surviving on the smallest possible quantity of resources produced in the most efficient way on the available land. The Indians and Chinese, he noted, ate primarily vegetarian diets based on rice and other local cereals. Europeans, by contrast, expended extensive resources in fattening up huge numbers of animals, often wastefully using land to grow animal feed rather than food that could have been more efficiently used for human nourishment.

But there was a catch in the efficient system of the Asians. Because they did not have any slack in the system, Malthus argued, there was famine every time they had a bad harvest: 'It is probable that the very frugal manner in which the [Indians] are in the habit of living contributes in some degree to the famines of Indostan,' he wrote. Malthus regarded luxuries like the wasteful production of meat as buffers against shortage. In extremely bad years, Europeans could avoid starvation merely by wasting fewer agricultural resources on them.[26]

Malthus had good reason to be familiar with such eventualities. Just two years before the first publication of his *Principle of Population* (1798), England had been afflicted by a scarcity of wheat following two consecutive bad harvests. This was compounded by poor yields across Europe and America and was thus not remediable by food imports, which Malthus in any case did not favour precisely because they were not dependable. Both Houses of Parliament and the Privy Council warned that if wheat were consumed at the usual rate, there would be none left before the next harvest: the only option was to use more efficiently what remained. It was time to start eating into those buffers and tightening the slack in the system. The Archbishop of Canterbury issued a letter calling on the wealthy to consume less in order to leave more for the poor. As one preacher, William Agutter, explained, this meant observing with special urgency Christ's commandment to 'Gather up the fragments that remain, that nothing be lost.' 'Waste,' he explained, 'proceeds from ignorance, ingratitude and unthankfulness, from luxury, and want of compassion . . . He, then, who eats more than is requisite . . . is guilty of waste. He heedlessly consumes what does himself no good, and what many really want.' Beyond straightforward wastage and over-consumption, Agutter suggested that keeping unnecessary numbers of animals constituted a waste of common food stocks: 'In times of general or particular scarcity,' he explained, 'it is necessary to omit some articles of food which may neither be luxurious or extravagant in themselves, but which would consume too much of the article most wanted; in which case it is wise and patriotic to restrain where we can.'[27]

At around the same time, the Society for Bettering the Condition and Increasing the Comforts of the Poor issued a report affirming a similar point: 'Whenever the means of subsistence are inadequate to the population . . . nothing, in short, *but increase of food, or improved economy and management in the use of it*, can supply the deficiency, or remedy the evil.' This meant limiting meat consumption and profligate wastage, through the 'increase of the most productive modes of husbandry; as of corn and potatoes in preference to fattened animals, and . . . by instructing the rich, as well as the poor, in a more economical use of food, and in a less wasteful application of the necessary articles of life.'[28] Fattening livestock and indulging in profligacy were ways of converting surplus into luxuries at times of plenty, and providing a dispensable cushion in times of dearth. Wasting resources and consuming unnecessary luxuries were not crimes: they performed the role of a self-regulating buffer, or homeostatic system, in the human agro-economy, arguably benefiting society by stimulating surplus production, which in turn protected it against extreme conditions. It was when resources reached their limits that over-consumption and waste became sinful.

In the Western world today we have a greater buffer against famine than Malthus or his contemporaries ever imagined would be possible. In favourable ecological conditions, producing and consuming all this surplus can be harmless or even useful; but there is a trade-off when the danger of over-exploiting resources is so great that *it* threatens to undermine food security. What if, like the inhabitants of Easter Island, our ecological limits are reached? Profligacy could become a lethal habit.

In the past other societies have reined in their wasteful habits in response to similar ecological limits, and this could hold valuable lessons for us. One of Marvin Harris's favourite examples was the ancient Indians who used to sacrifice cattle as an extravagant display of wealth and power. But cows – the source of milk, manure and farm labour – were worth more to Indian peasants alive than dead, and so at a time of population growth and agricultural hardship a grassroots rebellion erupted against cow slaughter. At first led by the break-away religions of Buddhism and Jainism, both of which railed against meat eating and particularly cow sacrifice, even the Brahmanic elite who had officiated in the cattle sacrifice eventually absorbed their message. From being a creditable display of wealth, cow slaughter became a heinous crime.[29]

A parallel development occurred on the Pacific island of Tikopia 400 years ago. There, farming people had lived for thousands of years in the densest possible populations on the available farmland. After the arrival of pigs with Polynesian migrants in around AD 1200, pork became a primary protein source and a central sign of status. But by 1600 the people of Tikopia realized that pigs ate too much agricultural produce and had become an unsustainable luxury. In a dramatic resource efficiency drive, a decision was made to kill every pig on the island.[30]

Today, rich countries channel surplus food supplies into farm animals, rubbish bins and their own overweight bodies. If there were a global democracy, among the first measures proposed by poorer people would probably be a cull of livestock fattened on cereals and a proscription of the unnecessary waste of food. On Tikopia and in ancient India, people did start doing what was 'good' and 'practical' for them by reining in their wastefulness, even if it took many years for this to be achieved.

As this point suggests, the problem with Harris's definition of benefit to 'society' is that it does not sufficiently separate competing interests within and between individual societies. Supermarket directors may profit from – and therefore have a rationale to instigate – the waste of agricultural resources. Similarly, rich nations may profit from excessive meat production and waste despite the fact that it is fuelled by unsustainable exploitation of the land and the sea. Even though there may be nothing approaching global democracy, however, there is more reason than ever to view society's interests in a global perspective. It is no longer rational for rich nations to deplete natural resources regardless of where they are in the world. Doing so harms those local environments and indigenous people, and it deprives others of food needed for survival, which is no longer morally tenable, if it ever was. Waste may still accrue short-term benefits for a few individuals or groups with vested interests, but for human society as a whole it is potentially catastrophic.

In the past, consuming local resources unsustainably could temporarily fuel the growth and muscle-power needed to overcome neighbouring territories – and this is still what we are doing by encroaching on tropical forests inhabited by peoples less populous and industrialized than ourselves. But it is increasingly evident that doing so threatens to upset the climatic system of the planet, which could have a devastating effect on our ability to grow as much food as we currently do. This time, when the whole planet has been over-exploited, there will be no neighbouring territories left to invade.

12. Adding It All Up and Asking ... 'What if?'

> Mr. Forester. All luxury is indeed pernicious ... but luxury,
> which ... destroys the fruits of the earth ... is marked by
> criminality
>
> Mr. Fax. At the same time you must consider, that ... The waste
> of plenty is the resource of scarcity ... The rich have been often
> ready, in days of emergency, to lay their superfluities aside
>
> Mr. Forester. What then will you say of those, who, in times of
> actual famine, persevere in their old course, in the wanton waste
> of luxury?
>
> Mr. Fax. Truly I have nothing to say for them, but that they
> know not what they do.[1]
>
> Thomas Love Peacock, *Melincourt* (1817)

In the UK, WRAP estimates that the total mass of food waste generated from farms down to people's homes comes to approximately 18–20 million tonnes.[2] This is an initial attempt to agglomerate waste from all sectors in the food industry in Britain, but its accuracy, as WRAP concedes, is very doubtful. Another attempt by researchers at Cardiff University and the Royal Institute of International Affairs suggested that 5 per cent of food is wasted at the agricultural level, 7 per cent in processing and distribution, 10 per cent at retail, and 33 per cent at the consumer level.[3] However, there is no account of losses in commercial and public service catering, nor an estimate of what proportions of the food supply go through these various channels, and the figure of 33 per cent loss by consumers is based on the WRAP study of household food waste and, as already discussed, exaggerates actual losses of food because it includes inedible organic waste such as tea bags and orange peels.

WRAP and other organizations commissioned by the British government are currently undertaking further research to arrive at a more accurate estimate for waste throughout the supply chain and may publish the results in 2009. But without a full-scale investigation involving physical examination of the waste coming out of shops, factories and farms (which will not be possible, partly because of insufficient funding and reluctance from food companies), the conclusions will still rely on self-reporting from the industry and will therefore be tentative.*

In the US, the Department of Agriculture (USDA) published a major study on food waste in 1997; it relied on old statistics, some of them collected as long ago as the 1970s, and it left out waste on farms and a lot of waste in manufacturing.[4] The author – well aware of the shortcomings of the available data – estimated that in 1994 American consumers, retailers and food services wasted 41 million metric tonnes of food, or around 27 per cent

---

* The UK government tenders for research into food waste in 2009 (FFG 0809 and 0810) make no requirement for applicants to make a physical examination of commercial and industrial waste arisings, as was shown to be essential for accurate assessments of household waste. Neither is this within the scope of the Defra-funded studies 'Greenhouse gas impacts of food retailing' (FO0405) and 'Evidence on the role of supplier–retailer trading relationships & practices on waste generation in the food chain' (FO0210). This latter research project, being conducted by Cranfield University and IGD, which is due to report in June 2009, is obtaining its data by interviewing retailers and suppliers from a selection of sixteen case studies (Cranfield University (2008) and David Whitehead, IGD, personal communication, 23 March 2009). WRAP is currently proposing to conduct some other research in collaboration with IGD which could involve some physical measurements of food waste, but this has not yet been agreed (Mark Barthel, personal communication, 20 March 2009).

of the total US food supply.[5] Every year since then, the USDA's Economic Research Service (ERS) has estimated how much of the country's food supplies are wasted and has published them in their 'Loss-Adjusted Food Availability' data sheets. Although they never documented why exactly, the losses on these sheets are in fact significantly higher than found in their study.[6] For example, while the study estimated red meat to be wasted at a rate of 1 per cent by retailers and 16 per cent by consumers, the data sheets have indicated for the past decade that retailers waste 7 per cent and consumers 30 per cent. In 2004 the amount of food that ended up in shops and restaurants for consumers to purchase came to a total of 3,900 kcal per person per day. From this, the ERS estimated that what remained after 'spoilage, plate waste, and other losses in the home and marketing system' was just 2,717 kcal per person per day – a loss of 30.3 per cent of all food between the shops and restaurants and people's mouths.[7]

However, as the ERS concedes, this may still underestimate waste.[8] When American adults have been asked how much they actually eat within a 24-hour period, the wide-scale surveys of 1994–5 found that the average adult intake was 2,002 kcal per person per day (1,800 kcal for women and 2,200 kcal for men). Marion Nestle, one of the foremost nutritionists in the US, suggests that because people under-report how much they actually eat, the likely average intake was in the region of 2,500–2,600 kcal. As Nestle points out, although there is no more recent good data, food consumption has almost certainly risen since then,[9] but as the surveys were conducted on *adults* only, rather than a cross-section of all ages, including babies, children and old people, it seems unlikely that average intake would be as high as 2,700 kcal per person per day, and therefore the amount of US food supplies wasted at just the retail and consumer level seems from these figures to be an almost incredible 30–50 per cent, or 1,200 and 1,900 kcal per person per day.[10]

More recently, Dr Timothy Jones conducted a study of food waste as part of a government-funded project at the University of Arizona in the 1990s and early 2000s. His results, which have unfortunately never been published in peer-reviewed journals, suggested that waste from consumers, food services and retailers was around 20 per cent higher than the earlier USDA study, with a total of 53.8 million tonnes of edible or once edible food wasted (29.3 million from consumers and 24.5 million from supermarkets, restaurants and convenience stores). His work was particularly interesting in that it actually measured the food bought by families over long periods of time, rather than relying, as WRAP did in the UK, on national average purchases. Jones's sample size was relatively small and localized, but if his figures were representative of national averages, they would imply that consumers in the US purchased more food than the USDA estimates is available in the whole country for both household purchases and food services put together.[11] In Jones's opinion, America could well have even more food than has hitherto been estimated, but it is no wonder that the USDA has not adopted his findings.[12] It must be conceded that the USDA food availability estimates do leave out quite considerable quantities of foods; for example, the import and export of processed foods do not always find their way onto the data sheets and could therefore add to the total food supply without being counted. Similarly, edible offal – an enormous potential source of very nutritious food in such a carnivorous society – is largely left out of the food supply data, so its wastage does not even figure. If nothing else, Jones's studies show that much more work needs to be done to find out how much food Americans have, and how much is wasted.

Jones also conducted some studies of US agriculture, and he estimates that around 10 per cent of all food grown on US farms is wasted, with a further 3 per cent lost in storage and transport. Following more limited studies, he estimates that manufacturers and processors waste about 10 per cent of the food they handle (for example, he claims that around 16 per cent of pigs died while being transported to slaughter after road journeys typically of 600 miles and that the pork industry also failed to market numerous edible portions of pork such as offal and trotters).[13] At the commercial catering and food service level, some convenience stores wasted 26 per cent of the food they handled; fast-food restaurants wasted 10 per cent of their food and full service restaurants wasted 3 per cent.[14] He estimates that, in total, food worth an estimated $136bn is wasted before it reaches the belly of the American consumer.[15] If waste at every step in the chain is included, Jones estimates that around half of all food in the US is wasted.

After several years of neglect, the topic of food waste is finally making it back onto the agenda within the USDA. Thanks very largely to the dedication and genuine curiosity of Jean Buzby at the ERS, money has been provided for four new studies on food waste. There is one study looking at the loss of some commodities between farm and retail level,[16] another one tracing the waste of fruit and vegetables at different stages in the supply chain, and another on losses of fresh fruit, vegetables, meat, poultry and seafood in the retail sector which calculates how much food goes into six large supermarket chains using supplier shipment data, and comparing that with how much actually gets sold. The results of this study were published in March 2009, but it did not significantly alter the ERS's existing estimates of food waste overall.[17] But the most interesting research of all is a study that measures food purchases in households and compares that with the food actually consumed in the home as recorded in national dietary intake surveys.[18] This could give the clearest picture yet on how much people buy, how much they consume, and how much they throw away.

It is only regrettable that the US central government has not provided the funds to make a thorough forensic study of the contents of bins in people's homes and in businesses, accompanied by surveys seeking to identify the reasons behind the decisions to waste food. This is the only empirical way of confirming how much food is being wasted, and how best to convince people to stop doing it. In the UK, the WRAP investigation into household food losses cost a mere £420,000; by March 2009 the follow-up campaign to try and persuade people to waste less had cost £4 million: it has already saved British consumers an estimated £300m.[19] The US is a nation of over 300 million people with the largest food supply per capita in the world – and half of it is being wasted. Currently there are only one and a half people in the ERS tasked with finding out why, and what can be done about it: this gigantic problem merits much more investment.

The figures based on observations of waste in the US are significantly higher than a figure that can be inferred from another set of interesting, though very approximate, data – the 'food scraps' content of Municipal Solid Waste (MSW), which includes any food waste collected from residences and commercial outlets (though not industrial manufacturing waste). By sampling portions of MSW, the US Environmental Protection Agency (EPA) comes up with very rough estimates of the proportion that consists of food, and in 2007 it came to 12.5 per cent, or 29 million metric tonnes.[20] Assuming that only 76 per cent of food scraps are composed of once-edible food (the rest being inedible organic matter),[21] this would suggest there are just 22 million metric tonnes of edible food in MSW – around half of the quantity one might expect from the studies by Jones and the USDA. However, the MSW figure does not include food waste that is composted in back yards, fed to animals, poured down the sink or washed away in sink shredding units, and some weight will be lost through evaporation, decomposition and leaching,[22] but it still means that either the sample surveys of MSW need to be made more extensive and accurate, or those measuring food waste in homes and businesses need to explain where all the unaccounted-for food waste goes.

The empirical studies suggest that rich countries waste around half of their food supplies, but at present the data is too scant for an accurate assessment. For the time being, however, there remains another way of estimating total food losses. The FAO publish food supply data on their 'food balance sheets' for each country. The reliability of the data is variable, but it serves as a useful corroboration of measured food losses and it allows us to estimate surplus and waste across the entire globe. The results are shocking. Vaclav Smil used FAO data from 1990 and found that of the 4,600 kcal per capita per day of edible food harvested globally, only 2,000 kcal on average were actually consumed (suggesting that 57 per cent of global food supplies do not reach human mouths). Of the 'lost' 2,600 kcal per person, a large part of this was due to feeding grain and pulses to farm animals, an input equivalent to 1,700 kcal per person per day, which yielded just 400 kcal of meat and dairy products for consumption (a net 'loss' of 1,300 kcal per day for each person on the planet).[23] A further 600 kcal were apparently lost between the field and the food industry, while the waste in distribution, retail, institutions and households came to the equivalent of 800 kcal per day for each person in the world, implying that in total over 30 per cent of the world's food supplies are actually wasted.[24]

The levels of waste at the retail and consumer levels, as Smil points out, are by no means uniform: they are much higher in rich countries, and drop as one descends the developmental ladder.[25] In all high-income countries, says Smil, per capita food supplies are now more than 1,000 kcal per day greater than actual consumption.[26] Just taking the waste at retail and consumer level, Smil calculated that if the rich world cut its food losses to 20 per cent, the annual savings would be equivalent to at least 100 million tonnes of grain,[27] which would be enough to satisfy the hunger of all the malnourished people in the world nearly four times over.[28]

As the example of the US illustrates, surveys of dietary intake invariably suggest that people in developed countries eat much less than the total amount of food available in the country. Empirical studies on food waste demonstrate that a great deal of the discrepancy between supply and intake can be attributed to food waste. However, because dietary intake surveys are themselves subject to enormous degrees of inaccuracy and have not in any case been conducted in many countries, I shall focus on the discrepancy between *nutritional needs* and *food supply*.[29]

As discussed in the previous chapter, supplying a population with 130 per cent of its nutritional needs is believed to provide an adequate buffer against food shortages, as long as people within a country have adequate access and entitlement to food. It should be noted that food supplies in shops and restaurants amounting to 130 per cent above nutritional needs actually hides much greater surpluses embedded in other inefficiencies and waste, particularly in rich countries: the amount of food available does not include the cereals and pulses inefficiently fed to live-stock, nor the diversion of food crops into other non-food uses, nor fish discards; nor does it include the proportion of harvests that farmers leave to rot in the field or feed to animals but which in bad years can be harvested for humans to eat. As Smil points out, 'there is no need for any margin [of 130 per cent] in the West, where we throw away roughly half of what we grow.'[30] The West's inefficiencies in these regards are largely due to cultural choices about the desirability of meat and the 'quality' of vegetable products. For now, I shall return to the issue of the surplus that is actually supplied in shops and restaurants available for people to eat. Anything significantly above 130 per cent has a good chance of ending up as waste: dietary intake surveys do not find that populations actually eat more than 130 per cent of their needs. The human appetite, and the size of our stomachs, is finite.

Although the size of a country's food surplus is strongly corre-lated with levels of affluence, some rich countries appear to supply their populations with a higher level of 'unnecessary surplus' than others, which at least gives an indication that being rich does not mean a country *has* to waste so much food – it is also determined by other cultural factors. The US's 3,900 kcal per person per day (USDA's 2004 estimate) means that it has the greatest food surplus of any nation, with 200 per cent of the energy requirements of the population.[31] According to FAO's 2003 data, the UK, Ireland, France, Belgium, Italy and Canada all supply 170–190 per cent of requirements; the Netherlands, Iceland, Finland and New Zealand between 160 and 170 per cent; Sweden and Australia between 150 and 160 per cent, while Japan falls below 150 per cent, providing a substantial range, despite all of these countries having similar levels of affluence.[32] Neither are high levels of surplus found exclusively in the very richest countries, but also in some intermediate ones like Egypt, Morocco, Tunisia, Mexico and Hungary, possibly because these countries have less well-developed retail and distribution systems which cause a great deal of waste.[33] A chart showing all these countries with food surplus plotted against income levels can be found in the appendix.

Suppose that the developing world were equipped to cut post-harvest losses down to the levels achieved by Western agriculture. Suppose meanwhile that rich countries learnt to cut waste by treating food more carefully. What would this mean for world agriculture? Using FAO data in 1994, the food economist W. H. Bender calculated that global food demand could decrease by 19.6 per cent if *richer* nations reduced their food supplies to 130 per cent of nutritional requirements and the *poorer* nations reduced their post-harvest losses down to the level of richer nations.[34] Bender worked this out by taking all the food supplies in any rich country that exceed 130 per cent of requirements; for the proportion of that 'unnecessary surplus' that consists of meat and dairy products, Bender assumed that the 'unnecessary surplus' livestock would no longer be raised and therefore the cereals and pulses used to feed them would not be needed either. This proposed improvement in efficiency still allows for 4 per cent post-harvest losses in procurement, storage and distribution in poorer nations and it still allows richer countries to make available 30 per cent more food than actually required by their populations, without changing the proportion of meat they already enjoy in their diets, and it is thus by no means imposs-ible.[35] Given the approximate nature of the FAO data, Bender's 19.6 per cent potential savings should probably be adjusted to 'something in the region of 15–25 per cent'.[36]

In fact, recalculating the potential savings using the most up-to-date FAO data for each and every country, it would appear that if *all* countries kept their food supplies at the recom-mended 130 per cent of requirements, and poor nations reduced their post-harvest losses, then 33 per cent of global food supplies could be saved.[37] This level of 'unnecessary surplus' would be enough to relieve the hunger of the world's malnourished twenty-three times over, or provide the entire nutritional requirements of an extra 3 billion people.[38] This does not include the savings that could be made if Westerners ate a smaller pro-portion of cereal-fed livestock products in their diet, which would liberate grains that are inefficiently fed to animals rather than people, and it does not include the potential savings from agricultural products currently wasted in rich nations before they enter the human food supply chain, such as potatoes rejected on cosmetic grounds.

Sticking with Bender's more conservative figure for now, and requiring only rich countries to reduce their unnecessary surplus food supplies, the potential for all the agricultural land that would thus be liberated is formidable: 19.6 per cent of the world's agricultural land comes to 294 million hectares of crop-land and 659 million hectares of pastureland.[39] Just 4 million hectares of good-quality cropland would be needed to produce the 27 million tonnes of cereals currently required to sate the hunger of the world's malnourished.[40] More land may be needed to feed the world's growing population, and no doubt much of it would be used to satisfy the global increase in meat consump-tion. But setting aside these demands for the moment, one option would be to use this land to tackle global warming and repair some of the damage we have already done to the planet.

We could, for example, grow willow coppice or miscanthus

to use as fuel instead of oil and coal.[41] Wood is an excellent and potentially sustainable source of fuel. Fast-growing species such as willow can be grown in cycles of just a few years, industrially harvested and burnt in centralized plants where the energy can be harnessed for power generation and heating in domestic or industrial applications. In many countries, Britain included, biomass is already used in power stations and burnt instead of fossil fuels in other heat-demanding industries such as steel or concrete manufacturing. The British government, among others, is promoting this as a viable renewable fuel, and there has been a great deal of research into the potential that this has for reducing carbon dioxide emissions. However, using agricultural land for dedicated biofuel and biomass crops is not always a good idea in current circumstances. Firstly, because there are no international protocols to prevent this from increasing demand for agricultural land, thus potentially fuelling deforestation elsewhere in the world and consequently releasing more greenhouse gases than it aims to save. And secondly, because such crops are, in some instances, currently in competition with people for food.[42]

But suppose that these objections were dealt with through effective global land-use governance that protected forests and natural habitats and ensured food supplies for the world's poor. If the land that could be liberated by saving 19.6 per cent of the world's food supplies were used to produce energy with willow coppice or similar crops, it could produce around 164 million terajoules of energy, which is equivalent to the heat used by 1.8 billion European-style homes.[43] Used in a range of industrial and domestic applications, this would eliminate a theoretical maximum of 35 per cent of all anthropogenic emissions from fossil fuels.[44] To these totals we can add the emissions saved by not producing food on all that land. Assuming that global food consumption is responsible for around 30 per cent of total emissions, eliminating 20 per cent of production would cut emissions by around 6 per cent.[45] In other words, reducing unnecessary food production and growing willow to use as fuel instead of oil and coal has a theoretical potential to cut greenhouse gas emissions by a staggering 40 per cent.[46]

Another inspiring result would be achieved by planting permanent forests on all the liberated land.* Owing to population pressure and rising demand for food, reforesting most of this agricultural land is an unlikely scenario (although net increases in forests are reported in fifty countries worldwide), and the land would have to be chosen carefully because some land is unsuitable.[47] But, if it were achieved, this would reinstate lost natural habitat and could provide a sink for a maximum of 50–100 per cent of man-made greenhouse gas emissions.[48]

These figures represent only theoretical scenarios and they are in no way a substitute for cutting emissions by reducing fossil fuel consumption. They are primarily intended to illustrate how valuable the world's agricultural land could be to us and how senseless it is to waste it. In practice, we are unlikely to achieve anything like this much. But, with such potential, even managing a fraction of this is surely worth a go.

* This is a fundamentally different proposition to popular 'carbon-offsetting' schemes whereby consumers are invited to emit greenhouse gases (by flying, for example) and then offset those emissions by planting trees or funding other projects. In this hypothetical scenario, by contrast, people are being invited to *reduce* their emissions by ceasing to waste so much food, theoretically liberating land which can then be put to better use.

PART III

Where There's Muck There's Brass

13. Reduce: Food is for Eating

Eat and drink: But waste not by excess, for Allah loveth
not the wasters.

Quran, 7:31

In the medieval city of Kashgar in far north-western China,
outside a pilau shop that looks like it has served Silk Road
merchants for centuries, I push aside my empty bowl, and burp
in satisfaction. Abdul, the owner of the restaurant, with whom
I have been discussing in basic English the viability of opening
an Uighur restaurant in London, pours me a second cup of
aromatic tea, and I sip it while answering his questions. 'You
could charge 45 Chinese Yuan for that bowl of pilau in London,'
I assure him – about six times the price in China. But as I am
talking, it dawns on me that his attention has shifted. He is,
I realize, peering into my bowl in dismay.

'Clean?' he demands. I look down in surprise. The pilau,
enriched with apricots and mutton bones, was delicious, and I
ate it all. But with his finger Abdul is pointing at four or five
stray grains of rice stuck to the side of the bowl and in the
crevices of the bone. Feeling thoroughly beaten at my own
game, I scrape up the last grains and lick the bone clean, enjoying
his look of satisfaction as I do so. Then I gesture to his colleague's
bowl, where some traces of sour yoghurt still remain. 'Clean?' I
ask, and Abdul roars with laughter as his sous-chef dutifully starts
cleaning his bowl.

The expectation that people finish every scrap of food set
before them is not the only cultural bastion against waste among

the Uighurs – a Turkic people with their own language and partially autonomous province of Xinjiang. As Muslims of central Asian ethnicity, the Uighurs regard themselves as radically different from the Han Chinese who run the country, and in some quarters there are deep strains of hostility towards the people they regard as colonial masters. The clash of cultures is particularly evident in their attitudes to food. Chinese hospitality consists of putting more food on the table than guests can eat. Tables in restaurants in China are left piled high with dishes that have hardly been touched at the end of a meal. A reported 81 per cent of Chinese restaurant clientele leave food on their plates.[1] The Uighurs, by contrast, consider it *haram* (forbidden) to waste food. When they kill an animal, each part of it is used; stroll around any bazaar early in the morning in Xinjiang and you will see the butcher who specializes in offal selling a trailer-load of cow, goat and sheep's hearts, lungs, kidneys, feet and heads. At a livestock market I visited twenty miles from Kashgar, a stall dedicated entirely to broiled cow hoofs fed queues of hungry peasants and townsfolk. Knowing what to do with every part of an animal is part of the Uighurs' cultural heritage.

This frugality extends right from the slaughterhouse down to the kitchen. To ensure that no-one takes more than they can eat, each restaurant offers its main dish in three different sizes: full, half or quarter, so no-one can accidentally order more than they want. If food in the home is left over, it is customary to offer it to neighbours or to take it onto the street for the poor. To leave food on your plate is an insult to the host, to the cook, to the farmer who grew it, and to Allah.

Why these contrasting groups – the Han Chinese and the Uighurs – differ so markedly in their attitudes to wasting food is unknown. Perhaps the pig-keeping Chinese always found a use for their scraps and so had less to lose by leaving food on their plates, in contrast to the pig-abhorring Muslims; perhaps the Chinese court splendour and hierarchy encouraged conspicuous consumption as a central part of the notion of hospitality. The Uighurs, meanwhile, must to some degree have been influenced by the Quranic injunction to 'waste not by excess, for Allah loveth not the wasters'. Perhaps these cultural attitudes have been determined by the physical environment – the Turkic people, having long inhabited arid zones, learnt to use every morsel the soil yields, in contrast to the relatively affluent Chinese, who lived in the abundant, fertile river basins of eastern Asia where food was less scarce. Or, conversely, the recent history of catastrophic famines in China may have fuelled a cultural backlash against frugality. Whatever the ultimate origins of this particular contrast, it is clear that the amount of food a society wastes is dependent on cultural attitudes. There are legal, fiscal and logistical measures that can be taken to reduce food waste – but their strength will derive from what society deems acceptable. In this sense, the solution to food wastage lies in our hands. If we felt, as intensely as the desert-dwelling Uighurs do, that food is a finite, invaluable resource to be cherished, our situation would be very different.

To experience just how different things could be, go to any landfill site in Britain, the US or countless other countries, and examine its contents. Among the mass of general detritus is an array of uneaten food. The smell betrays its presence first; look closer and it seems that a supermarket has been tipped in with the rubbish and left to fester. Some of it – the decomposing pasta, half-finished loaves, sprouting potatoes – has evidently come from restaurants and individual households. But there are also entire crates of food that have clearly never seen the inside of a shopping bag: eggs, oranges, cauliflowers in sprawling piles like a scattered bag of children's multi-coloured marbles. The whole world is represented here in one unaromatic nutshell: bananas from the West Indies, grapes from South Africa, rice from India or America. All of it has come from the earth, and to the earth it has been unceremoniously returned, now blended with plastic, paper and clapped-out furniture. Such a vast collection of rubbish is hard enough to assimilate: it is even more

difficult to think backwards to all the individual lives and processes that are amassed here. Multiplying this by the many thousands of similar heaps thrown away by whole nations is beyond
the imagination. That apple which had only one bite taken from
it before being discarded, that pallet of chocolate bars that never
made it beyond the factory – they all deserved a better fate than
ending up here, rotting into toxic effluent and methane.

The cornerstone to managing food waste – which any
Western nation could well learn from the Uighurs – is the waste
hierarchy, or 'pyramid of use': Reduce, Redistribute, Recycle.
First and foremost must be to stop creating surplus and waste in
the first place – thus avoiding the expenditure of resources and
strain on the environment. The second priority, when surplus
does arise, is to ensure that whatever can be eaten by people *is*
eaten by them, which means donating and redistributing it for
human consumption. Even when all avenues for this primary
use have been exhausted, food still has enormous potential, and
the question is how best to recover value from it – whether to
feed it to livestock, or to break it down in industrial digesters
to make heat, power and soil-enriching compost. The current
practice of sending food to landfill is the worst possible way of
dealing with it and constitutes one of the clearest instances of
our mismanagement of resources.

The environmental and social reasons for reducing food waste
outlined so far in this book may only change the behaviour of
some. Others' entrenched resistance to change may be discerned
in the negative responses to the call by the British Prime Minister, Gordon Brown, in July 2008 for consumers to waste less
food. The timing of Brown's announcement was unfortunate –
he was about to sit down to an eight-course feast at the G8
summit in Japan; his own popularity was at an extreme low; and
politicians had just voted in favour of keeping their contro-
versially generous personal allowances for furnishing second
homes. Furthermore, making consumer waste his principal
response to the food crisis gave the impression that Brown was

pointing the finger of blame at individuals, rather than all the innumerable other factors creating waste in the global food industry. There was little chance that his advice would be taken to heart. He was called fatuous, patronizing and hypocritical. One *Times* reader reflected a widespread sentiment with a comment on the newspaper's website: 'I waste some of the food that I buy because I'm happy to swap money for time and choice. This from the chancellor who has put us into so much debt my children may still be paying it off. How out of touch is this man?' Another more pugnacious writer added, 'Ummm. I think I remember paying for my stuff the last time I went to Tesco's. I paid for it. I can do what I want with it after that. Get it?'[2]

It would be nice to believe – despite evidence to the contrary – that people would not maintain such an attitude if they realized that throwing food away deprives someone else of eating it. But neither Gordon Brown nor his successors, nor colleagues in other countries should be discouraged: they should introduce firm policies to tackle the problem of food waste throughout the supply chain. But is it really possible to bring about change? WRAP has measured the success of its 'Love Food Hate Waste' campaign, and since its launch in 2007 has received innumerable endorsements from members of the public, politicians, supermarkets and the media, and it has apparently convinced 1.8 million households to make an annual saving of 137,000 tonnes of food worth £296m. By working with local authorities, retailers and organizations like the Women's Institute (WI), it has begun to bring about a meaningful change in attitudes.[3] A survey of householders reveals that the newly convinced army of non-food-wasters feel they now plan meals better so that food isn't wasted (37 per cent), they use up food that is already in the fridge before buying new food (31 per cent), they use their freezers more effectively (20 per cent) and they measure portion sizes more accurately to avoid cooking too much (22 per cent). However, other research suggests that 84 per cent of householders are still under the impression that they don't waste

significant amounts of food – so most people, it would seem,
still haven't got the message.[4] In other countries, even small
efforts in schools where children have been encouraged to examine more carefully what they throw away have resulted in a fall
in plate scrapings by 35 per cent.[5] We know from history that
cultures can adapt when necessity demands it. There have always
been puritanical voices urging frugality – it is just that their
prominence, or relevance, comes in cycles. When resources are
stretched, profligacy appears foolish; when there is abundance
the issue of waste falls away.

It was not so long ago that the West was more frugal than
today. A survey in Britain during the 1930s found that household
food waste comprised only 2–3 per cent of the calorific value
of food that entered the home.[6] In 1976 waste was apparently
only 4–6 per cent, and similar studies in America during the
1960s and '70s found wastage levels of about 7 per cent.[7] In
those leaner times, scraps were used up instead of being thrown
away; surplus was bottled or pickled to keep for winter months;
leftovers would be re-heated or incorporated into the next day's
meal and offal – even lung or udder – was a common repast.[8]
In the US during the Great Depression, American families
adopted the motto 'Take all you want, but eat all you take', and
on both sides of the Atlantic during the world wars shortages
made reducing food waste a matter of national security as well
as domestic necessity.[9] American, British, French and German
governments issued edicts ordering their populations not to
throw away food – in the First World War, Britain made it a
summary offence to waste any wheat, rye or rice and proposals
were made to make the waste of any kind of food a punishable
offence. Shopkeepers, individuals and institutions were taken to
court and fined when food was found in their bins.[10] In the
Second World War, the French government produced a film
encouraging people to rear rabbits on vegetable trimmings, and
Britain, too, designated food raised in this way at home as being

'Off the Ration'.[11] Propaganda posters emerged in both wars with slogans like 'Can all you can'[12] and 'A clear plate means a clear conscience: Don't take more than you can eat'.[13]

Older generations who lived through the Second World War remember rationing when everyone was more thrifty. Some might not go as far as my Scottish-born grandmother-in-law, who denounces throwing away milk even when it is ready to walk – 'Perfect for scones', apparently. She once scolded me as I was throwing away a tea bag. 'But Tristram!' she cried. 'You've only used it once!' It used to be common practice to eat an old joint cold in sandwiches and cook any remainder as stew; milk on the turn was made into cottage cheese; stale bread was used to make breadcrumbs for frying, or baked into the magnificent crusty stodge called bread-and-butter pudding. Sadly, despite WRAP's initial conclusions that older people waste less, further analysis reveals that they are now about as wasteful as the rest of us: bamboozled by over-sized supermarket ready-meal portions or large packs of fruit and vegetables that they simply cannot get through.[14]

It is no coincidence that during this period of recession Penguin has just republished the wartime classic *Keeping Poultry and Rabbits on Scraps* (1941),[15] while the WI, established in Britain in 1915 to teach women how to reduce household waste during the First World War, is now giving classes on how to use up leftovers. 'People want those skills,' said Ruth Bond of the WI. 'Apart from anything else, it helps them save money.'[16] Some restaurants have taken measures into their own hands: at the Obalende Suya Express, a Nigerian restaurant in east London, customers have to pay a £2.50 donation to Oxfam if they fail to finish their dinner. The Chinese restaurant Kowloon, on Gerrard Street in London's Chinatown, has a similar scheme, and the idea has caught on in at least one restaurant in New York.[17] Timothy Jones facilitated a project in Chicago, where restaurants reduced the size of all their portions, but offered to 'supersize' for free if people asked them to do so. This way people with large appetites did not feel ripped off, but much less food was wasted.[18]

If the hidden costs of wasting food become more tangible – through unhindered environmental degradation, economic recession, resource scarcity, or just high food prices due to excessive demand – people may learn to stop their counter-productive behaviour, as they have done in the past. There are signs that the current credit crunch is teaching super-markets, institutions and individuals the value of avoiding waste – 60 per cent of householders in Britain claim that the recession is encouraging them to shop more carefully and waste less.[19] But the question remains whether, this time round, the lessons will be learnt for good, or whether the principles will merely fall away again when things start looking up. Many people now understand the need to use fossil fuels and other resources more economically – through home insulation, double glazing, efficient boilers and recycling, which offer financial savings as well as reduced impact on the environment. These are long-term measures and indicate that people have accepted the need to conserve resources regardless of what we can personally afford. But food has hitherto remained a blind spot, and there are still huge resource-saving opportunities to be exploited.

Beyond the walls of our own kitchens, we can also exert influence over the food industry. Consumer power is the new face of democracy. We only vote for politicians every four years or so, but we vote every day with our money, and we can use it to bring about change, often much more rapidly than legisla-tion can ever achieve. The remit of a business is to make as much money as it can within the legislative framework of the country or countries in which it operates.[20] It is possible for companies to adopt ethical policies even when this is injurious to profits, but this happens very rarely. There are therefore two main ways to ensure that a business's profit-making activities do not harm the interests of society and the environment. First is to convince its managers and directors that it will make more money if it behaves in certain ways, and less if it behaves in others. This means persuading supermarkets that they will lose popularity if they fail to take significant measures to reduce wastage in their stores and all the way up the supply chain. The key is to prevent over-production and overstocking waste: which means telling them that we do not want to see any more bins full of food.

There are numerous ways they can achieve this without even hurting their profit margins. The fact that large supermarket and sandwich chains (such as Sainsbury's and Pret A Manger) advertise their donations of surplus food to charities indicates that they know this will win them public approval, and therefore customers. Other supermarkets have piecemeal relationships with so-called 'cast-off' shops, where their surplus can be sold at 10 per cent of its original value to, for example, emergency services staff and health workers.[21]

Tesco has recently responded to public demand by introducing a new carbon labelling scheme. A Carbon Trust-commissioned survey found that 67 per cent of all consumers said carbon footprints would influence their choice of product, which shows that real change could be achieved by putting more information on foods. This kind of labelling should take account of wastage – so if 20 per cent of a product is lost during its production, its carbon footprint would go up accordingly and people would be less likely to buy it, encouraging efficiency among retailers, manufacturers and farmers all the way up the supply chain.[22]

Reducing the price of products nearing the end of their shelf-lives is an important way supermarkets can cut down on their waste. Most supermarkets now do this to some extent, but none of them do it as much as they could. According to the British Retail Consortium, the main obstacle to this is simply because it takes too much staff time going round shelves reducing the prices. The head of technical services of one supermarket chain estimated that marking down cost £11m per year in labour and lost margins.[23] A system could be designed to overcome this problem while at the same time dealing with the fact that many consumers reach to the back of shelves to take items with the longest shelf-life, leaving older stock unsold. Supermarkets could announce that if an item is close to its use-by date, the till will automatically make a discount proportional to the remaining shelf-life.

Supermarkets claim (not without some dissent from personnel) that the widely used 'electronic point of sale' system has cut down on waste by recording sales, helping to predict regular demand and indicating when stock should be reduced on offer.[24] A new technology called Radio Frequency Identification (RFID) now enables companies to track stock during its entire journey through the supply chain, thus helping them to manage supply, demand and storage more efficiently. Despite its proven ability to save millions of pounds in a single year and its popularity in the US, however, there has been slow uptake in Britain and Europe.[25] In most cases relating to food and drink products, RFID is currently only applied at the pallet level during distribution into depots, and not at the product level, where it could continue to track product life (and other) information even after stock is broken down for delivery into stores and people's homes. A new generation of more cost-effective 'conductive ink labels' may make product-level RFID a possibility in the medium term with consequent savings on waste.[26]

I spoke to one district manager from the pan-European, German-origin discount supermarket chain, Lidl, who took the view that waste reductions could easily be achieved even without these advanced technologies, claiming that one store had dramatically cut down on waste simply by following company protocols more strictly and paying careful attention to exactly how much of each product it sells. 'Until two months ago,' he said, 'the local store manager was constantly overstocking because he simply wasn't focusing on the issue of waste. But just by being more careful, and by training those responsible for making orders, we've cut down on waste enormously.'[27] Discount chains like Lidl have relatively high incentives for waste reduction owing to their very tight profit margins – particularly given the typical socio-demographic profile of their customer base. But these and other instances of waste reduction initiatives in the industry do indicate that the food industry is responding to pressure from the public and the media. Seven years ago food waste hardly figured on the public agenda; now, every newspaper in Britain and many television and radio programmes have addressed the issue, and businesses are beginning to listen.

The second way the public can change business practices is to convince governments to take measures to entice – or force – businesses to be less profligate with the world's natural resources. Once the sources of waste have been identified, the way to alleviate the problem often appears relatively clear, even if it requires substantial political will-power to do so (which is only forthcoming when politicians are confident they have popular backing).

The first problem is the lack of transparency. We need the food industry to declare how much it wastes. As a first step in the UK – and other European countries where similar processes are observed – food waste tonnages could easily be entered into the solid waste reports that large manufacturers are already required to fill in under the Integrated Pollution Prevention and Control (IPPC) Regulations. This would immediately cover 30–40 per cent of manufacturing production in the UK, and in Defra's estimation it would take just two employees at the Environment Agency to manage the entire scheme at a cost of just £100,000 per annum. A similar system could also be extended to large food retailers, i.e. the supermarkets. Given the possibility of false reporting, it would also be advisable to have a mechanism for checking the data supplied by these companies to ensure they are not hiding the extent of their wastefulness.[28] A system similar to the enforcement of EC fish quotas could

work, whereby businesses would keep records of how much
they waste, and these figures must tally with what is actually in
the bin when the inspector calls.

Instead of legally binding targets, in 2005 the British govern-
ment and WRAP instigated a voluntary agreement called the
Courtauld Commitment by which companies opted to reduce
their waste. Original signatories included Alliance Boots, Asda,
Budgens, the Co-operative Group, Iceland, Londis, Marks &
Spencer, Morrisons, Sainsbury's, Somerfield, Tesco and Waitrose.
By 2009, another thirty-seven companies in the food industry,
including Heinz, Northern Foods and Unilever, had signed up.
Packaging is the principal focus, with an overall target of starting
to reduce the amount used by 2010. Government-funded research orchestrated by WRAP has now assisted several of these
companies in re-designing their packaging to reduce waste. On
the issue of actual food, signatories initially committed merely
to the vague target of 'identifying ways to tackle the problem of
household food waste'.[29]

Tim Lang, who is among other things the Commissioner of
the government's Sustainable Development Commission, concluded in 2008 that 'there has got to be a ratcheting up of targets,
whether through legislation, fiscal measures, or name and shame;
the Courtauld Commitment is a low hurdle and you jump over
it; what we need is a revolution.'[30] The 2008 Cabinet Office
report *Food Matters*, which laid out the government's food strategy for the twenty-first century, emphasized that the government would seek 'a new voluntary agreement to cut significantly
the amount of food wasted *in the supply chain* and in the home
– for conclusion by February 2009'.[31] The 'supply chain' element
of this, however, has been put on the back boiler. Food companies have stalled on making this commitment, postponing a
decision until 2010, and they have not indicated what this will
eventually be.[32] In January 2009 the Courtauld signatories –
including the major supermarkets and some manufacturers – did
announce their new initiative to discourage consumers from

throwing away 155,000 tonnes of food by 2010. However, there is no way of measuring their achievements, and the target will be subsumed into (rather than being additional to) WRAP's own direct achievements in household food waste reduction – not to mention the effect that the recession will have on people's level of profligacy regardless of other efforts since WRAP currently estimates the savings primarily by asking people what they are doing to reduce waste. The supermarkets promised to minimize food waste in people's homes by making adjustments to labelling, pack sizes, storage advice and the design of packaging so as to keep the food fresher for longer. Sainsbury's, for example, has implemented new storage guidance on fresh fruit and vegetables and has been running the 'Love Your Leftovers' campaign, which offers advice and menu ideas for using leftover food. Marks & Spencer has re-designed its meat packaging to use less plastic and keep meat fresher for longer. The manufacturer Warburtons introduced a new 600 g small-size loaf designed for single-occupant households.[33] However, the targets are still aimed at *household* waste, rather than eliminating the waste within their own companies or in their supply chains. Supermarkets can certainly advise people on how to avoid wasting food in their homes so this is definitely better than nothing, but customers will be forgiven for sensing an element of hypocrisy when they see supermarket bins full of more food than they would throw away in a whole year. Besides, for businesses that fail to meet any of their targets there are still no sanctions, remediation costs or penalties (which could be collected in a fund and awarded to the best-performing companies): if they really intend to meet them, why not agree to penalties for failure?

Meanwhile, a joint government and industry initiative, the Food Industry Sustainability Strategy Champions' Group on Waste (CGW), comprising representatives from Britain's food manufacturers, supermarkets, the government and other interest groups, voluntarily agreed to try and reduce packaging and food waste in the food manufacturing industry (not including the

retailers) by 3 per cent a year over five years from 2006. The
inclusion of food waste in the target was a much needed supplement to the EU Directive on Packaging and Packaging
Waste. But apart from being a weak target, there is currently no
way of monitoring progress, and baseline data for 2006 is lacking.
To measure such small changes requires accurate, robust waste
reporting. It is risible to suggest, as the CGW does, that two
thirds of the manufacturing sector could be meaningfully assessed
on the basis of self-reported waste data from 'a voluntary active
panel' of companies, when it is clear that the few companies
volunteering to join such a panel are likely to be unrepresent-
atively progressive about waste. And yet, the Group has firmly
opposed the introduction of regulations obliging food companies
to report their waste. All reporting should, according to them,
be voluntary and anonymous so as to avoid revealing information
that could benefit competitors – even if it helps all companies
achieve greater efficiency. Instead of each company being held
to account, the target was set for the industry *overall* so that
companies that fail to meet the target 'should not be disadvantaged'.[34] The CGW has also joined other organizations in moot-
ing a Waste Trading Scheme which would work like carbon
emissions trading. Waste permits would be granted, and then the
most efficient companies could sell excess permits to more-
wasteful companies.[35] However, this could only work if there
were a mechanism to measure and monitor waste properly and
it would have to avoid the pitfall of the European carbon-trading
scheme, which became a cash give-away to polluting companies
because of the way permits were allocated. Unsurprisingly, the
CGW project has lost momentum – no meetings at all were
held during 2008. It is true that initiatives work best with the
goodwill and co-operation of the industry and the public. But
these voluntary agreements currently lack teeth.

Some industry initiatives do claim to be yielding fruit. The
Food and Drink Federation (FDF), a trade association representing the interests of food and drink manufacturers in the

UK, issued a press release in November 2008 claiming that a survey of its members found that they recycled or recovered 82 per cent of the food and packaging waste created in factories and had 'prevented over half a million tonnes of food waste being created'.[36] Their recycling rates are impressive, but when they say they have 'prevented' half a million tonnes of food waste (which would include food such as all those millions of slices of bread thrown away by sandwich makers), they really mean 'diverted' it from falling into the legal definition of waste by feeding it to animals. This is clearly vastly better than land-filling it, but the term 'preventing food waste' should be reserved for the much greater task of actually preventing food from being wasted. WRAP has indicated that the FDF survey might be used to establish baseline data for the industry as a whole, but the FDF is the self-proclaimed leader in waste reduction and its relatively enlightened members – especially those who volunteered to respond to the survey – are undoubtedly unrepresentative of the industry as a whole, and furthermore the survey consists of yet more self-reported and anonymous data which cannot be verified.[37]

One way of getting round these problems would be to select a representative sample of food companies for a compulsory waste audit during which companies would regularly record their waste and would then be spot-checked for the reliability of their figures. Participating companies could be compensated for their pains by receiving priority assistance from the government's environmental consultancy service, Envirowise. The effects of this assistance could then be measured and held up as positive examples within the industry. Another way of dealing with companies' concerns about revealing their waste data for fear of benefiting competitors would be to make data reporting compulsory, but *aggregate* the data rather than releasing the data for each individual company, so that at least an overall picture of the industry can be formed and improvements measured.

The other main government initiative relating to food waste

is the landfill tax, which discourages the disposal of biodegradable waste into landfills. After a decade of setting landfill tax so low that it had little effect on the destiny of biodegradable waste, the UK raised its standard rate of landfill tax from the 1996 level of £11 per tonne up to £32 per tonne in April 2008, with the intention of raising it by £8 per year for at least two more years. On top of these taxes, there are also charges paid to the private companies collecting and disposing of the waste, so the total cost can be in the region of £50–60 per tonne or more.[38] This does not in itself significantly discourage businesses from wasting food. The cost to a business of wasting a tonne of food might be around £1,000 depending on the type of food; it is unlikely that paying an extra £32 in tax will dissuade the business from disposing of it.[39] However, the increased tax rate does at last appear to be proving effective in persuading businesses to stop sending their food waste to landfill, and to send it instead for anaerobic digestion or composting, which is just as well because failing to reduce the amount going to landfill could have very unpleasant consequences.

The EU Landfill Directive (1999/31/EC) required that in all member states biodegradable municipal solid waste going to landfill must be reduced to 75 per cent of 1995 levels by July 2006, to 50 per cent by 2009 and down to 35 per cent by 2016. Because the UK, Spain and Greece had such poor recycling rates initially, and landfilled more than 80 per cent of their municipal waste in 1995, the EU allowed them an extension of up to four years in which to catch up (meaning that the UK targets are for 2010, 2013 and 2020). Despite this allowance, in January 2009 the National Audit Office (NAO) warned that the UK is still 'at risk' of missing the 2013 target because Defra was too slow in implementing sufficient waste reduction and diversion policies.[40] Defra passed on the job of diverting waste from landfill to local authorities and set each one an annual allowance reflecting its share of the national targets. Since 2005, if a local authority fails to meet its target it has to buy additional allowances from local authorities that have exceeded theirs.[41] Local authorities have tried to build recycling infrastructure and divert waste mainly through government-funded Private Finance Initiatives (PFIs), which have suffered chronic delays and which Defra has not monitored sufficiently to ensure that they work. Difficulties in the financial market have made it even less likely that these schemes will be up and running in time to meet the national targets. Although the EU has not yet announced how much it will fine the UK for failing to meet the 2013 target, the government estimates that it may be in the region of several hundred million pounds, and Defra has said that it will levy a fine of £150 on local authorities for every tonne of waste that exceeds their allowance. These fines – which could reach £1m *every day* – will be crippling, and ultimately will have to be met by taxpayers. The NAO has warned that the landfill targets will not be met unless local authorities instigate more non-PFI projects (including composting for communities and households). But central government does not fund these so there is at present no requirement for local authorities to report on them and no effective way for the government to monitor their progress.[42]

In its pathetic struggle to meet its targets, the UK has been confirming its old reputation as the 'dirty man of Europe', as many other EU member states are either comfortably on track, or met their targets years ahead of time. Austria reached its final 2016 target at least fifteen years early after instituting a law that biodegradable waste, both household and commercial, had to be collected separately and composted, and they accompanied this law with extensive information campaigns in order to change the behaviour of the public. In 2005, Germany had reduced biodegradable waste going to landfill to a mere 5 per cent of 1993 levels by collecting biodegradable waste in separate bins provided to half of all households and sending it for composting or anaerobic digestion. Before being sent to landfill, any remaining municipal waste in Germany has to be incinerated or undergo mechanical biological treatment which treats waste so that it can either be applied to land or be used as a fuel. France, Denmark, the Netherlands, Italy, the Flemish region of Belgium and Sweden all achieved targets ahead of time by one or more of the following measures: banning the landfill of biodegradable material or unsorted household waste, collecting biodegradable waste separately for composting or anaerobic digestion, taxing landfill, encouraging home composting, and also by incinerating waste (usually with energy capture to harness the heat from burning).[43] Ireland is one of the only countries performing worse than the UK: there the amount of food waste going to landfill is actually increasing by 5 per cent a year. The only likely way that Ireland can achieve its targets, as the environment minister has conceded, is to introduce bylaws to enforce the segregation of food waste from both households and commercial premises.[44]

In general the EU landfill directive is having some success at diverting food waste from landfills, and therefore having a positive – though so far unmeasured – impact on global warming and pollution. However, there is still essentially no legislative framework aimed directly at discouraging the waste of food, either in the UK or in most other countries. In Britain and Europe, governments are using fiscal instruments to bring about change, but the problem is so acute that there is now a strong case for robust legislation, regulation or at least an adjustment to the current tax regime. Primary legislation could take up to five years to get on the statute book and may be so costly to the industry that governments would block it, and enforcing it could cost the taxpayer. On the other hand, food waste also creates a cost for consumers and the environment. Legislation may be more effective at dealing with this than the current combination of voluntary agreements and landfill taxes alone, and it would start to have an effect even before it reached the statute book because food companies would make changes in anticipation of the law.

As current trends suggest, if food were more expensive we

would not waste so much of it; but ideally it is not so much that eating food should be more expensive, but that *wasting* food should become so. This would be difficult to enforce on individual consumers, though the South Koreans have proved it to be possible – not by raising taxes but simply by making it illegal, in 2005, to send any food waste to landfill whatsoever. There, it is obligatory to separate all food waste in the home into designated bags which are then gathered up for composting or animal feed. The food waste bags must be purchased, so the more you waste, the more you pay. Currently the price is very small, but just making people measure their waste already causes them to think about how much they are throwing away and, in conjunction with public awareness campaigns, reductions have been achieved. Anyone who fails to separate their waste and ends up putting food scraps in their general garbage bag is also liable for a fine. But these fines are hardly ever issued because the rules are almost universally observed. My friend and translator in South Korea, Tehion Kim, told me he hated being made to separate his food waste – but he did it all the same. The country as a whole has 98 per cent separation of food in its waste stream. This is the mirror image of the US, where only 2.6 per cent of municipal food waste is recycled.[45]

Initially, making it expensive or more difficult for *companies* to waste food may be simpler, more remunerative, and easier to enforce than targeting consumers. Taxes should be used very carefully, but the food industry's pitiful offerings so far warrant their consideration. In order to create a significant disincentive, there could be a special tax on wasting food *irrespective of the disposal method*, reflecting the social and environmental costs of doing so. The landfill tax is designed to encourage the diversion of food waste from landfill; but there is currently no further incentive for diverting surplus food from *disposal* to *use* as food for people. There is a level of tax, somewhere above £32 a tonne, that would dissuade companies from wasting food so recklessly. Furthermore, rather than imposing the tax by weight (i.e. per tonne), which assumes that all food has the same value and environmental impact, it may be more effective if the tax were imposed on food according to its *value* or, when further research on food production has been conducted, on its environmental impact. Any revenue could be raked back into schemes to help companies reduce their waste (as currently happens with landfill tax revenues), or it could be compensated for by reducing tax on other less environmentally damaging items, though if the scheme were effective, the tax revenues would steadily decline. There is already a partial rationale for this within the existing food taxation regime in Britain as well as many other countries. Governments help to make food affordable for consumers: in Britain, for example, there is a zero rate of value added tax on staple foods, whereas there is a standard rate (usually 17.5 per cent, but temporarily held at 15 per cent during the recession) on luxury foods such as sweets, crisps and ice cream.[46] A zero rate makes sense for the necessities of life, but the portion of food that is thrown away is not a necessity; it is no less a luxury than ice cream and it is a good deal more harmful. This would at least provide an additional basis for imposing a tax on the portion of food that companies throw away. The administrative viability and success of this tax may be greatest if imposed solely on retailers, with the condition that they take responsibility for waste in their supply chains.

An alternative, not incompatible with the above, would be to take forward agreements such as the Courtauld Commitment, but make them compulsory, raise the targets, penalize companies that fail to meet them, and press for them to be adhered to internationally. Instead of 15 per cent food waste reduction over five years, as with the CGW commitment, it should be much higher – 50 per cent, for example – with the eventual aim of eliminating the waste of ready-prepared and fresh edible food altogether.

Failing this, governments could always reintroduce something like the laws that operated in the First and Second World Wars which simply made it illegal to throw away food. Draconian legislation is generally regarded as unpopular and unviable and certainly it is preferable to achieve social change without recourse to such measures. But the recent wave of smoking bans, which have been almost universally obeyed in the UK, Ireland, France, Japan and the US, or the previously unthinkable invasion of public privacy in the wake of the 'War on Terror', or the rapidly effected South Korean food waste laws, do suggest that people can be induced to submit to dramatic laws when there is either sufficiently widespread public acceptance of the need for such laws or sufficient effort on the part of legislators to convince the electorate of such a need. Businesses would certainly object to any legislation forcing them to change their wasteful habits, but if they fail to do so rapidly enough without such inducements, this approach – backed up by careful research and negotiation – could be justifiable.

14. Redistribute: The Gleaners

There must be a reason why some people can afford to live well.
They must have worked for it. I only feel angry when I see waste.
When I see people throwing away things that we could use.[1]

Mother Teresa, *A Gift for God* (1975)

The warehouse in Bermondsey, south London, has an aura of industrialized efficiency. A forklift truck shifts pallets of supermarket food into a chilled room, stacking the walls high with crates of Marks & Spencer's chicken breasts and Sainsbury's minced beef. Next door, workers lug boxes of pomegranates and apples, broccoli and lettuces, hundreds of loaves of bread and enough breakfast cereal to feed the five thousand. All of this food is in pristine condition, within date – and all of it has been donated as food industry surplus. Far from being scraps and leftovers, it is of the very highest quality, including some products that most people in Britain could not ordinarily afford to buy. From this central depot, it will be transported by the charity Fareshare to hundreds of community centres and homeless shelters whose beneficiaries rely on it for their daily bread.

Food redistribution charities have been in operation for half a century in Europe and America, and have learnt to match the abundance of surplus food with those millions of people in developed countries who do not have enough money to eat well. This way, surplus food is put to its proper purpose, and the poor are provided with a decent diet. There is so much premium-quality food currently being thrown away that there is no reason to give people anything substandard.

The limiting factor in food redistribution schemes is not the availability of quality food, but the availability of funds to redistribute it. Even when there are plenty of people offering food, charities lack the financial resources to make use of it all. It costs money to manage food depots, buy refrigerated vans and organize trained drivers and food handlers to collect and distribute the food, while others liaise with companies and charities to match supply with demand. Fareshare would be able to redistribute far more food if it had the money to do so, but so far the British government has failed to support it, even when it applied for funding under a recycling grant scheme it had helped the government to set up. When it received a £1m Lottery Grant in 2005, Fareshare almost immediately doubled the amount of food it redistributed annually. In 2008 it redistributed 3,000 tonnes of food to 25,000 people in 500 different community centres and other organizations, with a further 5,000 tonnes that either ended up being diverted into animal feed, anaerobic-digestion plants, composting, or other waste-recycling routes.

Why the Lottery Fund should have to pay to clear up the supermarkets' excess is not clear; instead, it is at least arguable that the supermarkets themselves should pay to redistribute their own surplus. In fact, this has already started happening to a small extent because Fareshare now charges most companies around £16 per tonne to relieve them of good-quality surplus food. The companies are willing to pay – and even deliver the food – partly because it saves them the £50 or more per tonne they would be charged in landfill taxes and disposal costs. These fees could in time help to make food redistribution a self-sustaining concern, and it would certainly become so if all suitable food companies took part and paid enough.

Meanwhile, supermarkets and other food companies in the UK are beginning to notice the growing public clamour for them to take the issue more seriously. In the past, supermarket food donations have had a somewhat token value – something to put in their Corporate Social Responsibility reports. But now some of them are taking an increasingly constructive role in the expansion of food redistribution (see chapter 2). In fact, Fareshare now has an agreement with all the major supermarket chains in Britain – with the sole exception of Morrisons, which Fareshare has given up pursuing.[2] The amount the supermarkets donate still represents only a tiny fraction of their overall waste; the trend is promising but movement is still far too slow.

Over-production surplus from manufacturers is among the most suitable kinds of food for charitable redistribution. It is typically well within its use-by date – even chilled goods like yoghurt will have around five days' shelf-life – and it is available in large quantities, very often on pallets which can be transported easily. Some manufacturers, such as Kellogg's in the UK and Kraft in the US, co-operate with Fareshare or Feeding America and actually deliver this valuable food to redistribution depots. From the larger food companies, Fareshare frequently receives full truck-loads, consisting of around twenty-six pallets or around 18 tonnes of food and other supplies that otherwise would have been wasted. These companies are beginning to do what all of them ought to do.

If all food companies rapidly decided to redistribute their surpluses in this way, there would be a case for governments to take a laissez-faire approach – relying on businesses to respond to public pressure for change. But given the food industry's history of neglect, the likelihood is that they will drag their feet. There is therefore a case for obliging food companies to make surplus food available to redistribution charities, and certainly this use for food should be prioritized over disposal to landfill, composting or anaerobic digestion.[3] Nor would it be unreasonable to expect retailers to pay for the full cost of redistribution, or at least contribute to it significantly. After all, the food has been created at a cost to the environment, and will otherwise be disposed of at a cost to the country and the climate.

If the tax on wasting food (regardless of disposal method) proposed in the previous chapter were adopted, this would encourage supermarkets to pay redistribution charities to take food away for people to eat, rather than incur the tax. At the moment, the redistribution charities are in competition with anaerobic digestion and composting plants, whose costs may prove to be lower, partly because they are receiving government subsidies, which gives them an unfair advantage over redistribution charities. As Fareshare's Maria Kortbech-Olesen puts it, 'These new recycling technologies are all very good, but the government does not look at the steps further up the waste hierarchy: the *use* and *re-use* options for human consumption are not being invested in.'[4] Although Defra's literature, including its Waste Strategy, talks about the need to observe the waste hierarchy, its fiscal policies do not reflect this. The policy of taxing food waste regardless of disposal method would be in line with the government's preference for fiscal coercion rather than direct regulation. The higher the tax, the more companies would be willing to pay redistribution charities in order to avoid it and the harder they would try to avoid wasting food in the first place. In other words, this tax would create a *market* for food redistribution, just as the landfill tax has created a market for anaerobic digestion and other non-landfill disposal methods. Redistribution charities would thereby perform a service to the industry.

The government would also reap benefits from effecting this change. It may have been solely the threat of fines under the EU Landfill Directive that forced Britain to use landfill tax to encourage businesses to divert waste from landfill, but there are also legislative incentives for the government to prevent food being wasted at all: doing so would help to meet the greenhouse gas emissions reductions which Britain, for one, is legally bound to achieve under its own Climate Change Act (2008).[5]

Possibly even more cost-effective than making supermarkets fund the redistribution of their surplus would be to encourage them also to give away surplus food in the actual stores themselves. This would help to quell the supermarkets' inevitable argument that taxes would make food more expensive for consumers. Recipients of this free surplus food could – if necessary – be limited to certain groups, such as emergency services staff, health workers, or those on state benefits. That would make it more expensive for supermarkets to waste food – because they might lose out on potential sales. They would react by avoiding overstocking and thus reduce the amount of surplus arising in the first place. Currently it is only understocking which is perceived as causing lost sales; thus managers are under pressure to overstock. Under this system, both understocking and overstocking would be pernicious to profits, and retailers would have a powerful incentive to try and get their food orders just right. These changes could be brought about through legislation, but there are signs that businesses could be persuaded to adopt it voluntarily as it does offer substantial PR benefits and could even increase the attractiveness of competing supermarket chains.

The supermarket Lidl, for example, recently trialled the policy of giving away surplus food. At Christmas time particularly, products it has over-purchased end up at the cash counter and are given to shoppers at the cashiers' discretion. I spoke to one of their district managers, who told me they did that for 'ambient goods' (foods kept at room temperature) as a way of avoiding the costs of waste disposal. However, they stopped giving away bread when they realized that people were just waiting to get it for free rather than buying it. 'It was affecting our sales,' the manager told me.

A simple modification could solve this problem: allow shoppers to take free surplus of up to, say, 20 per cent of the value of the goods they buy. This would prevent people from taking liberties, and would serve the supermarket's interests by attracting more people to the store. An organic-wholefoods shop near to my home does something like this with any fruit or vegetables that are blemished or over-ripe: instead of binning them the owner puts them into a 'free box' for customers to help themselves. This assures his customers that he is not profligate with the produce that comes through his store, and no doubt there are some who are attracted to it because they know there might be some freebies.

I hesitate in suggesting this to supermarkets, firstly because I am reluctant to help them do their marketing, and secondly because free and discounted produce might just end up in shoppers' own bins. This may be so, but the problem would be reduced by the fact that potential lost sales would create the disincentive to produce surplus.

Other initiatives, which have already been introduced piece-meal in some stores, include using products such as fresh fruit and vegetables approaching their best-before date for processing into other products, where possible within store. Fruit, for example, could be turned into smoothies and given or sold to customers through dispensing machines. Although the relatively small quantity of surplus produced in individual stores would make it too expensive to transport it to processing outlets off the premises, even this could be achieved through the increasingly popular practice of back-hauling surplus from many stores to central depots, where it can be collected more economically. As a last resort, for suitable products such as prepared ready-meals, meat, fish, bread and certain dairy products, freezing at this point could help with the logistics of redistributing it to charitable community centres and the like.

At present in the UK it is estimated that the total amount of food redistributed is unlikely to be more than 15,000 tonnes a year (between 1 and 3 per cent of retail food waste and a fraction of a per cent of all food waste generated in the supply chain),[6] though others estimate that the real figure could be half that.[7] In the United States, more than 1 million tonnes of food are redistributed,[8] helping to feed around 26 million people.[9] This means that for each person in the US, nearly fourteen times more food is charitably redistributed than in the UK – and yet advocates of the system in the US aspire to increase this more than threefold.[10] Although this has much to do with the American philanthropic approach to charity, in contrast to Britain's guaranteed welfare state, the corporations' willingness to co-operate is a valuable model for other nations. Fareshare makes the conservative estimate that if redistribution organizations could utilize just 25 per cent of suitable surplus food generated by retailers alone, British charities could multiply their capacity by more than eight times, to handle 125,000 tonnes of quality food a year.[11] They point out that this would create substantial savings for businesses and the UK economy as a whole, as well as protecting the environment. Although other European countries have a stronger history of redistribution, supported by the European Federation of Food Banks, there is still plenty of scope for expansion there too.[12]

This estimated potential expansion does not even take account of the capacity for donating food on a smaller, more informal scale from catering outlets such as restaurants and canteens. Maintaining the safety of cooked food means that restaurants cannot easily transport their meals, and for this reason it is seldom practised in the UK, though some of the gastro-pubs and restaurants of the Whitbread chain, such as TGI Friday's, Pizza Hut and Beefeater pubs have donated surplus to Fareshare. In the US, charitable food rescue teams regularly organize soup kitchens giving away restaurant surpluses. It would not even be impossible to introduce a system similar to that widely practised in India and Pakistan, where poor people are fed in the restaurant – either funded by donations from clients or given surplus food.

Nor does this include the potential for gleaning food from farmers' fields when the crop has become uneconomical to harvest. Organizations in the US such as Feeding America and The Society of St Andrew organize volunteers to collect good fresh fruit and vegetables from farms under schemes such as the Potato Project, the Gleaning Network and Harvest of Hope. No such system exists in the UK, but there is no reason why it shouldn't. In the past, surplus agricultural products in the EU that were withdrawn from the market to support prices were allowed to be redistributed through the EU Surplus Food Scheme. However, for twenty-five years until 1997, the UK Intervention Board chose to use surplus for animal feed and composting, and it was only with the 1996 Common Agricultural Policy reforms that donating surplus to charities for human consumption was explicitly given priority. Under the same reforms, it was decided that surplus staples such as butter and dried milk should be subject to subsidized export, largely to poorer nations.[13]

The United States has a legislative framework that is favourable towards food redistribution. Under the administration of President Clinton, a number of laws were passed to encourage corporations to donate food rather than discard it. The Good Samaritan law protects food corporations from liability in the extremely rare, indeed virtually unheard-of, scenario where someone falls ill from donated food. This gives corporations confidence, and in cases where this fear has merely been an excuse for their reluctance to give food away it has removed the pretext.[14] Large companies are also allowed to deduct the cost of the donated food against tax,[15] and companies also get a tax break on shipments of food if they transport charitable food on the return journey when the truck would have been empty anyway. It remains questionable why taxpayers and the state should forgo tax revenue to persuade corporations to redistribute their excess food, but it is certainly the case that these tax breaks and other incentives have been a spur to increasing food redistribution in America.

Instead of offering the carrot of tax breaks, the UK and Europe have opted to use the stick of raising taxes on sending waste to landfill. Although this needs to be adjusted to prioritize redistribution in the ways described above, it is arguably a more equitable approach than the American system, which can be seen as pandering to business interests, even if the measures are popular among the redistribution charities.

Perhaps the best model can be found in Canada, where a

system similar to that in the US prevails, but with a few key differences. Canada does not offer corporate donors tax incentives and the state has not historically provided financial assistance to help redistribute. Canada's welfare state system is similar to Britain's, so there is less chance that recipients will develop an unhelpful degree of dependence on food hand-outs – as can happen, arguably, in the US, with its combination of redistribution charities and the state-run 'Food Stamp' scheme. Instead, the Canadians emphasize the community basis of their work, by 'helping people to help themselves'.[16]

Food redistribution is a way of improving the diet of poor people. It ameliorates the problem of food waste but it does not solve it. In the UK alone there are an estimated 4 million people suffering from 'food insecurity' – not just the homeless, but pensioners, impoverished immigrants, single parents and many children. In order to reach them all, food redistribution would have to expand by 160 times, and even this many people would not eat all the surplus food that is currently wasted. The supermarkets' historic refusal to do even this little thing to alleviate hunger with their unimaginable quantities of waste is an illustration of the extent of their past neglect of social and moral concerns, a neglect that largely continues into the present.

One of the main hindrances in the expansion of the food redistribution system in the UK is the fact that we very rarely employ the system of 'food pantries' or 'food banks', a mainstay of redistribution networks in the US, where recipients can come in off the street and collect free groceries to take home and eat. Instead, organizations such as Fareshare insist on providing food only to community centres and homeless shelters that give meals to recipients *in situ*. The rationale for this is sound, in that it ensures that the food is prepared safely and comes with a more holistic welfare package, including contact with others in the centre and social workers. But there are also benefits to the food bank system. Being able to prepare food for the family in the home is a source of pride and can be an important part of family life. This way recipients can receive food in much the same way as ordinary shoppers – the only difference being the fact that they do not have to pay for it. As one recipient of food in the wake of the hurricane Katrina disaster said, 'They make you feel like you aren't begging.'[17]

The other major hurdle hitherto has been that food redistribution has had its sworn enemies. Tim Lang, who has for decades been a pioneering advisor on improving the environmental and social sustainability of Britain's food supply, has led a long battle against food redistribution charities such as Fareshare. He considers that giving food to the poor creates cycles of dependence and takes pressure off governments to do something more fundamental about the causes of poverty. In a BBC Radio 4 programme in which I was advocating food redistribution some years ago, he commented that giving surplus food to people was equivalent to 'treating the poor like pigs'. He still maintains this view.[18] He's wrong. Domestic pigs are kept in captivity and do not have a choice about the food they eat. People, by contrast, do. The scenario advocated by Lang, by which millions of tonnes of top-quality food go to waste but the poor are prevented from taking any of it because it would supposedly be injurious to their dignity, is the one that deprives people of choice, and of their dignity too.

It is true that there are issues which should be treated with caution. Certainly there need to be safeguards to ensure that food redistribution does not become a substitute for other welfare initiatives. Having taken a back seat in advancing food redistribution, the UK government cannot currently claim credit for it, or use it as an excuse to withdraw other types of welfare assistance, which is one potential argument against them getting involved now, though not, I think, a conclusive one. Lang rightly points out that people should not be allowed to fall into the position of needing food assistance in the first place. However, even if we had the best imaginable welfare state, there would still be people who would benefit from saving money on food and they would then be able to use that money for other things, such as heating bills, paying off debts, extras for their children, or holidays. Even if the whole population was well-fed, there would still be demand for free food, which can liberate a family's resources.[19]

But there is no need to take my word for it, because Fareshare biannually surveys the beneficiaries of food redistribution: 92 per cent of them say the food they receive helps them keep healthy; for 29 per cent of the clientele it's the only time they eat fruit and vegetables; 78 per cent say it makes their money go further; and 44 per cent find that coming in for the meal helps them to engage with fellow diners and access a wider range of available support services.[20] If, despite these evident benefits, there are still those who think it a bad thing to give free food to poor people, then instead of arguing against food redistribution altogether because it is patronizing, they should argue that surplus food should be given away to anyone, irrespective of their financial situation.

It is not just large organizations which can make a difference either. More-politicized groups such as Food Not Bombs act deliberately outside the law, practising freeganism – or 'dumpster diving' as it is known in the US – by giving away free meals to bystanders or community centres. The aim of this organization, which operates in many cities across the Western world, is not confined to feeding people; it is also highlighting the grotesque levels of waste and the mismanagement of resources under current trends of Western capitalism.

Some of the most inspiring and effective instances of food redistribution I have encountered have been informal groups and individuals, often working below the radar of official policy. In 2000, when I was moving house, I called the number on a scrawled note I saw stuck to a lamp post in Victoria advertising the services of a 'Man and Van'. When Kurt arrived, his eyes lit up as we hefted my various 'reclaimed' pieces of furniture out of the door, including the double bed my friend Sangam had made from wood reclaimed from skips, chairs, tables, a sofa and even an antique Persian carpet that had been left out on the street.

Kurt told me that most of his time was spent as a self-appointed volunteer collecting unwanted office furniture and computers and redistributing them to charities all across England – and even to Africa. For years, Kurt and his partner Carolyn had also been collecting surplus food from shops and restaurants and opening informal soup kitchens. They used to operate out of the back of their van in the City, where people from every walk of life – from bankers to homeless down-and-outs – used to come and eat their free food. Someone donated them the use of a shop which they ran just like an ordinary grocery store, except that everything in it had been donated, a lot of it was out of date, and no money changed hands. 'Anybody and everybody came there to get their free stuff,' Carolyn explained to me.

It gradually dawned on me that I had bumped into the freegan king and queen of London. Years later, I learnt that Carolyn – who is now known as Peace – had got into all this when she was herself homeless and living on the street. Once she had discovered where she could get enough food and clothes for herself, she soon found that shop-keepers and others were willing to give more than she could possibly handle, and so she started collecting for others. Carolyn became a bag-lady and amassed trolleys and rucksacks to carry around the goods she had been given, distributing them to other homeless. Then she met Kurt and they became a team.

Having spent thirteen years collecting and giving informally, Kurt and Carolyn have now gone official. They have taken out contracts with five major food outlets, have one volunteer assistant, and now collect and give away food on the street to 'anyone humble enough to come and get it'. Whenever I hear sceptics ask what possible difference mere individuals can make in the grand scheme of things, I think of Kurt and Carolyn.

15. Recycle: Compost and Gas

Riches, then, which benefit also our neighbours, are not to be
thrown away. For they are possessions, inasmuch as they are
possessed, and goods, inasmuch as they are useful and provided for
the use of men[1]

Clement of Alexandria (AD 150–c.215)

Many of us remember wading into ponds as children, watching
bubbles rise from the muddy bottom, and recoiling at the rank
smell. This gas – mostly methane, or 'marsh gas' – is the by-
product of microbes breaking down leaves and other organic
matter in oxygen-starved water. Billions of years ago, when life
on earth was still in its infancy, the microscopic organisms which
evolved from the primeval soup found two ways of harnessing
energy from the environment. One required atmospheric oxygen; the other could occur without it, anaerobically. In the early
phases of the earth's history, when the atmosphere consisted
primarily of water vapour and carbon dioxide, anaerobic conditions dominated. It was only after millions of years during
which photosynthetic organisms absorbed carbon dioxide and
pumped the earth's atmosphere full of oxygen that aerobic res-
piration took over. This was disastrous for those microbes that
could survive only in anaerobic conditions, and their niches
slowly receded.[2] However, there were many places on earth
where oxygen was still in short supply. In bogs, marshes and at
the bottom of stagnant ponds, anaerobic microbes still reign,
slowly releasing methane.

Methane is flammable, and it has been said that as early as the tenth century BC Assyrians were using it to heat their bath water. Sophisticated systems were developed by British imperial engineers in the nineteenth century to collect organic matter such as human sewage and siphon off the 'biogas' to burn in street lamps. China had such plants in place by the end of the nineteenth century, channelling in manure and pouring out gas, compost and effluent that nourished aquatic plants and fish. By the 1980s, the Chinese government had helped to construct millions of biogas digesters piping methane into rural homes while also creating a hygienic treatment of animal and human faeces.[3]

Today, anaerobic digestion is used to treat sewage and, increasingly, farm slurries across the industrial world. Methane from farm manures was recently found to be among the most environmentally beneficial biofuels currently available.[4] Increasingly, these systems are also being touted as one of the best ways of treating food waste.[5] In the developing world, they receive funding under the United Nations Clean Development Mechanism and have the potential to reduce deforestation if the gas collected is burnt as fuel instead of wood.

When food waste, manure, slaughterhouse by-products or other organic 'feedstock' is digested in large tanks, the gas can be burnt to produce electricity, and the hot 'waste water' used to provide heating to nearby industries or homes. Alternatively, it can be purified and pumped directly into the gas mains and used in people's homes for heating,[6] or it can be bottled and used instead of petrol or diesel – as it is in Switzerland, Germany and particularly Sweden, where a fleet of buses, taxis and a train line are currently run on gases from slaughterhouse waste. In effect, these are all being run on green, clean solar energy, stored in the food during the plant or animal's life and released as the organism decomposes.

Anaerobic digestion as a means of food waste disposal is preferable to landfill not just because of this kind of power generation, but also because when methane (CH_4) is burnt, the carbon in it binds to oxygen to form carbon dioxide (CO_2), which is a much less potent greenhouse gas.[7] One recent study estimated that avoiding landfill could save emissions of between 0.4 and 1 tonne of carbon dioxide equivalent per tonne of food waste.[8] Indeed, this carbon dioxide is not considered to contribute to global warming at all because the carbon released comes from plants that have recently absorbed carbon dioxide from the atmosphere – unlike the burning of fossil fuels, which releases carbon that has been stored for millions of years. The decomposed organic matter left over after anaerobic digestion can be used to fertilize fields or gardens, so potentially, though not always in practice, replacing industrial nitrogen fertilizers; furthermore, because digesters can be constructed in, or very near, urban areas, they could reduce the distance that urban waste needs to be transported, though at present the opposite is sometimes the case. It is for these reasons that many governments are now promoting anaerobic digestion as the solution to their food waste problems.

Because anaerobic digestion produces power, it is generally deemed to be better than straightforward composting. Composting does not harness the energy in the organic matter. It breaks it down aerobically, releasing carbon dioxide rather than methane, and produces compost, and it is therefore generally considered to be better than landfill. But it can also produce ammonia emissions or even methane if not properly aerated.[9]

The benefits of both centralized composting and anaerobic digestion are partly counteracted by using a specialized fleet of vehicles for door-to-door food waste collection, though in the future, when councils learn to use one vehicle to collect all kinds of segregated waste, this may no longer be such a big issue. According to WRAP, anaerobic digestion is also better than alternative methods such as gasification and pyrolysis, which are currently very expensive and require a lot of energy to heat the food to temperatures up to 800 °C to convert it into residual char and 'syngas' or oil which can then be used as fuel.[10] It is

also cleaner to use anaerobic digestion for separated organic
waste than either to incinerate it, along with other waste, or
to extract it from mixed municipal waste through mechanical
biological treatment or autoclaving systems in which waste is
treated under pressure by steam to allow the recovery of the
organic matter, which can then be spread on land or used as
fuel. Nascent technologies, such as enzymatic hydrolysis, which
uses enzymes to break down the food into sugars that ferment
into ethanol for fuel, may have a role in the future, but they are
currently uneconomic.[11] For these reasons, anaerobic digestion
is at present probably the best treatment method for most mixed-
food waste in Europe.

One significant disadvantage, however, is that anaerobic
digesters are expensive to construct and difficult to maintain.
Several governments in Europe have therefore decided to try to
boost the industry by supplying subsidies and grants. Germany
has been one of the most successful countries in promoting the
expansion of its anaerobic digestion capacity, largely in the form
of on-farm digesters dealing with both animal slurry and food
waste. Initially this expansion was fuelled by the high prices that
farmers could charge for allowing companies to dispose of their
waste, but they are also supported by the Renewable Energy
Sources Act (2000), which pays a fixed subsidy for every unit of
renewable energy produced.[12] Britain has been slow to follow
suit, but the cost of sending waste to landfill has now reached a
level where many companies and local authorities are deciding
to pay anaerobic digestion plants to treat their waste instead.

In addition to this indirect assistance, anaerobic digesters in
the UK receive direct grants for building costs and can now
apply for extra money through the Carbon Trust. In 2009 the
Welsh Assembly decided to make £26m available in capital
and procurement funds for anaerobic digestion plants. Most
importantly, they can sell the electricity they produce at an
elevated price under the government's Renewable Obligation
scheme, and in the future they will receive even more than

other renewables because anaerobic digestion fulfils the dual
purpose of waste treatment and green power generation.[13]
According to Joan Ruddock, the British minister for Climate
Change, Biodiversity and Waste, 'it ticks all the boxes environmentally. It produces 100 per cent renewable energy, and the
treated material can be returned to the land as a fertilizer –
offering us a virtuous circle. And the whole process reduces
greenhouse gas emissions when compared with other waste
processes.'[14]

As a consequence, there are many anaerobic digestion plants
in the planning pipeline in Britain, and it promises to be a
booming industry. Already, a survey in the UK found that of
604,883 tonnes of food waste produced by a selection of food
manufacturers, 66,239 tonnes were sent to anaerobic digestion
in 2006 – though it is unclear how representative these figures
are.[15] There are a number of different models of anaerobic
digester, and each one differs in the type of waste it handles and
how the waste is treated.

When I arrived by train in Ludlow, a small town in western
England, in March 2008, I knew I had reached the right part of
town from the sensation in my nostrils – a faint whiff of kitchen
bins. Complaints about odour have been a hurdle to food waste
treatment plants, so I knocked on a few doors and asked the
locals what they thought about it. All of them participate in the
government-funded recycling scheme by separating their food
into special bags, and everyone I spoke to seemed good-willed
about the plant on their doorstep. 'It is a shame about the smells,'
they said, 'but it has to be done.'

At the Greenfinch plant itself, Michael Chesshire, the brains
behind the project, showed me the huge heaps of bagged-up
food waste from the viewing gallery. A closer look gave an
intimate insight into Ludlow's kitchen bins – half-eaten bananas,
cabbage leaves and loaves of bread still in their plastic packaging
(which can contaminate the residual compost of some plants).
From the holding bay, the waste is lifted by conveyor belt into

a pulverizing macerator, pasteurized at 72 °C and thence pumped into the first of the digestion tanks. A month later it comes out the other end as a peat-like compost which is collected by farmers. The gas is used to generate renewable energy which Chesshire sells to Marks & Spencer at 'a very good rate'. (Unfortunately, the plant does not currently sell any of its heat, though it has plans to link up with a new eco-business park which would buy at least half its surplus heat during winter, and possibly even use some for cooling during the summer.) Chesshire is a great enthusiast of the system. But he concedes that plants like his can only be profitable if they handle over 10,000 tonnes of food waste every year – about three times the present volume.

The Holsworthy plant in Devon cost around £8m to build, but despite the government paying for half of this the business eventually folded. Furthermore, in the early days the government found that it had an overall negative impact on the environment because the greenhouse gas savings were outweighed by its emissions of sulphur dioxide and the spreading of excess nutrients onto the land.[16] The plant was subsequently bought by Summerleaze Ltd in 2005, which has boosted its income by moving from low-yielding cattle manure to food waste. This way, food companies seeking to avoid the cost of landfill pay tipping fees from £2 to at least £42 per tonne – and this currently brings in about 60 per cent of the plant's income.[17] In effect, therefore, the government's tax regime provides the financial backbone of anaerobic digestion, and studies in the US have also shown that anaerobic digestion needs to be funded by these 'tipping fees' in order to operate cost-effectively.[18] By itself, turning food waste into energy is not economical: it is a waste disposal service that must be paid for.

How much power does anaerobic digestion generate, and is it really the best way of treating all kinds of food waste? There are scores of scientific articles published across the world comparing anaerobic digestion with composting, incineration, landfilling and numerous other alternatives.[19] Governments have funded studies calculating which of these various ways of treating food waste are the most cost-effective and environmentally friendly. But there is one waste disposal method which is routinely and inexplicably left out of scientific comparisons. Unlike anaerobic digestion, it is financially sustainable without government funding; it actually makes money without charging businesses or taxpayers for the service; and it is arguably far more environmentally beneficial. It is also around nine millennia old. It is called swill.

Not all food waste is suitable for feeding to livestock, and currently animal by-product laws in Europe, Australia and some states of the US prevent using most food waste for this purpose. But there is still a great deal of extremely nutritious food waste either being sent to landfill, or increasingly to anaerobic digestion plants, which would be much better used as livestock feed, and this could replace the need to use so much environmentally damaging cereal-based feed. Giving food by-products to livestock requires less capital investment and processing than anaerobic digestion, and if you want electricity and heat too, you can channel the manure which results into an anaerobic digester. For comparatively low costs, you can get electricity, heat, rich digested fertilizer and huge quantities of sustainably grown meat.

Surprisingly, there is not one study anywhere, as far as I know, dedicated to comparing the environmental and financial benefits of anaerobic digestion with pig feeding, so I have had to use my own calculations to compare the two.[20]

The figures to assess the carbon savings from sending food waste for anaerobic digestion were provided to me by Michael Chesshire at Greenfinch. They are, as he says, tentative, but a comparison with several other attempts to measure similar biogas plants suggest they are accurate. The Greenfinch plant is not an ideal comparison, because under current European law the household kitchen waste it uses cannot be fed to livestock. However, there are commercial and industrial waste foods which would yield similar amounts of biogas but which could alternatively be fed to pigs, and thus the comparison should at least provide a useful indication.

Sending food waste to anaerobic digestion nominally replaces conventional fossil fuel energy sources with a renewable energy source, so the carbon saving is represented by the greenhouse gases that would have been emitted by generating the same amount of energy conventionally. On the other side, the carbon saved by giving food waste to pigs comes from negating the need to produce conventional pig feed made from grains and pulses, which requires tractors to be driven, land to be ploughed and agro-chemicals to be manufactured.

For every tonne of food waste, the Greenfinch plant yields 95 m³ of methane,[21] which generates 255 kWh of renewable electricity,[22] thus saving 110 kg of carbon dioxide emissions.[23] If the plant achieved its aim of using half of its hot water for heating, the total carbon dioxide savings would be 143 kg.[24] The savings that could be made by replacing conventional pig feed with swill, by contrast, would be 236 kg of carbon dioxide, or more than *twice* as much carbon saving as using the food waste to produce electricity in an anaerobic digester.[25] Even if you include the potential emissions savings from the currently unused heat from the anaerobic digestion plant, feeding general swill to pigs is still 63 per cent better than using it for anaerobic digestion.[26]

This does not include the fact that pig manure can be used to generate biogas.[27] More importantly, it does not include the avoided emissions from changes in land use associated with producing soymeal for animal feed, such as deforestation in South America. If you include land use changes, the emissions generated by a tonne of South American soymeal grown on deforested land and fed to animals in Europe soars from 700 kg of carbon dioxide equivalent up to 232,800 kg per tonne – or, if you allowed the impact of deforestation to be spread over twenty years' worth of harvesting soy on that same piece of land, then the emissions would be 11,600 kg per tonne.[28] In other words, supposing you had a tonne of food waste with a similar nutritional quality to soymeal, it could be between 26 and 520 times better to give it to pigs than to put it into an anaerobic digester.[29]

On top of these global warming advantages, sending food waste to pigs rather than anaerobic digesters is preferable because it can reduce the risk of other environmental problems associated with anaerobic digesters, such as acidification, eutrophication, and excessive nutrification of terrestrial habitats which can occur as a result of spreading digestate on land.[30] Furthermore, using food waste instead of cultivated animal feed would save around 1 million litres of water per tonne and it could liberate millions of hectares of agricultural land for better purposes.[31] Perhaps most significant of all is the avoided damage to the world's biodiversity currently caused by cultivating animal feed. Anaerobic digestion has none of these advantages.

That is the environmental benefit. The economic case is just as clear. The potential value to the economy of one tonne of food waste in an anaerobic digester would be £37.20.[32] By contrast, a tonne of food waste converted into pork would have a retail value of around £330.[33]

Anaerobic digestion may currently be the best option for most kinds of food waste which are not allowed to be fed to livestock, such as catering and household kitchen waste. But I am worried by the technophilic faith that people in governments, the food industry and the media have in its environmental benefits. There was one high-profile BBC documentary which gave people the impression that the benefits of anaerobic digestion even compensated for wasting food in the first place.[34] This gives the food industry too easy a route to make itself look green – and as a consequence virtually all the commitments by the large supermarket chains have concentrated on reducing the amount of food waste going to landfill, rather than the far more significant step of reducing food waste per se. The fact is that the energy reclaimed by sending food to anaerobic digesters is a tiny fraction of the energy expended in growing the food. (For example, putting a tonne of tomatoes through an anaerobic digester would recoup less than 0.75 per cent of the emissions released in producing them in the first place. From a global warming perspective, that means it is at least 130 times better to avoid growing the tomatoes than to turn them into gas.)[35] Supermarkets have claimed that anaerobically digesting surplus food 'wastes nothing along the way', while media stories have called it 'an environmentalist's dream', 'the Robocop of food waste' and 'too good to be true', assuring readers that it could 'provide the UK with fuel security'.[36] Estimates suggest that if the UK's household food waste was collected for anaerobic digestion, it would realistically still only produce between 0.5 and 1 per cent of domestic electricity demand.[37]

It is stunning how quickly the food corporations are now falling over each other to appear the greenest, with, for example, an announcement from McDonald's in 2008, saying it aims to send nothing to landfill by 2010, a commendable commitment. But we can now distinguish between 'slightly less bad' waste management measures, and actually good ones – between green measures and greenwash. McDonald's – whose 'green credentials' were hailed in the press when they made the landfill announcement – is conducting a waste incineration scheme which in March 2009 still only covers less than 1 per cent of its stores. It hopes to expand this, but as its environment manager explained to me, the target of zero waste by 2010 is 'not set in stone'.[38] The supermarket chain Asda received accolades and reams of media coverage for making a pledge to send no waste to landfill by 2010. But it is planning to do this primarily by sending its food waste for incineration, composting or anaerobic digestion, rather than something more useful. Sainsbury's announced in 2008 that it intends to send all of its annual supply of 60,000 tonnes of food waste from the back of its stores to anaerobic digesters and other non-landfill disposal routes by the end of 2009, no less.[39] They do give some surplus to charity, and some goes for pet food; but if they separated their (unpackaged) bakery, fruit and vegetable waste, it could all go much more beneficially to livestock. The Holsworthy anaerobic plant takes cream and whey which pigs would love to eat, and the Biogen plant near Luton is sent industrial food wastes like bread that would make suitable pig feed. Anaerobic digestion enthusiasts are even talking about trying to compete with farmers who buy fruit and vegetable by-products for their livestock. This is only economically possible because of the government's current policy of subsidizing anaerobic plants. In the absence of comparable assistance for pig farmers who want to use this food waste, the government's 'green' policy is skewing the market and arguably having a negative impact on the environment.

One reason why the savings in greenhouse gas emissions offered by swill-feeding are ignored by European governments is the problems with the way the calculations have been set up under the Kyoto Protocol. You might think that the emissions released in the production of a sausage eaten in a European country ought to count against it. But, in fact, when European consumers pay South Americans to burn the Amazon rainforest to grow soy to feed the animals which will be turned into European sausages, the emissions are counted against Brazil, and the emissions from shipping the soy across the Atlantic are counted against no-one at all.[40] So although switching from commercial feed to swill-feeding would reduce carbon emissions globally, much of that reduction would not count in the European country's favour.

Biogas plants are attractive to people who think that hi-tech solutions are always the best answer. Their carbon 'savings' are immediately recognizable and accountable under the Kyoto Protocol; they show on paper that emissions have been reduced and they help to meet renewable-energy targets. But in this case a lack of research and clear-headedness has led governments and businesses, even with the best will in the world, to make the wrong decisions.

16. Omnivorous Brethren: Pigs and Us

Dearly beloved brethren, is it not a sin,
to peel the potatoes and throw away the skin?
For the skin feeds the pigs and the pigs feed us,
Dearly beloved brethren, is it not thus?

British folk rhyme

For 9,000 years, humans have lived alongside domestic pigs. Traditionally, pigs consumed human refuse, and humans ate their flesh. So useful was the pig that people domesticated it in regions as far apart as the Philippines, western Europe and Africa – right across its natural range of habitat. As far as the pig species was concerned, having some of its number regularly killed and eaten turned out to be a worthwhile price for the benefits of being given plentiful food and shelter. Indeed, pig domestication probably originated when pigs entered human settlements to forage, rather than as a result of humans actively going out to capture them. As the relationship between pigs and humans evolved, people took them wherever they went. Ancient Polynesians carried them in their canoes when they sailed off to colonize new islands; Koreans penned them beneath their household privies to clean up everything from kitchen waste and animal viscera to human excrement. So effective at reducing outbreaks of disease were these pig-privy structures that the ancient Chinese buried terracotta figurines of them with their emperors. Pigs cleared away garbage that would otherwise be breeding grounds for harmful rodents, insects, pathogens and parasites, and converted the unwanted waste into nitrogen-rich manure and delectable meat proteins and fat.[1] Today, with a few notable exceptions, nearly anywhere there are people in the world, you will also find pigs.

Right up until modern times, waste recycling remained one of the pig's primary functions in human societies – sometimes more important than as a source of flesh.[2] Pigs were often fattened in herds in woodland, foraging for acorns, beechmast and chestnuts; but in agricultural and industrial areas, where forests had been felled, their fodder was primarily by-products from the human food chain. In Europe, this role in life was regarded with a mixture of admiration and revulsion. On farms across the continent, any food that would otherwise go to waste was left out for the pigs. In the seventeenth century, when the pig population of England alone reached 2 million, the author Gervase Markham remarked that the pig:

is the Husbandman's best Scavenger, and the Huswives most wholsome sink; for his food and living is by that which will else rot in the yard . . .; for from the Husbandman he taketh pulse, chaff, barn dust, man's ordure, garbage, and the weeds of his yard: and from the huswife her draff, swillings, whey, washing of tubs, and such like, with which he will live and keep a good state of body, very sufficiently.[3]

Echoing Markham in 1707, the agricultural improver John Mortimer added that 'Swine are very advantageous to the Country-man . . . in that they feed upon what would otherwise be of no use or advantage, but would be flung away.'[4] Pigs thus turned the useless into something useful. In other parts of the world, in Goa, South Korea and even in metropolitan centres like Delhi, pigs are still free to rummage in dumps and latrines for organic garbage, which they convert into flesh.

Industrial-scale pig-rearing first became possible where large concentrations of waste food were available in one place. The most common sources were from large millers and dairies – which produced quantities of protein- and fat-rich whey – and from breweries which had a constant need to dispose of spent hops and grain. Owners of such institutions were advised either to sell their waste to pig farms, or better still, to get a pig herd of their own. In the 1720s, Daniel Defoe wrote in his *Tour through the Whole Island of Great Britain* that the main bacon-producing areas in England were the dairy counties of Wiltshire and Gloucestershire: 'the hogs being fed with the vast quantity of whey, and skim'd milk, which so many farmers have to spare, and which must, otherwise, be thrown away.'[5] In 1748 a Swedish tourist in England noted admiringly that:

the Distillers keep a great many, often from 200 to 600 head, which they feed with the lees, and any thing that is over from the distillery: . . . In the same way, and with the same object, a great number of pigs are kept at starch factories, which are fed and fattened on the refuse wheat.[6]

This was still the case in the 1830s, when one writer noted that pigs were 'especially valuable to those persons whose other occupations furnish a plentiful supply of food at a trifling expense'.[7]

Until very recently it was widely recognized that feeding pigs anything other than waste or woodland forage was economically or ecologically unviable. The other most important domestic animals that humans had kept for 10,000 years – cattle, sheep and goats – were useful because they were ruminants and could digest cellulose in grasses, leaves and shrubs, which were otherwise useless to people. The pig, by contrast, being a monogastric (single-stomached) omnivore like humans, needed very similar kinds of food to people. Pigs are relatively efficient, but they will still turn only about a fifth of what they eat into meat – hence the events on Tikopia and also, according to Marvin Harris, the reason why the ancestors of Jews proscribed eating pigs after prehistoric deforestation in the Middle East had removed the pigs' natural forage and shade: God instituted a

resource efficiency drive and instructed his people to eat any
grass-loving ruminant that 'chews the cud', but not the pig
because 'It cheweth not the cud.' (That is what the scriptures
say, not the commonly repeated notion that pork can cause
disease, which is also true of other kinds of meat.)[8] What was
true then was equally so millennia later. At the end of the
nineteenth century the palaeontologist Nathaniel Southgate
Shaler warned that 'In the earlier state there was no cost for his
keeping; in the latter, except in so far as he could be fed from
the waste of a household, he is an expensive animal.'[9]

It was the dramatic agricultural changes during the nineteenth
century, principally in the United States, that overturned tradition. When European settlers made it over the Allegheny
mountains, they found themselves in the vast plains of what
would become the Midwestern states. Huge savannahs could be
utilized for grazing cattle or growing maize. But at such a
distance from the main centres of population these commodities
were worth virtually nothing; it was said that maize 'was of so
little value as to be substituted for wood as fuel'. At this price it
was economical to let pigs loose into maize fields to reap the
whole harvest themselves, fattening on cereal crops before being
transported to the coastal metropolitan centres of Cincinnati and
later Chicago. But even here, before the advent of refrigerated
railroad carts in the second half of the nineteenth century, pork
was so abundant that packers often used only the prime cuts,
disposing of the remaining carcasses in the river.[10] As one weekly
paper commented in 1908, 'A large proportion of the pigs . . .
used to go to waste. They did not think about caring for the
many little portions that now are gathered up and used for food,
or used for grease, or for soap.' It was no doubt to burnish their
sullied image that Chicago meat-packers later boasted of their
thriftiness for making use of every part of the pig 'except the
squeal'.[11]

From this time onward, pigs in Europe and America were
increasingly raised on cheap grains. However, in times of

scarcity, such as the two world wars, this extravagant practice had to be curtailed. Towards the end of the First World War, giving to animals any cereals designated for human consumption was made a fineable offence to reduce 'the competition of pigs with human beings for cereal food', and instead only traditional by-products and waste were permitted.[12] The same thing happened again in the Second World War, when Britain's food imports were disrupted. As the historians Stephanos Mastoris and Robert Malcolmson have shown, the number of pigs in Britain declined by 60 per cent from 1939 to 1945. In June 1940, one article in the *Journal of the Ministry of Agriculture* pointed out that Britain would have to revert to traditional methods:

During the past sixty years we have come to rely to an ever-increasing extent upon imported feeding stuffs; now we must learn to make shift with reduced supplies; and we can get some help with our problem by studying the systems of feeding and management that prevailed up till the time (about 1880) when the flood of cheap grain and other feeding stuffs began to arrive from overseas.

Relearning the old ways of rearing pigs in Britain was promoted by the innovative Small Pig Keeper's Council, convened in 1940 to encourage the sharing of pigs in associated 'pig clubs' and the collection of kitchen swill for their fodder. King George VI gave his backing to the project by organizing a photoshoot showing him joining a pig club; and a leaflet issued in May 1940 declared it government policy: 'To-day, those who can keep and feed pigs largely on waste foodstuffs from kitchens, gardens or allotments will be doing a national service in addition to assisting themselves. They will be helping to increase the nation's meat supply, and to save shipping.'[13]

As part of the war effort, the horse stables in Hyde Park in London were converted into a piggery tended by the police. Swill buckets stationed on street corners were collected on rounds and given straight to the urban pig herd.[14] For children who grew up during the war, the swill bucket is still a vivid memory; for many, it has become an icon of thrift to contrast with the profligacy perceived to prevail in the modern world. 'Nothing ever got wasted,' remembers one civilian of wartime Chesterfield in Derbyshire, England: 'We used to have a pig bin [in] which we would collect all the food waste, for example potato peelings, stale bread, in fact anything . . . They used to collect the bins in lorries, not unlike the dustbin lorries we see today.'[15]

The Germans, who had analogous food supply problems during both wars, were no less alert to the wastefulness of feeding grains to livestock, and the savings to be made from organized swill-feeding. The local authority in Stuttgart was typical: in the First World War it purchased a 740-acre farm where pigs were fed on an officially organized kitchen waste recycling programme. During the difficult winter of 1916–17, the pork was sold at a significant profit which was then used to feed 7,500 of the city's schoolchildren.[16] During the Second World War, the Nazis, like the British government, tried to stamp out the uneconomical practice of rearing animals on grain rather than by-products. Franz Wirz, a member of the Nazi committee on Public Health, argued that it took about 90,000 calories of grain to produce just 9,300 calories of pork and Hermann Göring said that farmers who fattened animals on grain were 'traitors'.[17]

After the Second World War the drive for agricultural intensification – achieved by increasing irrigation, breeding higher-yielding crops and injecting fossil fuels (and thus nitrogen fertilizer) and state subsidies into agriculture – once again made cheap grain available in abundance, and prices have dropped steadily ever since (until very recently).[18] Total supply of grains increased by 43 per cent between 1980 and 2004 and international prices halved in real terms between 1961 and 2006. Between 1964 and 2004 the world's annual soy harvest grew from 29 million to 200 million tonnes, and most of it is now used for animal feed after the oil has been extracted.[19] By the

1980s, swill-feeding in Europe and America was in serious decline, as cheap meat, produced with cheap subsidized grain, flooded Western markets. From the 1960s to 1994 the number of pigs fattened on swill in the US fell from 130,000 a year to less than 50,000.[20] A handful of larger swill-feeding operations managed to stay in business, and in England one of those farmers from the old school is John Rigby. Rigby took on pig farming from his father and grandfather, who started swill-feeding after the Second World War, when farmers were paying up to five shillings for a binful of swill. With the availability of cheap grains, compounded by the spread of foot-and-mouth and classical swine fever and depressed prices from slaughterhouses that were suspicious of pork tainted with fish by-products, swill-feeding became less and less competitive.[21]

Swill-feeding came to an abrupt end in 2001. In that year, the British government concluded that the catastrophic foot-and-mouth outbreak originated on a farm feeding swill to pigs. It transpired that the farmer had not been observing the law on boiling food waste for an hour to kill off pathogens such as the foot-and-mouth virus, and the untreated waste he allegedly fed his pigs may have contained illegally imported infected meat. The government's response, on 24 May 2001, was to ban swill-feeding with the Animal By-Products (Amendment) (England) Order 2001. The European Union quickly followed suit. Nervous administrators in the United States scrutinized events closely, and Texas, once a thriving swill-feeding zone, brought in similar laws. In Australia too, swill-feeding is illegal.[22] Before 2001, institutions were paying John Rigby £7 per tonne to take food waste away – and he would sell it on as animal feed for £120–160 per tonne after mixing it with meal, yielding a peak annual profit of £750,000. In 2008, Rigby was making a loss on his pigs and he is considering getting out of pig farming altogether.[23]

In 2007–8 a sudden rise in the price of agricultural commodities, including animal feed, meant that in the twelve months running up to July 2007 the price of cereal pig feed rose by 40 per cent and was destined to increase further. In September 2007 animal feed-grade wheat was selling at around £170 per tonne, twice that of the previous year, and maize was selling at 75 per cent more than the 2006 average. In nominal terms, these were the highest prices ever recorded; and even in real terms they were the highest since the sharp spikes during the crisis of the 1970s.[24]

Livestock farmers, particularly pig farmers, went out of business because they were paying more for their feed than they receive for their pork. For intensive pig-rearers, feed represents around 60 per cent of total costs.[25] In 2007 the UK pig industry estimated that farmers were losing £20 for every pig they raised; the industry has shrunk by 40 per cent already and this trend is bound to continue. Despite the agricultural subsidies that are designed to keep farmers afloat, grains are becoming too valuable to feed to animals. Even as prices come down, governments around the world are realizing that an industry so dependent on cheap grain is vulnerable to inevitable fluctuations and insecure global food supplies. This provides a new economic incentive to reconsider the ban and revert to the former practice of swill-feeding.

Modern science has not completely forgotten the virtues of swill. Research has shown that, on average, swill from catering outlets can contain a healthy 20 per cent crude protein and over 25 per cent fat. That is more than sufficient for pigs fattening for slaughter, and swill generally contains enough minerals and vitamins, and more than the recommended level of different amino acids (with the possible exception of lysine, which can be supplemented). A team of American researchers found that feeding pigs purely on swill could lower growth rates from 1.8 lb per day to 1 lb a day, but a respectable growth rate of 1.45 lb per day could be achieved just by adding 25 per cent cereals to the swill to bring down its water content.[26] In this scenario, the swill-feeder's pigs would take an extra quarter of the time before they get to slaughter-weight, but for each day the farmer would be saving up to three quarters of the cost of his feed and possibly charging extra money for relieving companies of their food waste. Even better weight gains can be achieved by crushing, drying and mixing waste with grains and pulses to form feed pellets. Some feed manufacturers, such as Enviro-Feed Corporation in New Jersey in the US, have developed a system in which the swill is both cooked and dried by a single blast of hot air, which complies with the legal cooking requirements for those states in the US where it is still permitted and also converts the swill into easily transportable non-perishable pig feed. This is an energy-intensive process which usually counteracts many of the environmental benefits of using waste food, but instead of using fossil fuels to provide the heat, they burn food waste.[27] A study in South Korea showed that replacing half of the conventional corn-soy diet of finishing pigs with a food waste mixture composed of restaurant waste, bakery waste and poultry litter aerobically digested by bacteria and vacuum-dried to a moderate moisture content reduced feed costs by 33 per cent: in fact, the food waste mixture was so palatable that the pigs fed on it had larger appetites than those on conventional food.[28]

In Britain, it is estimated that an annual total of up to 1.7 million tonnes of restaurant, supermarket and industrial food waste which used to be fed to pigs had to find a new home after 2001.[29] Some of it is fed to pets; most of the rest ended up going to landfill, while the swill-feeders went out of business. To their continuing dismay, the British government refused to compensate them as they did other farmers affected by the foot-and-mouth outbreak, even though they had recently been encouraged to invest millions of pounds in equipment for swill processing. Other countries, such as Germany and Austria, had much bigger swill-feeding industries; their governments helped them to change their businesses into anaerobic digestion plants.

In the panic of 2001 the British government had to be seen to be doing something and had a strong incentive to claim they had found the source of the infection. A temporary ban may have been justified, but making it permanent was a hurried, drastic and unnecessary policy decision. The only problem with the previous law requiring farmers to boil their swill was that it was not properly enforced.[30] But the new law is not enforced either, and I know many farmers who regard it as so excessive that they regularly break it, including a vet tasked with upholding the law. As I have found while rummaging through their bins, most of the well-known supermarkets and other retailers break the law too, by occasionally throwing raw meat away with other trash destined for landfill, whereas the law stipulates that it should be sent for specialist treatment because if it ends up in landfill it could come into contact with livestock. Even the official way of dealing with some animal by-products means that they end up being composted on farms and spread on the land where there is still a potential risk of disease outbreak.

The non-sequitur in the British government's support of the ban was demonstrated by the then Under-Secretary of State for Environment, Food and Rural Affairs, Ben Bradshaw MP, who said in the House of Commons that 'experience demonstrated that no matter how stringent the legislation, there will be those who will do not [sic] follow the rules ... There is only one conclusion to be drawn: the Government made the right decision in banning the feeding of swill to pigs.'[31] If, as Mr Bradshaw points out, there will always be rule-breakers 'no matter how stringent the legislation', how is it true that the only possible measure to deal with this is to make the legislation more stringent?

Dealing with this issue is fraught. My uncle was a Pennine shepherd, and his sons practised the ancient craft of stone-walling in Cumbria, which was the county worst hit by foot-and-mouth in Britain. The flocks of sheep that had been nurtured for centuries were remorselessly culled and incinerated. In the wake of the disaster, they emigrated to New Zealand, and tales of woe live on in their community. (Ironically, my cousin John is now raising pigs on swill, which is still legal in New Zealand.) No-one wants another outbreak of infectious livestock diseases. But the matter cannot be looked at from only one point of view.

Defra officials concede that the policy against swill-feeding was made without taking into account the environmental or economic costs and benefits, and was considered solely in terms of animal health. One policy advisor said to me on the telephone: 'I would say there would be a market for any research that showed the value of food waste to farmers. Policy makers and politicians would be interested. If one assesses a risk to be x, but the benefits are y, and there was good evidence of the benefits, you would be bound to look at that.'[32] Yet no study has yet been commissioned to examine it.

The foot-and-mouth outbreak cost the UK economy an estimated £8bn, led to the culling of 6 million animals and caused untold suffering to farmers and others dependent on the rural economy.

What, by comparison, are the costs of banning swill? The manufacturing and processing sectors of eighteen European countries produce an estimated 222 million tonnes of fish, dairy, animal, fruit and vegetable waste each year.[33] It would, of course, be much better if we stopped wasting so much food in the first place, but even if this was achieved we could still use the unavoidable by-products as animal feed. Still more waste is produced by the nine other members of the EU for whom data is lacking, and there are also many millions more tonnes from retailers, caterers, farms and even households – sources which have in the past been used to raise pigs, and in countries such as South Korea and Taiwan still are. Industrial food waste that has not been handled under the same roof as meat already can be fed to livestock, but supposing that just 44 million tonnes more food waste could be fed to pigs if the law were changed, then this could be used to produce 3 million tonnes of pork with a retail value of £15bn every year.[34] There would also be the enormous saved costs of disposing of the waste under the current regime. (Since the ban, the cost of disposing of waste food has shot up by 40 per cent;[35] and one food manufacturer in England said it saved disposal costs of £126,000 a year by sending its permitted food waste to pigs instead of anaerobic digestion.[36]) Millions of hectares of land currently used to grow cereals and pulses for livestock in Europe could be liberated, and supposing that just 3 million tonnes had an equivalent nutritional value to soymeal (which it certainly does), and this was used instead of deforesting tropical forests to raise livestock, the potential emissions savings of using this feed would be up to 700 million tonnes of carbon dioxide.[37] These are only approximate figures, but they give some indication of the forgone value of swill. What is needed is a thorough, government-funded study into the costs and benefits of maintaining the ban.

It is important to distinguish the ban on feeding food waste to pigs and chickens from the earlier ban on feeding animal by-products to ruminants with the Animal By-Products Order (1999). This prohibition was passed following years of enquiry into BSE (mad cow disease), believed to have been caused by feeding to cattle the brains of sheep suffering from scrapy.[38] Being herbivorous, cattle are not evolved to eat animals, and the BSE crisis was seen by many as nature biting back against unscrupulous modern farming practices.[39] Feeding swill to pigs, however, is entirely different. Pigs and poultry are both naturally omnivorous. In the wild, pigs will forage for meat – even devouring other dead pigs – as well as roots and leaves, and science has shown that properly heat-treating swill kills all pathogens, rendering it safe for pigs.[40]

European legislators have at last begun to re-examine these regulations and have now decided that feeding pig by-products to chickens might be permissible. This is good, but it does not go far enough. One of the concerns voiced in the media when the EC made this announcement in 2008 was that Muslims would no longer want to eat European chicken meat because it had been raised using pig offal, which is *haram*. I was in Pakistan at the time and talked to my Muslim friends about it. Without exception they rejected the idea, pointing out that chickens – everywhere in the world – have always greedily pecked at innumerable things forbidden for Muslims to eat, including insects, lizards, carrion and faeces. Yet chicken is eaten throughout the Muslim world.

There is now international recognition of the environmental imperative to lift the ban on feeding food waste containing animal by-products to pigs and chickens. The FAO has looked at the environmental damage caused by the ban and in 2006 called for Europeans to reconsider their position:

In various contexts, food wastes and agro-industrial by-products could contribute substantially to the feed supply, and by the same token release pressure on land. There is also an ecological interest in recycling the nutrients and energy embodied in food wastes and by-products, instead of disposing of them in environmentally damaging ways.

The FAO points out that, before the ban, the EU fed about 2.5 million tonnes of animal by-products to livestock each year, which in nutritional terms equates to 2.9 million tonnes of soymeal. Following the ban, EU soymeal imports increased by almost 3 million tonnes between 2001 and 2003. Much of this came from Brazil and other South American countries, where the Amazon rainforest is being destroyed to supply ever-increasing demand. The UN concluded that:

Soybean expansion and shipment creates environmental impacts in terms of biodiversity erosion, pollution and greenhouse gas emissions . . . The need to address such tradeoffs is likely to become increasingly acute, and policy decisions in this area will be critical to the environmental and social sustainability of the sector.[41]

No-one wants to see pyres of diseased animals being burnt in the fields of Europe again – but the fires raging in the Amazon rainforest only seem less terrible because they are further away. Managed safely, with proper enforcement of animal hygiene laws, swill-feeding can prevent both. Currently, valuable food waste is haemorrhaging at every link in the food chain. If we closed up those leaks, food waste producers would no longer have to pay such high disposal costs, farmers would be able to produce meat more sustainably, and pressure on the world's food markets would be reduced. A 2009 report by the UN argued – perhaps optimistically – that the food thus liberated 'could feed all of the additional 3 billion people expected by 2050'.[42]

There are advocates of a reintroduction of swill-feeding even in Europe. The Conservative Mayor of London, Boris Johnson, argues that it would be 'the right thing' to revoke the ban[43] and his colleague, Peter Atkinson, MP for Hexham, points out that the old swill-feeders 'had well-established businesses and provided a useful service in recycling waste from catering units and food manufacturers'. Tim Lang agrees that 'it's mad: that's what pigs are good at; the law should be reconsidered. We are putting into landfill what should be consumed.'[44] As Lord Haskins remarked to me, 'At the time of the crisis there was no alternative to the ban, but there is now a case for lifting it.'[45] Bringing back swill-feeding will require convincing EU member states and the European parliament which will be arduous but this process should begin immediately. In the future, particularly if cereals become even costlier and consumers start going without meat as a result, reality may finally catch up with our bureaucrats and legislators.

In the meantime, however, there is still a great deal of food waste which the law does allow farmers to give their livestock. Fruit, vegetable, bakery and dairy waste are all permitted 'former foodstuffs'. However, the law dictates that any food which has even come under the same roof as animal by-products has to be treated as contaminated. A factory that produces sausage rolls cannot give its waste bread to farmers, even if the production chains are completely separate. Only if the bakery institutes an official plan to ensure complete separation and has this approved by the local authority can the bread waste be put to good use.[46]

One might think that in this time of environmental consciousness, high feed prices, and urgent attempts to reduce the amount of biodegradable waste going to landfill, everyone is doing what they can to approve as many of these separation plans as possible and get the food wastes to livestock. Unfortunately not. The bureaucracy, fear, lack of awareness and relaxed attitude to environmental degradation prevent many industries from separating their permitted and banned waste foods, to the loss of both farmers and the food industry. Many food industry managers are unaware of (and do not have the time to find out about) the savings they could make by giving their waste to farmers rather than paying for it to go to landfill. Furthermore, in Britain many of the Animal Health officials responsible for approving the separation of permitted foodstuffs do not even understand the regulations themselves and issue incorrect and misleading advice on what can and cannot be fed to livestock.[47] The British government currently spends a lot of money threatening farmers and food outlet managers with two years in prison for breaking the swill ban. Instead, they should be collaborating with the industry to ensure that all legally permitted food wastes are sent to livestock. Sharing information between farmers, local authorities and the food industry would probably be enough to achieve this.

There are many instances of industries and farmers forming such mutually beneficial relationships. The great brewer local to me in Sussex, Harveys of Lewes, takes its spent lees to the nearby Plumpton Agricultural College for cattle. Our local cheese maker, which pumps out 10,000–15,000 litres of whey every week (a comparatively small operation), lets the local organic farmers take it away for their pigs. As one American farmer who feeds his pigs on bread, organic goat's milk whey, excess milk, cottage cheese and cheese trim testifies, 'All of these are what are termed pre-consumer wastes that are excellent foods for the pigs . . . Recycling these food items into pork reduces the flow into the waste stream, helps the creamery and bakery and saves us money on feed. The pigs love it all.'[48]

But, too often, such opportunities are missed. One food manufacturer in Luton currently operates two sandwich factories: at one they spend £65,000 a year sending their waste to an anaerobic digestion plant, despite the fact that it contains tonnes of bread which could be fed to livestock.* Their other factory produces only vegetarian sandwiches for the Linda McCartney range, and thus, instead of paying to have their waste taken away, they *sell* it to a pig farmer for £25 per tonne.[49] Researchers funded by the Austrian government found that tens of thousands of tonnes of nutritious foods, from organic pumpkin seed cake to buttermilk, brewer's yeast and okara (a by-product of tofu production), that could legally have been fed to pigs were being needlessly wasted.[50] One study in Britain found that 94 per cent of the fruit and vegetable waste currently being sent by food manufacturers to landfill could have been fed to livestock.[51] The supermarket chain Asda, in the UK, is already trialling a scheme whereby bakery waste is separated at source so that it can be fed to livestock, but it says it is currently unable to do so owing to the 'robust system'.[52]

This system prioritizes avoiding the manageable risk of disease outbreak over the definite consequences that feeding cereals to livestock has on the planet and the world's poor. During the Second World War, we knew that feeding pigs with cereals would deprive our compatriots of food. As one writer in wartime Britain said, 'it is better to keep five men alive on barley meal than one comfortably nourished on pork.'[53] Then we decided

---

* Subsequent to my enquiries about the logistics of separating this bread for animal feed, the company now plans, from late April 2009, to stop sending off all its waste bread for anaerobic digestion and to send it all to feed pigs and cows. This is much better than land filling, but it is still a waste of food that could be consumed by people.

to feed pigs swill instead. Why do we fail to come to the same conclusion now, just because the people we are depriving of food are brown-skinned and live a few thousand miles away?

17. Islands of Hope: Japan, Taiwan and South Korea

In Japanese there is a Buddhist concept known as *mottainai*, which
encourages . . . us to be grateful for the resources we have, to be
respectful of them and use them with care. It also calls for us not
to waste.[1]

Nobel Laureate, Wangari Maathai

My first experience of Japanese food (in Japan, that is, rather
than in London's sushi-filled bin-bags) was as wonderful as it
was fortuitous. Having completed my research on South Korea's
food waste laws, I took a train diagonally across the country and
boarded a boat bound for the port of Hakata in south-west
Japan. As we arrived in the dock I was asking my way to the
train station for my onward journey to Tokyo when a woman
called Yuriko Kawasaki overheard me and offered her advice. I
could try to get a train which, if I was lucky, would arrive at
around midnight in Tokyo; I could stay at a hotel in the nearby
city of Fukuoka and wait until the morning; or I could come
home with her.

Before I got the wrong idea, she added that she lived with
her relatives who used to be a 'host family', which meant they
took in foreigners from all over the world. There would be no
question of my paying, she said. I would be their guest.

When we arrived at her home, Yuriko's fiancé and her
brother ushered me into the traditional wood-and-paper walled
house and begged me to stop towering and sit down. Having
handed me a glass of home-brewed saké, Yuriko's mother,
Takeko, began laying the dinner before us: dishes of sashimi, a

bowl of cooked squid, a pickled cucumber, a platter of cold sausage and a terrine of home-made seaweed jelly. The family said a traditional Japanese grace, *hadakimasu*, and we started eating. But Takeko kept bringing more and more things to the table, until I became overwhelmed by their complexity and excellence. Taking chopstickfuls from the shared dishes, and dipping them into my bowl of soy-sauce, wasabi and chilli, I tucked in hungrily, enjoying the variety of tastes and textures – until Yuriko's father Jingo pointed out that these were merely the hors d'oeuvres. Takeko then came in from the kitchen with a meat and potato stew and several dishes of marinated fish. Finally we were given a bowl of rice with chopped vegetables 'to fill up on' and a bowl of miso soup. Each thing was eaten in small quantities, but the variety made the meal one of the most lavish I can remember.

Intrigued by how a family of modest means could sustain such gastronomic variety without inordinate expense, I asked them how they avoided wasting food when they had so much of it. As each dish was served, Yuriko's father, Jingo, proudly gave me explanatory footnotes on its source and origin, laying particular emphasis on freshness, and explaining how each thing could be used so as to get the most out of it. The rice and vegetables he grew himself in his own fields with 'a minimum of chemicals'; the fish came from Takeko's brother, who was a fisherman; and the seaweed they gathered from the seashore. All of it came from within a few miles of the house and was barely a day old before it was consumed.

Pointing to ancestral photographs on the wall, Jingo said that his father and grandfather had farmed the same fields as he did, built the wooden house we sat in, and had died in the Second World War 'defending' the land the food was grown on. (I resisted joining in as his Thai daughter-in-law protested at this gloss on Japanese military history.) Gesturing to the four-generation family sitting round the table, Jingo said he would do the same for them.

I told him he reminded me of my father, who used to boast to his guests that the vegetables they were eating were 'sushi' – so fresh they were still wriggling on the plate – and that he too felt strongly about the land his brothers died defending in the same war. Jingo became excited and pulled me over to the family computer to show me his fields on Google maps and to make me do the same for him. On the aerial photographs his lush paddy appeared as small dark green plots, squeezed into fertile valley bottoms between forested hills and urban sprawl. I had never looked at my family's fields on Google before and was astonished to be able to pick out every tree around my pig pen. The family asked me what was grazing in the fields, and I tried to explain that farming on such a small scale, on poor acid soil, had become uneconomical, that the only harvest we have is of wild deer which we cull to keep the soaring numbers down, that we manage the land primarily for wildlife conservation, and that since my father died we no longer have cattle in the fields. 'That is *mottainai*!' announced Onsri, Jingo's daughter-in-law (the best English-speaker in the family). 'Mottainai?' I asked. 'What is that?' The family glanced at each other and took a deep breath; then they proceeded to elucidate the many permutations of this concept in Japanese culture.

'Mottainai' cannot be translated, but it indicates a condemnation of wastefulness and squandering, and implies an endorsement of thrift and frugality. The word is used for anything from darning socks to scraping the last grains of rice from the bottom of a bowl. 'At school,' said Yuriko, 'we are taught that leaving anything uneaten after a meal shows ingratitude to farmers and that, if we do, the *mottainai monster* will get us.'

I confessed to the family that I had come to Japan precisely to investigate mottainai, and that I would be glad of anything they could tell me. Jingo immediately launched into a litany about life in Japan and how it has changed since his childhood (spoken in passionate broken English, with Yuriko and Onsri giving supplementary translations over his shoulder).

After the war, Japan went through a period of extreme food shortages, when a bowl of rice and some seaweed was the most one could hope for, and there was never any deliberate waste of food. As Jingo saw it, things changed after the 1964 Tokyo Olympic games, and by the 1980s Japan was undergoing a massive financial boom. It was to gourmet food above all that the newly affluent Japanese turned to enjoy and exhibit their wealth. Many saw this as a renaissance of Japanese culture, but it also led them to behave in ways which ran counter to their former traditions. 'Nowadays,' Jingo waved impatiently, 'kids have everything, and waste so much.'

He divided up the width of the table with his hands to represent the epochs of world history: 'for the past 4 billion years humans and other living things have thrived on this planet; but in the past century' – which he marked out as the last two centimetres of the table – 'people have spent almost all the world's resources.' Jingo pointed at his baby grandson: 'He will have to go back to living like we did after the war, planting rice by hand and using cattle to pull the plough. We have wasted so much there is nothing left for the next generation.'

The next morning I woke up early and came down at six, to find Takeko in a motherly frenzy around the kitchen, for today there were to be more guests and a big meal. As soon as I appeared she beckoned me into the car and we set off on a hair-raising half-mile drive down to the seashore, where her brother had been unloading his catch into the harbour. He had kept her aside a beautiful big sea-bream for sashimi and a crate of smaller glistening red fish. On our return, we passed Jingo wheeling along his bicycle, clutching a huge white radish and a bunch of other vegetables. Back at the house, their three-year-old grandson was watering the pots of rice-seedlings in the yard.

Takeko gave me the job of gutting the small fish which she then grilled and served for breakfast alongside dishes of salmon, a bowl of miso soup and cups of green tea. After fumbling through the challenging process of picking apart a whole fish with chopsticks, I caused more laughter by insisting on continuing to help Takeko in the kitchen – with her flowery lacy apron tied round my middle. My first job was to scale the bream and gut it into Jingo's compost bin, then extract the bones, pour boiling water over the fillets, dunk them in iced water, slice them thinly and lay them back onto the fish's skeleton in a dragon-like scaly array. The bream's head and tail were propped up with cocktail sticks, and the reconstituted fish was laid on a thick bed of 'cooling' grated radish and placed in the fridge.

At this point a friend of theirs strolled into the yard carrying a dead chicken. '*Tataki!*' everyone exclaimed, and cajoled him into swapping his fowl for some fish. After some friendly chat and a good peer at the enormous foreigner in the kitchen, he wandered off again, and Takeko handed me the bird. I did not believe it at first when the family explained that they would be eating this chicken raw, as tataki, the meat equivalent of sashimi. 'But surely raw chicken can make you seriously ill?' I protested. Yuriko explained that this chicken had been raised by their friend and killed by him fresh that day, so it was perfect for tataki: surely I knew that? I watched as Takeko filleted the breasts and grilled them for a few seconds, just to sterilize the surface of the meat where it may have come into contact with germs. The inch and a half of flesh inside the breasts, however, remained cold, pink and absolutely raw. The meat was then cooled rapidly in more iced water, and I was told to slice it thinly and lay it out like the sashimi. There were separate side bowls for the slices of raw liver.

Soon the lunch guests arrived, and the purpose of their visit became clear. They were a television company who, ten years earlier, had filmed the Kawasakis for a programme about host families, and now they had returned to make a sequel. With the cameras rolling, we all sat down to lunch, and Japan's television audiences were soon being regaled with the story of the back-packing Englishman who had turned up the night before to learn about mottainai and had just prepared all the sashimi and tataki on the table. After the coiffured presenter had complemented my 'cooking', I tasted the raw chicken and was astonished at how succulent and tender it was. Onsri had also made a delicious Thai seafood curry; and Takeko served more seaweed jelly and a huge bowl of sweet rice covered in vegetables and salmon eggs. Finally, when everyone had almost had their fill, she brought in a small grilled fish for each of us. With the plethora of dishes it was impossible to finish everything, and there was still a feast left after the guests had gone.

I tried to work out how each thing could be re-used and made into something new by the conscientious Kawasaki family. The raw chicken would be cooked up into a stew; the bones and heads of the fish collected to make soup, and leftover fish curry eaten that evening. Quite a juggle to manage such a variously stocked fridge, but just about feasible on a domestic scale.

But if the Kawasakis' tastes were representative of Japanese culture, how could the country's 126-million-strong urbanized population maintain such extraordinary standards of freshness, abundance and variety on an industrial scale? Here in the Kawasakis' home with the family closely connected to both land and sea, and with the different generations bound together by the inherited experience of growing food, there is a natural avoidance of waste. But in an affluent city, far from the sources of production, how could such delicacies be made available without large amounts of them losing their freshness and being wasted?

The answer is that they cannot. The Japanese predilection for high-quality, extremely fresh food results in enormous levels of waste. A study in 2005 found that the food wasted in Japan every year has a value of approximately ¥11 trillion (about US$100bn, or £55bn), which means that in monetary terms, Japan wastes food worth £437 per person per year. The total mass of food waste per person per year is 151 kg (including inedible stuff like citrus peels), which is actually much less than the British average

of 223 kg per person per year – though the difference may be
as much to do with different methods of calculating data as it is
to do with different quantities of actual waste.[2] Each year, Japan
creates around 19 million tonnes of food-related waste, though
the government defines only 6 million tonnes of it as edible.[3]
Of the total, 11.3 million tonnes comes from the food industry
and the rest is from households – these figures do not even
include waste from fishing and agriculture.[4] Japan thus apparently wastes as much food as it produces from its entire agricultural and fishery industries.[5] According to the Japanese Ministry
of Agriculture, Food and Fisheries, this represents one quarter
of Japan's entire food consumption,[6] but if the food dumped
into Japanese bins had been saved, it would have been worth
enough to feed more than 160 million people. In other words,
for every three Japanese people, four more people could be
fed just on the value of their waste.[7] Despite the extremity of
this problem, or rather because of it, the Japanese government
has instituted an extremely promising food-recycling system,
proving that things can change with surprising rapidity when
political will is galvanized.

Although the value of the food thrown away in Japan is
high, the Japanese do not waste as much as Western nations by
'over-eating'. I was struck by this as soon as I started going to
restaurants in Tokyo, where portion sizes are slightly smaller
than in Britain and much smaller than those served in American
diners. This no doubt helps to explain the very low levels of
obesity in Japan. The only really fat people I spotted while
travelling around were Sumo wrestlers. The Japanese are proud
of this fact, claiming that, being 'disciplined', they avoid the
pitfalls of Western 'gluttony and laziness'. The men are especially
boastful of the slim waistlines of the nation's women (though
arguably this can be no less pernicious than the size-zero men-
tality of the Western fashion industry). The nation's food supply
provides an average of 2,548 kcal per person per day (before
losses from waste),[8] which is below the world average of 2,808

kcal and, even taking into account their under-average height, is still far below the American average of 3,900 kcal.

However, while the Japanese keep their bellies empty, they still manage to make their bin-liners bulge. Ironically, it is their love of good food that often leads them to waste so much. The inhabitants of Japan's densely populated cities may lack the luxuries of other developed nations – they are self-conscious about their relatively tiny apartments and crowded public transport systems. But, perhaps in place of these hardships, the Japanese have cultivated their enjoyment of gastronomic pleasures to an almost unparalleled extent. As I experienced with the Kawasaki family, Japan's fridges and freezers are stuffed with dainties, and in middle-class urban families that array of local products is joined by garrisons of imported luxuries, from French cheeses to Mediterranean salads. The evidence of high-quality eating is everywhere in Japan, in people's homes and on every urban street. Tokyo has more Michelin-starred restaurants than any other city in the world, even Paris. As if to make the point, the Japanese have come up with a special term for funnelling income into food: *kuidaore* – 'eating until your finances collapse'.

The Japanese love of all things raw introduces many logistical problems which can result in waste. Sushi, sashimi, tataki, raw eggs and the vast array of other specialities the Japanese regard as central to their lifestyle all rely on extremely rapid food supply chains to get the food from the ocean or farm to the consumer in the minimum possible time. Japanese life revolves around this urgency. Tokyo is a city that never sleeps, but if there can be said to be a beginning of the day, then it occurs at around 4am in Tsukiji, home to what is probably the largest fish market in the world and the nerve-centre of Japanese existence. Early one morning, I made it down to the market in time to sneak my way into the wholesale section – technically closed to tourists – where garrisons of tuna fish, plundered from the world's oceans, lay thawing on the concrete floor, seeping blood into drains, eyes sunken and mouths open as if still gasping for water. From there, stall–holders drag them off to dissect them skilfully with long blades, saws and pincers. In five minutes the six–foot beasts have been carved into scarlet blocks and laid out on ice, ranked in order from the fattiest most expensive cuts downwards. Fork–lift buggies zip around with the latest deliveries, and between them dodge the restaurateurs and householders who come in their wellies, filling bamboo baskets with the freshest purchases – from flailing crustaceans and spiky sea anemones to deep–sea fish with enormous iridescent eyes. Over in the 'live' wholesale market, fish thrash around in bubbling blue tanks as men dip their hands in to grab anything from eels to octopuses, all destined to be served, hearts still pumping, onto the plates of the piscivorous Japanese. By nine in the morning, most of the day's work is done and – as if deserting their funerary shoal at Tsukiji – the fish fan out across the city in trucks, vans and baskets lugged onto the early–morning metro, heading for the nation's kitchens.

Their brief journey from the ocean is very nearly over, for no Japanese will keep fresh fish for long. Browsing Tokyo's supermarket shelves, I rarely saw tuna, salmon or mackerel with a use–by date more than a day beyond the date of purchase, and they had a maximum of two days' shelf–life. Even fish for cooking are rarely kept in stock for more than two days. Shop–keepers do their vociferous best to sell off produce at the end of the day, but, inevitably, a lot fails to reach its target before its freshness is compromised in the eyes of the picky Japanese.

The very week I arrived in Japan, the country's culinary world was recoiling from a scandal which had struck at the heart of Japanese pride. Senba-Kitcho, the 71–year–old president and third daughter of the late founder of one of the nation's most famous family restaurant chains, announced with her head bowed in disgrace that the business was closing its doors. At the end of 2007, branches were found to have re–served food which customers had left untouched, including fruit jelly, bamboo–leaf–wrapped sushi, and the much–loved abalone sea snail. They also illegally extended the eat–by dates of delicatessen products such as confectionaries and salted pollock roe. When an employee blew the whistle, there were murmurs about mottainai and cutting down on waste, but the damage had been done and the chain was on the road to closure.

The furore was the climax of a purge in the food industry during which several high-profile cases of violating use-by dates were uncovered. One famous meat-packing company had to close after police alleged they had been altering expiration dates for the past twenty-four years. A well-known cookie manufacturer with a 300-year history was likewise busted after falsifying production dates for a period of thirty-four years.[9] Finally, in November, McDonald's joined the hall of shame when a franchise running four Tokyo outlets was accused of illegally using yoghurt and milkshake ingredients past their expiration dates and, contrary to company policy, selling salads more than twelve hours after they were produced. Instead of throwing them away, employees allegedly stuck new labels on unsold products and sold them off as quickly as they could.[10]

The fact that there was not one food-poisoning case linked to any of these violations of Japan's ultra-rigorous food safety laws could have been taken as evidence that the laws themselves were unnecessarily strict. There have been suggestions in Japan of relaxing these to reduce waste: the prominent publication *Yomiuri Shimbun* ran a five-day series of articles on food waste from June 2008 and conducted an experiment in which food canned twenty-five years ago was eaten by reporters, who concluded that, although it tasted terrible, it did them no harm.[11] But the majority of people in Japan have reacted in a more predictable fashion with condemnation of the organizations involved, and a statement from the government announcing that they would consider tightening the rules still further.

Japan may one day have to decide whether these culinary preferences can be sustained, and already in the face of the global food crisis some are preparing people for this eventuality. 'The time will come,' says Akio Shibata, the director of the Marubeni

Institute and one of Japan's foremost experts on food supplies, 'when the Japanese people will realize that they will not have the quality, taste and prices of food they have come to expect.'[12]

Some corners of Japan are enjoying a revival of traditional ways of thinking about food, which are as much ideological as they are gastronomic. In the swanky Ginza district of Tokyo – home to luxury department stores and cocktail bars – the Chakodamari restaurant specializes in the centuries-old *hakozen* meal. 'With a hakozen meal,' explained the owner, Mitsuru Owada, to the *Japan Times*, 'people should not leave anything over, so we serve moderate amounts of food.' The restaurant even asks diners to emulate Zen Buddhist monks who pour hot water into their rice bowls after a meal and use a piece of pickle to wipe up any leftovers. As a spokesperson from an associated NGO said, 'If people stop leaving food, which then becomes waste, research has shown that Japan's self-sufficiency in food would increase to 70 per cent from its current 40 per cent.'[13]

The most spectacular sources of waste in fact come from the sectors of the food industry where traditional Japanese gastronomic customs have been eroded. Convenience stores have taken hold, and their numerous ready-meal 'lunch boxes' are wasted in staggering quantities. Japanese ready-meals tend on average to be far tastier than the salty slop served up as microwave meals in Europe and America. There is always a wide choice of noodles with vegetables and sauce, trays of sushi, or meat-filled dumplings, all relatively freshly made. The problem is that the shelf-life tends to be only a few days. Love them or loathe them, European microwave meals are hermetically packed in air-tight containers, sterilized at the point of manufacture and pickled with a brew of preservatives so that they become virtually indestructible by age and thus sport shelf-lives of around two weeks (though this has come down in response to people's distrust of additives). Disgusting though this may seem, it does reduce the proportion that is wasted. The result is that convenience stores and supermarkets in Japan are responsible for 6 million tonnes of the total 19 million tonnes of food waste generated in Japan.

Like most developed countries, Japan used to send its food waste to landfill, but being a small island it quickly ran out of space and started incinerating garbage to reduce its volume. As far as burning paper and plastic is concerned, this has some significant advantages: it creates electricity from what would otherwise have been useless (the disadvantage is that burning plastics produces harmful dioxins and other toxic chemicals). Burning food, however, uses up energy because around 80 per cent of it is water. As more people began recycling paper, the proportion of food waste in general garbage increased, making it ever costlier to burn. Thus, in addition to wasting millions of tonnes of valuable food, Japan's throwaway habits were creating mountains of problematic waste. The government demanded that the food industry sort this problem out, rather than dumping it on the nation as a whole.

Japan is also deeply worried about its vulnerable food supply. But unlike most other countries in the world that are tackling these two issues – food supply and food waste – the government has brought in dramatic new legislation which finds the solution in the problem.

The keystone in the Japanese government's approach to tackling the problem is its Food Waste Recycling Law which was passed in 2001, obliging food businesses to recycle 48 per cent of all their food waste by 2006.[14] By 2005 the industry had leapfrogged the mandatory level and was recycling 59 per cent of commercial and industrial food waste.[15] The latest revision to the law stipulates that businesses should raise this to an average of 66 per cent by 2012.[16] It is thus businesses who are leading the way in altering Japan's waste profile. The government chose to target them largely because it is simpler for companies to separate food from non-food, and much easier for recycling plants to collect and process it than it would be to organize a separate system for millions of individual householders.[17]

Japan's Food Waste Recycling Law differs in critical ways

from those introduced in Europe and America. Firstly, the Japanese law deals specifically with food, whereas European laws are
aimed at biodegradable waste in general, including waste from
parks and gardens. And secondly, in Britain at least, the greatest
effort has gone into coercing councils, and thus indirectly householders more than businesses, to improve their recycling rate.
This does have to happen too, but it can be frustrating for
consumers to carry their rubbish to a recycling bank, only to
pass by a pub on the way with a dustbin full of hundreds of
bottles that will not be recycled.

The key fact which is obvious to Japanese legislators, industri-
alists, farmers and scientists but which has eluded most Europeans
and some Americans is that the best thing to do with surplus
food that cannot be eaten by humans is to feed it to animals.
As Dr Tomoyuki Kawashima, Head of the Functional Feed
Research Team at the National Institute of Livestock and Grassland Science, explained to me, 'It is important to maintain the
cascade system. Don't just throw all food waste into a biogas
digester or compost heap (let alone a landfill site). First use the
surplus food to raise livestock. Then use the animals' manure to
make biogas, and finally use the digestate for compost. Otherwise
you're wasting the food waste's greatest potential.'[18]

Japan currently imports 90 per cent of its concentrated, high-
protein, high-carbohydrate livestock feed. Apart from the
environmental impact of growing grains and pulses and then
shipping them thousands of miles across the ocean, this causes a
continuous haemorrhage of foreign currency and now threatens
to tip Japan into a permanent trade deficit. Why continue to
import so much when 19 million tonnes of food waste are being
thrown away? The Food Waste Recycling Law's guidelines
themselves spell out that 'Since it is the most effective way to
utilize the nutrition or calorific value of the recycled food,
besides contributing to [Japan's] self-sufficiency ratio for feed,
it is important to make processing feed [from food waste] a
priority.'

It is regrettable that the government has not done more to reduce the amount of food that is wasted in the first place, and this is no doubt because it realizes that the desire for abundant fresh food is so deeply entrenched in the culture that any measures seen to be compromising this would be unpopular. But at least it is making the most of the waste that does arise. The unassailable logic of recycling uneaten food into animal feed has been heroically promoted by one man at the Ministry for Agriculture, Fisheries and Food: Itisaki Shimadu. Talking to Shimadu-san – surrounded by his main allies, Mr Satoshi Motomura and Ms Eri Utamaru – is to be in the presence of a technocrat-turned-evangelist. He talks in statistics and flow charts, but his eyes glisten with the passion of a zealot.

'Recycling Japan's food waste into animal feed,' he explains, 'will save Japanese farming, decrease the nation's food dependency, and sort out the waste disposal situation. Japanese people think they eat mainly rice, but what few realize is that they actually eat more imported maize – in the form of corn-fattened animals. That maize is becoming increasingly expensive, now that the Chinese eat more meat themselves.' While pig farmers in other parts of the world are being bankrupt by the high cost of animal feed, Japan's pig farmers are being given a cheap, environmentally friendly alternative. As Masaru Shizawa, chief director of the Japan Pork Producers Association and owner of 3,000 sows, opined, 'the rise in price of feed grains could strike a blow against Japanese livestock production in the future. In the meantime, however, food waste recycling should be promoted.'[19]

Shimadu-san explains that 'currently 37 per cent of the food recycled in Japan is converted into animal feed – about 2.5 million tonnes – but increasing that level is a great way to alleviate the self-sufficiency problem.'[20] He thrusts a brochure into my hands which outlines successful food waste recycling schemes dotted around the country, and insists that I must visit each of them to see for myself. In fact, I have already been to one or two, but he hands me the business cards of several CEOs and directors and I undertake to visit some more.[21]

What I find out over the coming weeks completely changes my impression of food waste recycling. I had been to many factories dealing with food waste before, but the Japanese have taken it to a new level of technical and sanitary accomplishment.

First, I visited Takeshi Tanami, manager of the two-year-old Alfo factory, situated on a peninsula jutting out into the ocean from Tokyo city. Built symbolically on top of a landfill site, the Alfo factory is part of the Tokyo Super Eco-Town Project which seeks to make the capital responsible for its own waste rather than continuing to dump it on neighbouring prefectures. Alfo's special innovation is to use cooking oil to deep-fry all the waste food, sterilizing and drying it into a convenient animal feed. The entire factory is consequently saturated with a thick sweet smell, something between a fish and chip outlet and a sweet shop. After passing through the 10-tonne-capacity frying pan, the feedstock is pumped through a mind-boggling series of gargantuan machines which stretch a hundred feet into the air, metallic pipes twisting and turning over three storeys; gases are purified and deodorized, oil is squeezed out of the food and re-used – before the final product drops out onto the warehouse floor as an odourless brown powder resembling coffee grounds. It is an impressive system using state-of-the-art machinery. But as Tanami-san is the first to point out, it does not come without problems.

Firstly, Alfo collects its food from restaurants, hotels and shops in plastic sacks, which means that large expensive machines have to filter them out again. This is never 100 per cent effective, and the other pitfalls of taking food waste from restaurants are exhibited on a table in the processing room – mangled forks, spoons and knives – which occasionally block up their crushing machine. Secondly, until recently, commercial feed mills who incorporate Alfo's food waste into conventional feed have been unwilling to add more than 3 per cent to their feed because they are worried about public fears of using waste food, and they euphemistically label it as 'food by-products'. Furthermore, they have only been willing to buy it for ¥10,000 per tonne, which is less than half what they would pay for a commercial feed with an equivalent level of protein.[22]

The other main trouble faced by Alfo, which Tanami-san complains about at length, is that the Tokyo local government gives such generous subsidies to the city's general garbage disposal system that it undermines the economic viability of recycling, and as a consequence the Alfo factory is only working at half its 140-tonne-per-day capacity.[23] Another problem with Alfo is that the deep-frying process used by the factory requires huge amounts of energy; it is favoured because it converts wet food waste into dry powder that can be added to ordinary pig feed, but one academic study showed that deep-frying is the only method of recycling food that emits more greenhouse gases than incineration.[24] It is also very expensive: to break even, the factory needs to process 70 tonnes of food waste every day, and it is difficult to see how, with a capital investment of ¥2.8bn (a third of which was paid for by the government), it can even begin to make a profit.[25]

Hiroyuki Yakou, the CEO of Agri-Gaia System Co. Ltd, was the next waste food gastronomist on my list. He was a sleek businessman with the fresh face of a missionary, eager to enlighten me. Agri-Gaia specializes in dealing with surplus packed lunches from the ubiquitous chain Seven-Eleven. It turns the highest-grade food waste into animal feed, and the rest goes to make compost. Instead of the homogenous brown substance produced by Alfo, Yakou-san proudly lays before me a menu of glass jars each containing a different kind of desiccated waste food, still clearly identifiable as noodles in one jar, vegetables or salad in another. I joke about what they must taste like – but he looks back at me with a serious expression: of course he has tasted them to test their quality. Separating each of the ingredients means that Yakou-san can sell each one, with its known nutritional value, to animal feed mills which then have the confidence of making feed almost entirely from his waste.

When Yakou-san leads me to the viewing gallery of his factory I cannot believe my eyes. A production line of fifty workers dressed in food-handling white gowns, caps, gloves, masks and boots huddle round conveyor belts disgorging onto them the contents of a warehouse-full of unsold packaged meals. I guess that there would be enough lunches here for 40,000 people. The cleanliness and streamlining of the process seems out of proportion with the unfussy habits of their end-consumer. Yakou-san sees me looking bewildered and remarks, 'The idea is to take the model of a lunch-box factory and reverse it.' Instead of loading ingredients into trays and fitting them all into packages, these employees are unpacking each tray by hand and sorting the ingredients into separate containers. Only this way, says Yakou-san, can 100 per cent of the packaging be removed, and the most value extracted.[26] Meanwhile, he adds, the pipes hanging from the roof suck up odours and methane, pressurize the combustible gases and use them to generate energy. Still in disbelief, I ask how on earth selling pig feed can come near to making back the costs of employing fifty workers in a factory which cost ¥3bn (half of which was paid for by the government). Yakou-san has heard this question before. 'What we are doing,' he says, 'is changing the image of food waste recycling. Thanks to government red tape and local opposition, it took six years to obtain permission to build a factory here. There are lots of mafia people in the waste industry,' explains Yakou, 'so the law is based on the idea that people are bad when they are born. We need to change all this before food waste recycling can take off.'

Government restrictions also hamper where they can collect waste from, and thus, having opened a year earlier, Agri-Gaia is still only working at one sixth of its 250-tonne-per-day capacity. But Yakou is passionate about the factory and speaks to me in great detail about its potential: 'In fact, I don't think the material we pick up from shops is waste at all,' he says, 'it is a valuable product. Thanks to President Bush's biofuels policies, food prices have rocketed. It is an economic rule that things go to whoever can pay the highest price: so if rich people can buy food, even if they're going to turn it into fuel, they can out-price poor people. People in Africa and Mexico need that food to survive. Now Japan needs to do its bit. Pigs and chickens are evolved to eat everything. It was only humans who had the idea of feeding them solely on corn. If animals are given the nutritious, varied food that was made for humans, they will produce better eggs and the meat will taste nicer. On blind tests, nine out of eleven people preferred the pork raised on our waste feed over that produced using conventional feed.' As far as Yakou-san is concerned, the key is to educate the public to see feed made from food waste as a positive thing. He stresses that Seven-Eleven get good publicity for recycling their food waste. Pork producers, too, ought to get good publicity for using it.

After the capital-intensive super-factories of Agri-Gaia and Alfo, I was relieved and delighted to visit the Odakyu Food Ecology Centre in Sagamihara city. An hour and a half from Tokyo, this factory is overseen by the inspiring figure of Kouichi Takahashi and everything about it bears the hallmark of sound thinking; wherever you look, economic and ecological interests appear perfectly aligned. The Centre is owned by Odakyu, one of the most renowned department store chains in the country and the owner of several other businesses, including train services. Following a model proposed to them some years ago by Takahashi-san, Odakyu delivers waste food from its supermarkets, restaurants and train lines to the factory to be turned into pig feed, and then buys back the pork to sell as a premium-quality eco-product in its own stores. This organic-retail loop is the holy grail of food waste recycling. Adopted by several projects in Japan, it ought to serve as a business model across the world. It reduces food miles, eliminates a waste disposal problem, provides sustainably produced meat, and creates valuable publicity for retailers.

The whole system hinges on Takahashi-san's brilliant and simple way of processing the waste with minimal fuss. The factory cost between a seventh and a tenth of the Alfo factory, and it may be no coincidence that it was the only food-recycling facility I visited which received no capital assistance from the government.[27] Not only does it draw in food waste from Odakyu outlets, but it also takes waste from a constantly growing list of eighty hotels, food manufacturers and another supermarket chain, ECO's. Instead of contaminating all their food waste with plastic bags, as Alfo does, Takahashi-san shows me the plastic dustbins which participating companies fill with their various kinds of waste. The Director of ECO's, Toshiro Myoshi, told me that his staff unpack the supermarket waste before sending it for recycling, arguing that it is best if everyone in the supply chain understands where the food is going and that 'it really doesn't consume that much time.'[28] Each bin that comes to the Odakyu factory holds different types of food – high-protein soya waste in one, carbohydrate-rich noodles, rice and dough in another, and fruit and vegetables in another – ensuring that the final feed product has a guaranteed nutritional content with a protein component at the perfect level of 15–17 per cent, depending on the farmers' requirements.

Each bin has a barcode on it, and as it passes through the system, a computer records the weight. This means that the nutritional content can be calculated automatically. It also provides the data for billing the companies that are charged according to the kind of waste they produce, though even the higher rates are half the cost of incineration.[29] The third benefit of the weighing system is one that Takahashi is just beginning to develop into a separate consultation facility in its own right. 'When food companies are sent the bills for the wasted food, they see exactly how much of each product they waste each day. That means that they can actually improve their own efficiency.'[30]

Once off the scales, the bins are tipped into a vat and as the food is drawn along a conveyor belt a quick check by one or two employees ensures there are no significant contaminants, such as chopsticks or labels. The food is then churned up and pasteurized at 90 °C for five minutes, which ensures that any pathogens dangerous to livestock are exterminated. Although they do not take in much meat waste (most of their protein comes from soya), Takahashi insists that 'in my personal belief it is OK to feed animal protein to pigs if it is properly sterilized. But the problem is that farmers don't like it in Japan because they know about the disease outbreak that happened in Europe.' But Takahashi argues that there is not much choice in any case. 'If farmers continue to buy commercial feed for livestock, they will go bankrupt.'

The next step after sterilization is the process which promises to revolutionize food waste recycling, not only in Japan but all over the world. Instead of using vast amounts of energy to dry out the liquid feed, or creating logistical headaches by quickly taking it to the nearest pig farm before it goes off, the churned-up swill is inoculated with *Lacto-bacillus*, a bacterium much like that we use to turn milk into yoghurt. This preserves the swill and even enhances its nutritional value, so it can be stored for up to two weeks without going bad.[31] The factory has only been running for three years and is thus still operating at half its 40-tonne-per-day capacity. But Takahashi insists that now demand for cheap animal feed has increased, new contracts are coming in every month, and farmers are queuing up to buy their feed.

However, even Takahashi sells his feed for half the price of conventional feed.[32] 'I could raise the price,' he says, 'but I am not trying to develop a system for making profit, it is a question of changing society. There are numerous invisible benefits which accrue from doing this business. Firstly, Odakyu can comply with the Food Waste Recycling Law, secondly, they profit from the sale of premium pork, and thirdly it improves their Social Responsibility image. Even if it went slightly into the red,

Odakyu would still continue it for ethical reasons. But the fact is that with the rising food prices, they are expecting to start making a profit earlier than anticipated.'

Having shown me round his excellent factory, Takahashi-san offers to drive me to the nearest pig farm, which receives a few tonnes of Odakyu swill every day. A short journey brings us to the farm of Kamei-san, a smiling ruddy man, with patched jeans and thick tousled hair. You can hear the richness of the soil in his voice, and he glows with satisfaction as he tells me how he converted from conventional feed to Takahashi's fermented waste. His pigs grow as fast as they did before: the only difference is that he now pays half what he used to for conventional feed.

I take a look inside, and although it is well-ventilated and sunny, I am dismayed by the crowding of the 400 pigs into tiny pens. There simply isn't room, in this peri-urban farm where Kamei-san also grows vegetables for his shop, to let the pigs roam free. But the farm's other credentials are impeccable. As it is situated close to its feed source and the market for its produce, the food miles and carbon footprint must be as low as any industrial farm I have ever been on. The entire feed–food loop is completed within a radius of about five miles. And this is not because the farmer or his customers are particularly environmentally motivated – but just because it makes economic and logical sense.

Kamei-san takes us to his shop and sits us down with a can of chilled coffee. He and Takahashi start to explain the problems they still face as food waste users. Firstly, the government has been giving special subsidies for farmers using unsustainably produced imported conventional feed to stop them going bankrupt. This gives them the competitive edge and means that American feed grains like maize are subsidized twice: first by the American government's farm subsidies; then by the Japanese government. The other hurdle faced by Kamei-san is the all-powerful Association of Agriculture, which sells commercial feed to farmers and then buys back the pork. If a farmer does not want to buy their commercial feed, and instead turns to 'eco-feed', the Association boycotts his meat. That is one of the main reasons why Odakyu buys the pork direct off farmers, to give them the protection and confidence to shift to eco-feed. 'Fifty years ago,' says Kamei-san, 'farms were small, with a few pigs: there was no mottainai. Now we have huge farms, loads of pigs and we import feed from the US, and mottainai is everywhere. It's a really bad system.' Before Takahashi and I leave, Kamei-san presents me with something which it is almost impossible to buy in Europe: a tray of sustainably grown pork.

Takahashi now wishes to show me the nearby Odakyu supermarket. He takes me past the garbage bins and thence to the shop counter which sells the recycled-feed pork. 'Yoghurt-pig from Kanagawa' reads the sign emblazoned above an array of fresh pork products. The sign shows the food-cycle loop flowing in a circle from the supermarket, through the recycling centre, to the farm and back to the shop as pork:

The advantage of Yoghurt-pig is that it is healthy, tasty and safe. These pigs are fed on lacto-fermentation feed, and compared to standard pork it contains 10 per cent more unsaturated fatty acids and 20 per cent less cholesterol. It is tender, delicious and juicy. Because the pigs are so well-maintained, they are not given antibiotics, so the pork is very safe and healthy.

Here is commercial sense combined with education of the public: a bold attempt to encourage consumers to know about the food they eat, and to buy what makes ecological sense.

The one thing that worried me in Japan was the apparent lack of concern over the risk of disease outbreak. Japanese law requires that swill should be heated at 70 °C for thirty minutes or 80 °C for three minutes, which studies show is the perfect level to ensure sterilization without destroying the nutrients in the swill. It is a far better law than the old British one which stipulated that swill had to be boiled at 100 °C for an hour – by which time a lot of the nutrients in the food have been destroyed. But the problem in Japan is that there is little or no system for enforcing the rules. Large food waste factories have rigid systems for sterilizing the waste. But there are hundreds of smaller farmers who collect food waste directly from restaurants. These small-scale operators may go for years without seeing the face of an inspector and therefore may be tempted to save on energy bills by side-stepping the law. I asked everyone I met about this problem and never once received a reassuring answer. Shimadu-san at the Ministry blinked at me through his glasses and told me he had heard of the ban on swill-feeding in Europe; but he had apparently not linked it with foot-and-mouth disease and he and his colleagues laughed at my concerns.[33] The manager at Alfo had never been inspected by the government and there was no obligation for him to report back to them on the results of the microbiological test results he conducts. Dr Kawashima at the National Institute of Livestock and Grassland Science justi-fied Japan's approach by admitting that although 'there is a risk', the benefits of recycling food outweigh them. 'Think about the known risk to the environment of producing all the unsustainable feed these animals would otherwise be fed on.'

However, they all agreed that Europe and those American states which have banned swill-feeding are handicapping themselves. Dr Kawashima told me that 'pig farmers in Europe are now asking us to encourage the European Commission to change the law. Once consumers know how important it is to stop wasting this valuable resource, things will have to change. The mistake Europe made was to mix together the BSE and foot-and-mouth disease problems when they made legislation. BSE required the complete ban of feeding animal by-products to ruminants but if you just feed the waste to pigs, all you need to do is sterilize the food. Indeed, it's true: people are starving in the world because rich countries are buying crops to feed to their animals. This would be alleviated by using food waste to feed animals instead.' Takahashi-san at the Odakyu

Food Recycling Centre was invited to Europe by a delegation of pig farmers from Denmark and Ireland who were keen to bring back swill-feeding in Europe using Takahashi's *Lacto-bacillus*. 'The situation in Europe?' said Takahashi after he had taken me on a tour of his food-recycling system. 'It's crazy . . . I think so.'

If the Japanese food-recycling system sounds rigorous, it is noth-ing to the draconian eco-legislation brought in by both Taiwan and South Korea. There, no food waste of any sort – whether from households, restaurants, supermarkets or factories – is allowed to go to landfill. In 2005 both Taiwan and South Korea introduced laws pointing to the same conclusion: burying food is madness. Instead, they now collect it separately, feeding some to pigs and composting the rest.

Like Japan, Taiwan is a small island, with little space for landfill; South Korea – on the end of a peninsula cut off from the rest of Asia by its hostile northern neighbour – is in a similar situation. Islands are often a test case for developments that happen on a continental scale: limits are reached more quickly here, but the forces are the same.

In the narrow jumbled streets that weave between the glisten-ing skyscrapers of Seoul, South Korea's capital, I heard many of the same stories that I was told in Japan. I visited factories that converted sacks of succulent restaurant waste into pellets of pig feed. I saw some of the 250 other plants that deal with Korea's daily output of 13,000 tonnes of food waste. I spoke to food technologists trying to train restaurant staff not to put plastic toothpicks (which stick in pigs' throats) into their food waste bins. I met government employees trying to wean the Koreans from their love of gastronomic variety that exceeds even that of the Japanese; every meal in Korea comes with an unlimited array of side dishes – fried sprats, fermented fish, octopus tentacles and *kimchi*, the spicy Korean version of sauerkraut – with the result that it is physically impossible to finish everything served (not

least because if any one of your dishes is emptied, waiters refill
it and react with astonishment if you ask them not to).

They may not be ready to give up their free *kimchi*, but one
thing all Koreans do, virtually without exception, is comply
with the law on waste recycling. This is not – as many would
have it – because they are an inherently obedient people, always
kow-towing to their authoritarian leader. On the contrary, I
walked among crowds who daringly reclaimed the streets of
Seoul in protest against the government's trade agreement to
import American 'mad cow' beef. The protestors were as stub-
bornly uncontrollable as any street marchers I have ever seen,
despite the presence of thousands of riot police trying to
shepherd them out of the road.

Koreans obey the waste recycling law largely because they
have resigned themselves to the reality: that sending food into
landfill is against their own interests and that of the planet they
live in. They know about the disease outbreak among livestock
in Britain, and they know about the ban on swill-feeding – and
they conclude, as a result, that Europeans are blithely continuing
their reckless, self-interested exploitation of the planet in the
manner that has characterized them for centuries.

I shall give the last word to Sung-Heon Chung, a smartly
presented, grey-suited professor at the Animal Research Centre
at Konkuk University in Seoul, who spent a hot spring afternoon
guiding me around Korea's waste-recycling system and lecturing
me on my place in the planet:

'It is a sin to interrupt the food cycle, and if people continue
to ignore the right way for selfish reasons, there will be environmental disaster. There is a pyramid in the natural food chain,
with an order of priority, from humans to animals; then plants,
and bacteria at the bottom. Anything that can be fed to humans
should be. Food waste is non-digested organic substance and the
pig's stomach is the natural medium for fermenting it. Only
then should their faeces be turned into gas or spread on a field.
There is a limit to how much nature can provide us with.

Europe, especially England, should be almost apologetic to the rest of the world, because England's political and other decisions in particular have been guided towards economic profit instead of environmental sense. If they continue to violate the laws of life, nature is bound to get its revenge.'

18. Action Plan: A Path to Utrophia

Everything in excess is an enemy to nature.[1]

Hippocrates

In an ideal world, there would be no freegans, because there would be no wasted food. All systems generate some waste, but, in the case of food, 'waste' should become a misnomer. If we viewed the food cycle in a holistic way – as used to be essential and is becoming so again – we would realize immediately that much of our wastage is avoidable. Throwing away food at one point in the system means we have to grow more in another. Instead, the two could cancel each other out.

The reason why this is not happening already is because environmental costs incurred in one part of the world are not passed down the food chain. Globalization of the food industry has moved faster than global governance or social consciousness. As the Kyoto Protocol recognized, international agreements need to put a price on environmental damage to encourage us to use resources more responsibly. This holds for food as much as for oil.

Food is not just a commodity but a vital interface between people and the earth. *Homo sapiens* is a terrestrial mammal; land is our greatest asset, and our fate depends on how we treat it. We need to grow food, but the costs involved mean that we have to do so as sensitively and efficiently as possible. Businesses and individuals have a responsibility to use resources carefully – to avoid reckless wastage and to recognize that wasting food does not just hurt profit margins, but also people and the planet.

In the imaginary land of *Utrophia* – a place of *Good-Eating* – farmers would sell all their potatoes regardless of shape or size. The chef would buy surplus ripe tomatoes from the wholesaler to make into that day's meals. Supermarkets would redistribute surplus food to people in need. All unavoidable organic waste would go to feed either animals or the soil. And the general public would learn to respect the food which sits in their fridges – to buy what they eat and to eat what they buy.

Consumers: Stop wasting food.

By throwing away food you are paying for: the wanton degradation of the environment, the starvation of people on the other side of the world, pollution in your local landfill site, unnecessary global warming, water depletion, soil erosion, habitat destruction and deforestation. Completely eliminating food waste in your home is achievable. It would be easy for everyone to cut waste down to well below 10 per cent, *today*.

The process starts even before you get to the shops: write a shopping list while still at home so you can check exactly what you still have in the fridge or cupboard. Think about what meals are likely to be eaten. In the shop, avoid being seduced by marketing devices to make you buy things you aren't going to consume.

When cooking, measure food portions to avoid preparing more than desired. Use up leftovers. Take them into work for lunch the next day rather than buying a sandwich. Re-heat or convert into new dishes the next day and save yourself the bother of cooking from scratch. Leftover meat, fish and vegetables make good soups, stews, curries or sandwiches. Buy the size of bread loaf the household will get through before it goes stale. Freeze surplus bread, or dry it out and turn it into breadcrumbs. Eat your crusts! They are at least 10 per cent of a loaf, so throwing them away is equivalent to throwing away 10 per cent of the arable land used to grow them. If you don't like crusts, don't eat bread. Potatoes, carrots, parsnips and other vegetables rarely need peeling, whether as chips, mash, roast, fried or boiled.

Treat *best-before dates* with extreme scepticism – they aren't telling you the food is harmful after that date: if it looks and smells fine –. eat it! *Use-by dates* are generally calculated with wide margins of error; don't risk food poisoning, but do ensure that anything such as raw meat you decide to eat after its use-by date has been properly refrigerated and thoroughly cooked. *Sell-by dates*: ignore them entirely, they're irrelevant.

If you think you don't have a food waste problem, try measuring over the period of a month or so exactly how much food you throw away, and then improve on it.

Eat more offal and less meat; eat fish from well-managed stocks and avoid eating species threatened by over-fishing and the problem of discards. Buy knobbly fruit and vegetables wherever you can, for example directly from farmers. This maximizes the efficiency of agricultural production; it can be cheaper; and there's a good chance that they will contain lower levels of pesticide residues and other toxic agri–chemicals.

Home composting, wormeries or separate organic bins for municipal collection are better than throwing biodegradable waste like orange peels, carrot tops, tea bags etc. in the general rubbish bin. BUT, it is not virtuous to throw *food* – whole lemons, bananas or whatever – into a compost bin. The value of the compost is a tiny fraction of the resources that went into growing the food and getting it into your home.

Parents: Children learn how to treat food from the earliest ages. Try to encourage children to finish the meals they have been served. Kids are very open to being told about where their food comes from – tell them stories about the land and the people that grow their food and it will help them to appreciate its value.

Governments: Initiate well-managed public awareness campaigns to change public behaviour. WRAP's 'Love Food Hate Waste' campaign and the wartime efforts in Europe and America are examples. Money spent is recompensed by saving resources, reducing the costs of waste management, increasing efficiency and enhancing national food security.

Impose mandatory food waste reduction targets on food companies: a starting point would be a 50 per cent reduction in five years with the eventual aim of eliminating all possible food waste (not including inedible food by-products). Alternatively, impose the waste reduction target on the governments, so they are obliged to introduce policies that will bring about that reduction. Where appropriate, provide assistance in improving efficiency through research, advice and infrastructure.

Introduce a tax on wasting edible food regardless of disposal method, to be imposed on food companies. This would create a fiscal incentive to prioritize giving surplus food to redistribution charities rather than other disposal methods.

Food companies should be made to report their waste arisings, specifying how much of it is food.

Avoid farm subsidies based on output which encourage the production of surplus food. American farm subsidies currently do this, as do EU export subsidies to a lesser extent.

Fund research and development in food technology to extract maximum value from food by-products and co-products.

When intervening in food waste management, governments should be very careful to avoid disrupting the natural waste hierarchy of reduction first, then redistribution, then livestock feeding, and only after these have been exhausted, promoting alternatives to landfill. Britain, as a bad example, has offered fiscal assistance for composting and anaerobic digestion and has not offered similar help to food redistributors or livestock farmers.

Ban sending food waste to landfill and take measures to ensure that adequate alternative disposal methods are developed.

Lift the ban on feeding swill (in the EU and some US states) to pigs. Instead, feeding swill should be encouraged or made mandatory. Start with commercial and industrial food waste and then follow the example of South Korea and wartime Europe by collecting household and catering food waste, instigating an effective awareness campaign to ensure people do not contaminate it with unsuitable material. Ensure the swill is properly heat-treated.

If Europe does not trust its farmers to treat food waste properly then it could at least allow commercial feed factories to do so. Large industrial outfits can easily be regulated and checked to ensure they are complying with the law. If governments still feel unsafe, they could allow food waste to go only to designated farms where contact with other livestock is restricted, or eliminated altogether by having a slaughtering facility on site.

Failing an end to the ban, instigate a nationwide scheme covering all commercial and industrial food companies whereby permitted former foodstuffs, such as bread, pastry, fruit, vegetables and dairy products, are segregated from banned meat products and collected for livestock feed.

To address the current anti-competitive fiscal bias in favour of sending former foodstuffs for anaerobic digestion rather than livestock feed, a scheme could be devised that remunerates farmers for the far higher carbon savings of feeding it to their animals. Under Kyoto's Clean Development Mechanism, swill-feeding projects in developing nations could be approved, so farmers could receive €18–30 per tonne of saved carbon dioxide emissions (though this should rise in the future). Carbon-saving agricultural projects within Europe are technically permitted under the Joint Implementation system; but owing to administrative hitches, not one scheme has yet been approved.[2] Facilitating this would make it more profitable for pig farmers or feed manufacturers to collect food waste. Direct government funding of such projects would also help: Japan, for example, pays pig farmers one third of the costs of installing feeding systems to take food waste.

Supermarkets: Stop throwing away food, except in extraordinary circumstances. Adopt this target voluntarily, and if not then submit to regulation. Give away any surplus food that cannot be sold. Any further permissible waste should be segregated under a local-authority-approved plan, backhauled to central depots and given to livestock farmers or feed manufacturers.

Remunerate managers on the basis of the accuracy of their ordering in order to avoid waste.

Submit to an enforced code of practice (as proposed in Britain) to protect suppliers (both manufacturers and farmers) from unfair practices such as volatile last-minute order changes, take-back clauses, dumping the cost of surplus on suppliers, exclusivity clauses or anything that prevents suppliers from selling surplus to other willing buyers or donating it charitably.

Reform use of best-before dates on fruit, vegetables and bread and inform the public on the meaning of the dating system.

Manufacturers and processors: Any of the waste problems in the supply chain would be alleviated by tackling the above problems with the supermarkets. Dramatic improvements can also be achieved through adopting best practice, already adopted by many companies and yet to be extended to others. Wherever possible, consider lengthening the shelf-life of products.

Canteens: Unnecessary waste is currently caused by the expectation that each person likely to dine must have a choice of dishes, so each dish has to be made in excess. Without reducing people's choice, canteens in schools, hospitals and other institutions could reduce this waste by asking or requiring diners to choose *a day or two in advance*. No choice is lost, just the expectation that the choice should be made spontaneously. Some restaurants already take orders in advance: for example, the exclusive restaurant Mildmay Hall, at the Glyndebourne opera house.

Full-service restaurants: Everyone has a different appetite and serving sizes could be adjusted accordingly. Make standard dishes smaller and offer to 'supersize' for free. Some restaurants have found that customers respond well to being told that a surcharge will be added to the bill for food left over on plates, especially if the reasons are explained and proceeds are given to charity: particularly useful for those restaurants offering 'all you can eat' deals. The best-run restaurants already use up leftover ingredients. Inventive chefs can turn odds and ends into complimentary appetizers. Discounts for ordering meals in advance can also be offered, as for canteens above.

Fast-food restaurants, convenience stores, sandwich vendors: Much of the food wasted in this sector arises from the need to keep food ready, prepared and immediately available at all times. For this reason, accurate forecasting of demand is essential in reducing losses. However, these stores – many of them comparatively small outlets – suffer from poor stock management by untrained branch managers.[3] Much of this can be dealt with by training staff adequately, keeping them in the job by giving decent employment terms and fostering best business practices to avoid waste. Any surplus at the end of the day should, wherever possible, be given away or donated to redistribution charities, if necessary at a charge to the company.

Fishing: It is up to governments and fishing industries to agree sustainable fisheries policies. Discards should be banned; human markets found for edible but currently unwanted fish; extensive marine reserves established and no-catch rules enforced within their boundaries; and modern fishing equipment and techniques that avoid by-catch made mandatory.

Farmers: Wherever practical sell direct to consumers: it can increase profits significantly and can cut the amount that is graded out of harvests by 30–90 per cent.[4] In the UK, work

with unions and the government to achieve the proposed out-
lawing of the supermarkets' unfair practices that cause unwanted
crops to be disposed of; outside the UK, press governments for
similar regulation. If no market can be found for surplus or
rejects, consider letting in gleaners, whether individuals or
established organizations.

Afterword

Reducing food waste should become one of the highest priorities on the environmental agenda. It is uncontroversial; it has the potential to curb deforestation and hold back global warming; it is relatively painless and easy. As the United Nations has warned, reducing waste may be essential in a world where the human population is set to increase from 6.7 billion to 9 billion by 2050,[1] and at a time when a quarter of the world's projected food production may be jeopardized during this century as a result of climate change, water scarcity, invasive pests and land degradation.[2]

But what is the point, exactly, of conserving resources? Are we aiming to reduce each person's consumption so that more people can enjoy a fairer slice of the cake? Is our aim to maximize human populations, and to fit as many people on the planet as possible? Are we trying to preserve resources so that future generations can survive? Or are we striving to protect the natural world?

It is generally assumed that conserving resources like food would both be good for people – because it would allow those suffering from food poverty access to more – and good for the planet because it would reduce demand for agricultural land. But there are a number of variables that may need to be dealt with before these benefits can be reaped.

If we stopped wasting so much food, perhaps all the resultant surplus would be equally distributed among the world's poor. Alternatively, the liberated surplus might be fed to animals to satisfy the soaring global demand for meat. In this case, more people would have access to richer, meatier diets but there would not be so great a reduction in the demand for agricultural products and the world's poorest might still go hungry. Furthermore, increasing food supplies by being more efficient might simply fuel population growth. More food might just encourage families to raise more children. Depending on one's perspective, any of these outcomes could be regarded as positive.

But arguably the most pressing concern in terms of the lasting health of the planet and all its inhabitants – both human and non-human – is the encroachment of our activities into natural habitats, particularly forests. Without protecting these, our planet has little chance of regulating its climate. In order to protect forests, we need to do more than just conserve resources such as food, land and fuel. There also needs to be robustly enforced national or international protection. The ongoing international negotiations to replace the Kyoto Protocol in 2012 are exploring ways to reduce emissions from deforestation and degradation in developing countries – by getting rich countries to pay poorer countries to conserve their forests on the grounds that their role as carbon sinks is essential to the survival of all of us.[3] This would reverse the current practice, which is for richer countries to pay others to destroy their forests for timber, food and fuel. (However, it would be risky to make forest conservation a mere addendum to climate change mitigation because forests have a value above and beyond their role in maintaining the atmosphere. They are also home to many of the world's species, some of which are undoubtedly of more use to people extant than extinct, and all of which, arguably, have their own inherent worth. They are also home to marginalized groups of peoples whose rights and investment in the land need to be guaranteed.) What is needed is an internationally funded total ban on destroying any more virgin forest, and corresponding controls on the consumption of products that drive deforestation. Better still would be a global system for governing appropriate use of the world's available land: the Earth is our common heritage, and we need to manage it more wisely.

In addition to protecting natural habitat directly, we need to look at the ultimate causes of increased demand. Waste is one. But at least as important is population growth. Throughout history it has been the aim of humans, like all other organisms, to propagate their species. Convincing people to forgo this would probably be futile, and luckily it is not necessary because curbing population growth is the most plausible way of ensuring long-term survival.

Education and contraception are the most constructive approaches. But in some cases there is an urgent need also for financial incentives and even legislation. This is controversial and in some places very unpopular. But the argument needs to be won. Any sacrifice in terms of personal fulfilment incurred by limiting the number of children per family is more than compensated for by the benefits of ensuring that future generations are not stranded on a planet that cannot support them. The alternative would be the likely outbreak of wars and famine.[4]

Many look at China today and point fingers at its failure to curb the use of coal-fired power stations and resource consumption. But it is practically alone in having tackled the ultimate environmental problem of population growth (and in any case per capita consumption in China is still a fraction of that in Europe or America). Its one-child policy has dramatically cut population growth to 1.3 per cent in 2001.[5] It may seem draconian, but it is surely better than a China with twice the current population.[6]

However, there is another fundamental environmental measure that China did not introduce. While its population has grown less rapidly, each of its inhabitants have taken up more 'room'. Like most people in the developing world, the Chinese want their share of modern consumer goods – richer food, more meat, more cars and more houses. While population growth has slowed, the number of households increased by 3.5 per cent each year from 1990 to 2005 – which is the equivalent of introducing a new nation with as many households as Russia onto the planet every single year. As Jared Diamond has observed, if China achieves its aim of reaching the level of consumption enjoyed by rich countries, this 'will approximately double the entire world's human resource use and environmental impact. But it is doubtful whether even the world's current human resource use and impact can be sustained.'[7]

Conserving food might therefore do nothing more than slightly mitigate the impact of the growing demand for it and other resources in China and elsewhere. If rich nations choose to share the responsibility for the impact of their resource consumption, this will require a plan of 'contraction and convergence', whereby they agree to curb their own per capita consumption in the interests of planetary survival. Reducing waste is one way of contributing to this essential process – but it is only one. Most of the others will be far more difficult to achieve.

APPENDIX
Graphs, Tables, Maps and Data

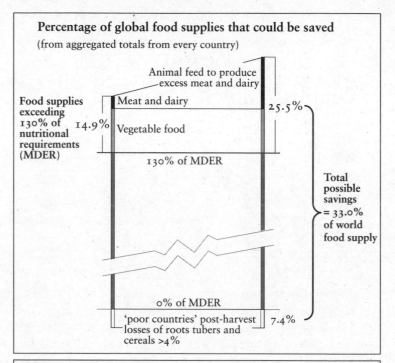

Percentage of global food supplies that could be saved

(from aggregated totals from every country)

Animal feed to produce
excess meat and dairy

Food supplies exceeding 130% of nutritional requirements (MDER) 14.9%

Meat and dairy

Vegetable food   25.5%

130% of MDER

Total possible savings = 33.0% of world food supply

0% of MDER

'poor countries' post-harvest losses of roots tubers and cereals >4%   7.4%

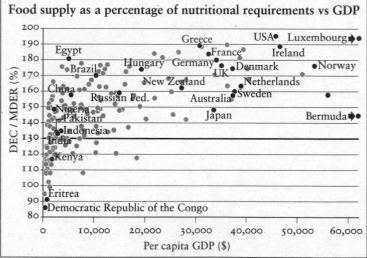

Food supply as a percentage of nutritional requirements vs GDP

DEC / MDER (%)

Greece   USA● Luxembourg➡●●

Egypt   France   Ireland

Hungary Germany ●   ●Norway
Brazil   UK Denmark

New Zealand   Netherlands
China   Sweden

Russian Fed. ●   Australia

Nigeria
Pakistan   Japan   Bermuda➡●

Indonesia
India

Kenya

Eritrea
●Democratic Republic of the Congo

Per capita GDP ($)

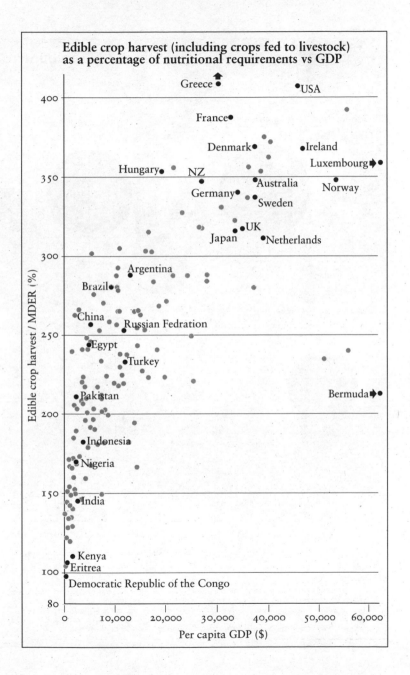

Edible crop harvest (including crops fed to livestock) as a percentage of nutritional requirements vs GDP

Food surpluses and waste in rich and poor countries

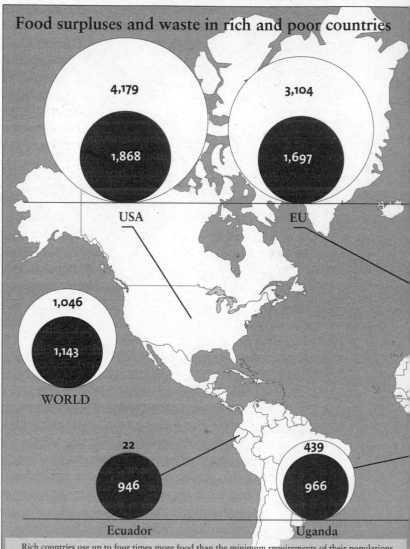

4,179

1,868

USA

3,104

1,697

EU

1,046

1,143

WORLD

22

946

Ecuador

439

966

Uganda

Rich countries use up to four times more food than the minimum requirements of their populations (after adding/subtracting imports and exports): surplus is either fed inefficiently to livestock, causing a net loss in food calories (outer ring), or it is wasted in the supply chain, or eaten in excess of dietary requirements (inner ring). The greatest food surpluses are in the US, followed by EU countries, while Japan has a more efficient food system. Poor countries have much smaller food supplies: fewer arable crops are fed to livestock, and less is wasted in the home.

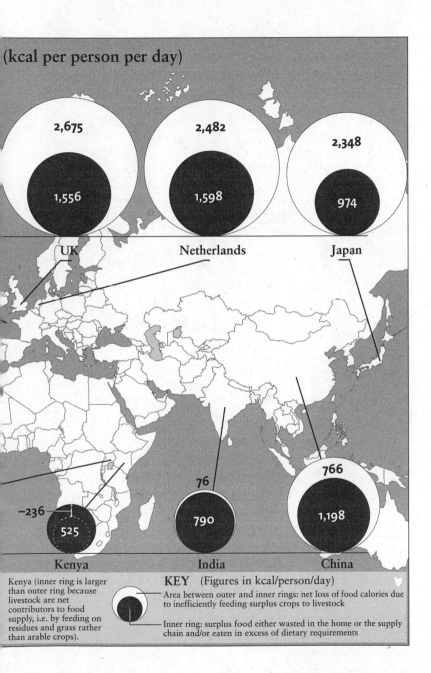

(kcal per person per day)

2,675

1,556

UK

2,482

1,598

Netherlands

2,348

974

Japan

−236

525

Kenya

76

790

India

766

1,198

China

Kenya (inner ring is larger than outer ring because livestock are net contributors to food supply, i.e. by feeding on residues and grass rather than arable crops).

KEY (Figures in kcal/person/day)

Area between outer and inner rings: net loss of food calories due to inefficiently feeding surplus crops to livestock

Inner ring: surplus food either wasted in the home or the supply chain and/or eaten in excess of dietary requirements

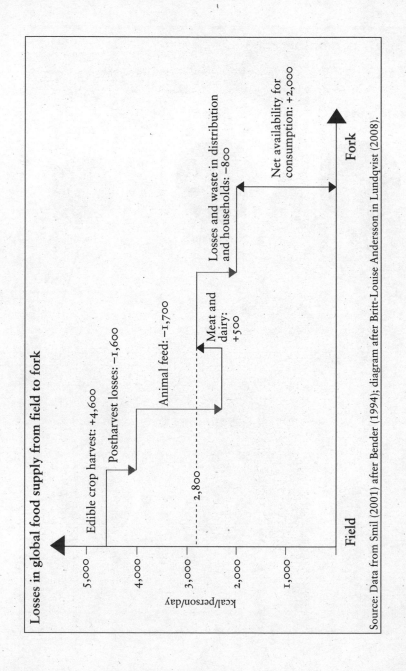

Losses in global food supply from field to fork

Net availability for consumption: +2,000

Fork

Losses and waste in distribution and households: −800

Meat and dairy: +500

Animal feed: −1,700

Postharvest losses: −1,600

Edible crop harvest: +4,600

kcal/person/day

5,000

4,000

3,000 ----- 2,800

2,000

1,000

Field

Source: Data from Smil (2001) after Bender (1994); diagram after Britt-Louise Andersson in Lundqvist (2008).

Feed to food in the US and India: total national inputs and outputs

The US diverts much more of its arable harvests to produce roughly the same amount of meat and dairy products as India (though it produces much more per person). (Area represents kcal/year to scale)

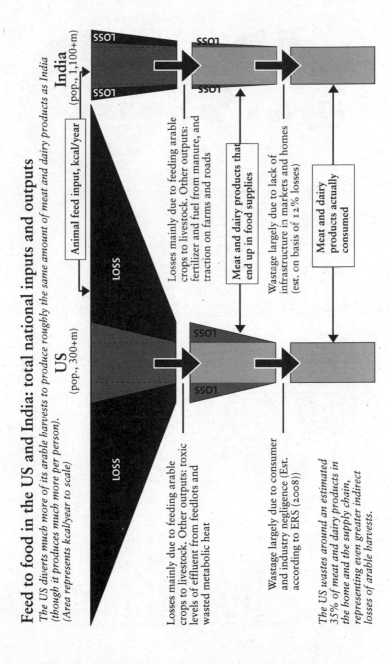

US (pop., 300+m)

India (pop., 1,100+m)

Animal feed input, kcal/year

Losses mainly due to feeding arable crops to livestock. Other outputs: toxic levels of effluent from feedlots and wasted metabolic heat

Losses mainly due to feeding arable crops to livestock. Other outputs: fertilizer and fuel from manure, and traction on farms and roads

Meat and dairy products that end up in food supplies

Wastage largely due to consumer and industry negligence (Est. according to ERS (2008))

Wastage largely due to lack of infrastructure in markets and homes (est. on basis of 12% losses)

Meat and dairy products actually consumed

The US wastes around an estimated 35% of meat and dairy products in the home and the supply chain, representing even greater indirect losses of arable harvests.

US food waste in households, food service and retail

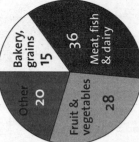

Proportion of waste by weight (%)

Bakery, grains 15 · Meat, fish & dairy 36 · Fruit & vegetables 28 · Other 20

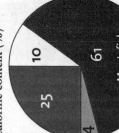

Proportion of waste by caloric content (%)*

Meat, fish & dairy 61 · 25 · 10 · 4

Commodity	Food wasted in US retail, food service and households *tonnes per annum*	Calories in food wasted *billions of kcal*	People this could have lifted out of hunger @ 250 *kcal/day (see Ch. 5)*	Environmental burdens	
				Emissions from producing food wasted *tCO₂e*	Land required to grow food wasted *ha*
Grain products	6,619,838	17,729	194,287,867	5,560,664	648,744
Fruit total	5,132,938	2,627	28,787,959		
Vegetables total	7,233,106	4,107	45,009,561		
Fluid milk	7,906,702	3,953	43,324,392	8,381,104	948,804
Other dairy products	3,164,767	7,626	83,577,266	19,304,505	2,185,416
Red meat	2,202,682	5,275	57,807,854	28,928,552	3,281,996
Poultry	1,241,503	2,495	27,347,084	5,710,915	794,562
Fish and seafood	290,758	427	4,675,684		
Eggs	1,127,650	1,703	18,660,284	6,202,073	755,525
Dry beans, peas, lentils	162,842	576	6,317,393		
Tree nuts and peanuts	133,812	775	8,490,646		
Caloric sweeteners	5,380,150	16,230	177,868,126		
Fats and oils	3,069,511	27,595	302,409,925		
Total	43,666,258	91,119	998,564,041		
Indirect waste of arable harvests used to produce wasted meat and dairy		87,851	962,755,588		
Total direct and indirect waste		178,970	1,961,319,629		

Source for food waste tonnages: Kantor et al., (1997)

Fruit and vegetables represent a comparatively large proportion of the total mass of food wasted, but in terms of caloric content, land use and environmental burdens, meat and dairy products are far more significant.

*Including indirect waste of arable harvests to produce wasted meat and dairy

UK food waste in households

Proportion of waste by weight (%)

- Bakery, grains 20
- 9
- Fruit & vegetables 47
- Other 24

Proportion of waste by calorific content (%)*

- 27
- 14
- Meat, fish & dairy 35
- 24

* Including indirect waste of arable harvests to produce wasted meat and dairy

Food type	Food wasted in UK homes (avoidable + possibly avoidable) tonnes per annum	Calories in food wasted billions of kcal	People this could have lifted out of hunger @ 250 kcal/day (see Ch. 5)	Environmental burdens	
				Emissions from producing food wasted tCO2e	Land required to grow food wasted ha
Bakery	895,583	2,415	26,469,617	752,290	87,767
Dried food and grains	164,969	365	4,003,783		
Fruit	637,585	326	3,567,209		
Salad	295,915	168	1,836,208		
Vegetables	1,587,183	918	10,062,787		
Dairy	187,181	358	3,925,616	640,103	73,317
Meat & Fish	301,776	633	6,936,089	1,969,307	251,198
Confectionary	62,500	329	3,605,918		
Drinks	147,500	42	459,967		
Condiments	153,085	390	4,275,825		
Desserts	54,300	100	1,090,608		
Mixed meals†	704,939	1,438	15,762,088		
Other	180,726	214	2,341,639		
Total	5,372,342	7,696	84,337,354		
Indirect waste of arable harvests used to produce wasted meat and dairy		2,597	28,457,668		
Total direct and indirect waste		10,293	112,795,022		

Source for food waste tonnages: WRAP (2008a)
†Ready and prepared meals

Supermarket waste statistics set against gross takings to show relative efficiency

Supermarket	M&S	Co-op Food	Morrisons	Waitrose	Asda	Sainsbury's	Tesco	TOTAL
Year	2007–8	2007–8	2007–8	2007–8	2005	2007–8	2007–8	
Total store waste, t p.a.		65,268	187,733	43,853	251,429	229,172	487,000	
Diverted, recycled, %		47	72	49	65	63	70	
Diverted, recycled, t p.a.		30,605	134,378	21,513	163,429	143,994	342,000	
Total waste to landfill, t p.a.		34,663	53,395	22,340	88,000	85,178	145,000	
Landfill waste (food, %) (bracketed numbers, extrapolations*)		29	[70]	60	69	70	[70]	
Food waste to landfill (reported, estimated), t p.a.*	20,000	10,100	37,377	13,404	60,720	59,625	101,500	302,725
Total food wasted t p.a. (incl. est. 23% raw meat, milk and other fluids diverted from landfill)	20,000	12,423	45,973	16,487	74,686	73,338	124,845	367,752
Consumer spending £,000s (12 wks to 2 Nov. 2008)	NA**	840,465	2,395,019	760,250	3,447,163	3,200,392	6,236,651	16,787,340
Index of food waste per £ spent in store (i.e. efficiency, 100 = average performance)		73	99	108	107	114	99	

*See Ch. 2, for assumptions used to calculate amount of food waste present in total landfill
** M&S considered large clothing store with grocery arm, therefore not included in grocery market share statistics

Sources: Interview Rowland Hill, M&S, 11 Nov. 2008; John Lewis Partnership (2008); Asda (2008); Julian Walker Palin (n.d.); Sainsbury (2008), p. 38; Tesco (2008); Garner (2008).

Notes, Sources, Method and Data for the Appendix

FAO (2003a) contains the Food Balance Sheet (FBS) for each country and region. The entry for 'food supply' (also known as 'domestic energy consumption' (DEC)) represents the food *available* for consumption in shops, restaurants, etc., not what people *actually eat*. World food supply is 2,791 kcal/person/day or 17.4trn kcal/day. In **Food supply as a percentage of nutritional requirements vs GDP**, the food supply (DEC) of each country was divided by the Minimum Dietary Energy Requirements (MDER) of each country, showing how much more than nutritional requirements each country has in its food supply, and this percentage was plotted against GDP for each country, following the method in Bender (1994). Note that the figure for the United States' food supply in FAO (2003a) is slightly lower than the 2004 figure in Hiza and Bente (2007) used elsewhere in this book.

The percentage of world food supplies that could be saved was found by taking each country in turn. If all countries with food supplies over 130% of MDER brought their food supplies down to 130% of MDER, this would save 14.9% of world food supplies (415 kcal/person/day or 2.6trn kcal/day). Meat and dairy represents an average of 22% of food supplies in those countries with food supplies over 130% MDER, so the proportion of this 'unnecessary surplus food energy' that is animal food = 559bn kcal/day. It is necessary to calculate the amount of edible arable crops that went into producing this portion of 'unnecessary surplus' above 130% of MDER.

Each country has a different method of animal husbandry which determines how much cereal, oilcake and other animal feed is used to produce the animal food in that country. This 'feed conversion ratio' was calculated for representative countries by taking the amount of each kind of animal feed used in that country, attributing to each feed crop its known calorific value to arrive at a total feed input in kcal. For the output, the total calorific contribution that animal food as a whole made in that country was calculated, minus imported produce, which was dealt with separately.

The FAO (2003a) FBS 'animal feed' column includes crops such as cereals, tubers and fishmeal that are fed to livestock. However, the FBS does not include oilcakes such as soymeal – a major feedstuff for livestock globally – in the 'animal feed' column. Although the extracted oil has its own entry on the FBS for each of the oilcrops, the oilcake, a co-product of vegetable oil production, is placed into the 'food manufacture' column, even though almost all of it does in fact end up as animal feed. Because oilcakes are not listed as commodities on the FBS, the 'import' and 'export' columns on the FBS do not properly record the import and export of these animal feedstuffs. Therefore, to correct the FBS for oilcakes, individual trade records for each country were incorporated to establish how much oilcake was actually used as animal feed within the country. Additionally, when calculating the total contribution of farmed animal products to food supplies it was necessary to deduct the energy contribution of *caught* fish (because these are not produced by feeding cereals, etc. and would therefore distort national conversion ratios). This was assumed to be 68% of total fish products, as per FAO (2009), p.4. Imported animal products were treated as having been produced under average conversion ratios of US/EU production.

Adding all the individual countries together, the amount of food energy that went into producing just the proportion of animal food in each country's 'unnecessary surplus' (above 130% of MDER), is 2.4trn kcal/day. Add this to the contribution of vegetable products only to the 'unnecessary surplus' and it comes to 4.4trn kcal/day or 25.5% of world food supply. That is the total direct plus indirect unnecessary food surplus above 130% of MDER. Add to this the figure of 7.1% of global food supplies saved by reducing post-harvest losses in poorer countries to 4% (Bender (1994)). Total potential savings globally = 33% or 5.7trn kcal/day.

Calorific values of oilseed meals were based on quoted digestible energy (DE) in McL. Dryden (2008), Table 5.7, p. 64 (pigs) (MJ/kg to kcal/kg using a conversion factor of 239). World production, import and export data for oilseeds and oilseed meals from USDA (2009), pp. 6–7, and Mattson *et al.* (2004), pp. 15–18. For the US, Mexico, the UK, Netherlands, Greece, India, Uganda and Egypt, the direct feed conversion ratio was calculated within those countries and applied to them. For other European countries (both EU and non-EU, including Russia and its satellites), the EU ratio was applied. For other non-EU OECD countries an average of the US and EU ratio was applied. Then, for any remaining developing countries, an average of the ratios for India, Mexico, Egypt and Uganda was applied.

MDER and food supply data from FAO (2008d) and FAO (2003a). FAO statistician Cinzia Cerri supplied MDER estimates for developed nations: western Europe, from 2000 to 2007, the minimum is between 1,900 and 1,910 and the maximum 2,000 kcal/person/day; in North America the minimum is 1,950 and maximum is 1,980 kcal/person/day; in Oceania (developed) the minimum is 1,940 and the maximum is 1,950; in developed Asian nations MDER is 1,810–1,890 kcal/person/day (Cinzia Cerri, FAO, personal communication, 18 December 2008). Population statistics from FAO (2005); GDP from CIA (2009). Many thanks to Simon Inglethorpe for doing much of the number crunching for these calculations.

For **Edible crop harvest (including animal crops fed to livestock) as a percentage of nutritional requirements**, the same method of feed conversion ratios was used to arrive at a total for each nation, except this time the feed conversion ratios were applied to the whole food supply, rather than just the portion over 130% of MDER; this represents the total food energy used to produce the food supply of each nation. **Food surpluses and waste in rich and poor countries** adds together the 'food supply' and the 'waste' column in the FBS and subtracts from it the country's MDER. This shows how much extra food energy is wasted either in the home or in the supply chain and/or eaten in excess of dietary requirements. The second figure (represented by the outer ring) shows how much food energy is *lost* in feeding arable crops to livestock (i.e. total inputs of feed minus output of meat and dairy products).

Feed to food in the US and India: total national inputs and outputs shows the total annual input of animal feed to produce meat and dairy products both domestically grown and imported in the US (566.4trn kcal) and India (109.9trn kcal). From these inputs, the two countries' outputs of meat and dairy products differ enormously, with the US having a feed input to meat and dairy output conversion ratio of 5.1:1 and India having a ratio of 1.4:1. The US ends up with just 111.2trn kcal of meat and dairy products in the food supply, of which an estimated 72.4trn kcal are consumed (the rest is wasted), while India has 77.6trn kcal in the food supply, of which an estimated 68.3trn kcal are consumed.

UK food waste in households: WRAP (2008a) lists 173 individual foods. Each one was given its calorific value from FSA (2008b) and added into a total for each food type (e.g. fruit, vegetables); for each food type there in an entry for 'others' which was given an average calorific content from that type. For 'possibly avoidable' food waste Andrew Parry (WRAP, personal communication) supplied tonnages per food type, plus a list of the most wasted individual foods which covers 76% of all 'possibly avoidable' food wasted. These were given their true calorific values; for the rest in each food type, an average calorific value was calculated from the calorific values of the remaining (non-listed) items. 'Environmental burdens' of meat and dairy production are taken from Williams *et al.* (2006) and figures given for *on farm* emissions of *avoidable* food waste only; for bakery, Nielsen *et al.* (2007), avoidable and possibly avoidable, all treated as ex-retail fresh wheat bread. For indirect losses, the UK feed conversion ratio, as calculated above, was 3.62:1.

For **US food waste in households, food service and retail**, tonnages were taken from Kantor *et al.* (1997). Calorific content per tonne for each food type was calculated by taking the total tonnage per type in UK household avoidable food waste and, using the total calorific content of that tonnage, finding the average calorific content per tonne for that type, except for fish, poultry, eggs and fluid milk, which were all calculated individually. Caloric sweetener given as per FAO (2003a). 'Environmental burdens' from Williams *et al.* (2006) for poultry, eggs, milk and red meat (average of pork, lamb and beef); other dairy (average of other dairy products) and figures given for *on farm* emissions only; for grain products, Nielsen *et al.* (2007), all treated as ex-retail fresh wheat bread. The US average feed conversion ratio, animal feed: animal food = 5.09:1, as above.

Abbreviations

OECD	Organization for Economic Co-operation and Development
AD	anaerobic digestion
BBC	British Broadcasting Corporation
BMSW	biodegradable municipal solid waste
bn	billion
BPC	British Potato Council
CGW	Champions' Group on Waste (FISS)
CO_2e	carbon dioxide equivalent
DEC	domestic energy consumption
Defra	Department for Environment, Food and Rural Affairs (UK)
EA	Envrionment Agency (England and Wales)
EPA	(Ireland) Environmental Protection Agency Ireland
EPA	(US) US Environmental Protection Agency
ERS	Economic Research Service of the USDA
EU	European Union
FAO	Food and Agriculture Organization of the United Nations
FBS	Food Balance Sheets (from FAO 2003a)
FDA	US Food and Drug Administration
FISS	Food Industry Sustainability Strategy (UK)
FSA	Food Standards Agency (UK)
GDP	gross domestic product
GHG	greenhouse gas
Gt	gigatonne (one billion tonnes)
GWh	gigawatt hour (one billion Wh)
GWP	global warming potential
ha	hectare
IPCC	Intergovernmental Panel on Climate Change
IRIN	Humanitarian news service of the UN Office for the Coordination of Humanitarian Affairs

kcal	kilocalories
kWh	kilowatt hour
m	million
MAFF (Japan)	Ministry of Agriculture, Forestry and Fisheries (Japan)
MAFF (UK)	Ministry for Agriculture, Fisheries and Forestry (UK) – superseded by Defra
MCS	Marine Conservation Society
MDER	Minimum Dietary Energy Requirement
Mha	million hectares
MP	Member of Parliament
MSC	Marine Stewardship Council
MSW	Municipal Solid Waste
N	nitrogen
OSPAR	Oslo/Paris convention for the Protection of the Marine Environment of the North-East Atlantic
t	metric tonnes
TJ	terajoule (one trillion joules)
trn	trillion
TWh	terawatt hour (one trillion Wh)
UK	United Kingdom
UN	United Nations
US	United States of America
USDA	United States Department of Agriculture
WFP	United Nations World Food Programme
WRAP	Waste & Resources Action Programme

Cider-pressing in Sussex, England, using apples that otherwise go to waste, gathered from the local area through the internet sharing site 'Freecycle'.

The Tale of Three Lettuces. These three identical Romaine lettuces were purchased at the same time and stored for ten days, from left to right
1) at room temperature
2) in the fridge
3) in a glass of water like cut flowers.

'Sheep's organs', a traditional dish of lung, liver, stomach and intestines, served in Kashgar, Xinjiang province, Western China.

Chicken feet and pigs' trotters for sale on a station platform in China.

...and what the British do with their chicken feet and other animal remains: a heap of composted poultry by-products maturing next to a field of grazing cattle. One bullock is grazing right up against the heap.

Unwanted fish being discarded from an industrial trawler.

UK Second World War poster.

UK Second World War poster.

First World War poster, 1917, Ministry of Food (UK). Similar posters were issued as far away as New Zealand in recognition that the waste of internationally traded commodities, such as wheat, contributed to shortages many thousands of miles away. The message was reissued as a postmark at the end of the Second World War.

First World War poster, United States Food Administration.

Unsold Seven-Eleven lunch-boxes being separated and converted into pigfeed in the Agri Gaia factory in Sakura city, outside Tokyo, Japan, 2008.

Food waste collected on the streets of Seoul, South Korea for composting and pig feed, June 2008.

Food waste being converted into pig feed at the Odakyu Food Ecology Centre in Sagamihara city, near Tokyo, Japan, 2008.

Kamei-san and his pigs, fed on food waste recycled at the Odakyu Food Ecology Centre, June 2008. The pigs are sold as eco-pork in the Odakyu department stores.

Notes

Introduction: The Effluent of Affluence

1. To feed a population that grew from 0.6 to 5.5bn between 1700 and 1993, global cropland increased from 270 to 1,450m ha through habitat conversion, and total agricultural lands including permanent pasture covered 4,810m ha or 35% of the earth's land area (excluding Antarctica): Goklany (1998).

2. Between 1980 and 2003 cropland in Latin America increased by 15%, in sub-Saharan Africa by 22%; and in land-scarce Asia by 12%: Steinfeld *et al.* (2006), p. 12.

3. UNEP (2009).

4. Food waste has been defined (Roy (1976)) as 'a potential source of food that has knowingly been discarded or destroyed. Waste includes inedible materials that could yield food through utilization as animal feed.' He also defined food loss as 'a potential source of food that has been inadvertently destroyed or spoiled', and relative waste as 'the inefficient use of food resources in feeding, say, cereals to livestock instead of directly to people': cf. Singer (1979).

5. Borger (2008a). The World Bank stated that 'the grain required to fill the tank of a sports utility vehicle with ethanol . . . could feed one person for a year' (World Bank (2008c)); Crooks and Harvey (2008); Wolf (2008); RFA (2008); Runge and Senauer (2007).

6. According to the authoritative International Food Policy Research Institute (IFPRI), maize ethanol produced in the US may cut GHG emissions by 10–30% compared to petroleum; but ethanol produced from cellulose or sugarcane (from Brazil, for example) could reduce emissions by 90%. But when either contributes to deforestation by increasing demand for agricultural land its impact is negative: IRIN (2008b); RFA (2008).

7. IFPRI estimated 94.9m t diverted to biofuels in 2007–8; other estimates suggested 60m t (Reuters (2008)) and Mitchell (2008) estimated 86m t. Estimates on the contribution this had to food price inflation varied from 3% (the US government) to 70%; IFPRI thought this was responsible for 30%: Wolf (2008); IRIN (2008b); IRIN (2008c).

8. Smil (2001) estimated that the equivalent of 100m t of cereals could be saved if rich nations reduced their waste to just 20% of supplies, and that 150m t are lost in post-harvest operations in developing world countries. In ch. 12, I show that the grand total throughout the world may be twice that.

9. BBC Radio 4 (2008b).

10. FAO (2008b), p. 2.

11. Wolf (2008); IRIN (2008b); IRIN (2008c).

12. Hinchcliffe (2005), cf. Sustain (2009).

13. Safeway (2009); Timothy Jones estimates 68m, personal communications, 2008. Feeding America aims to feed 30m.

14. The EC definition of 'food poverty' in this context means an inability to afford a meal with meat, chicken or fish every second day (EC (2008a); EC (2008e)). In other contexts, 'food poverty' is defined as 'the inability to acquire or consume an adequate quality or sufficient quantity of food'. 'Food insecurity', a term often used in the US, suggests a 'limited or uncertain availability of nutritionally adequate and safe foods'; cf. Hawkes and Webster (2000), p. 1; Leather (1996).

15. EC (2006), and see further discussion below.

16. GWP of methane over 100 years: Forster *et al.* (2007), p. 212.

17. Ackerman (1997).

18. Under EU law, waste is defined as 'any substance or object which the holder discards, intends to discard or is required to discard' (Waste Framework Directive (European Directive 2006/12/EC)) and has historically been considered an *external* to economic activity and discarded from it. Emissions standards and taxes on waste aim to *internalize* waste into the economic cycle by putting

a price on it (Lacoste *et al.* (2007), pp. 4–6). See Chalmin and Gaillochet (2009, forthcoming).

19. Lundqvist, Fraiture and Molden (2008), pp. 5, 31, launched on 14 May 2008 at the 16th Session of the United Nations Commission on Sustainable Development, outlining steps to achieve a 50% wasted-food reduction globally.

20. Ackerman (1997).

21. FAO (1981), Concl. 5.

PART I: Perishing Possessions

1. Liber-ate

1. *New International Version.*

2. Locke (1690), v. 37–8, 46.

3. BBC Radio 2, *The Jeremy Vine Show*, 18 December 2003.

4. BBC (2005).

5. BBC TV, *The Politics Show*, 19 October 2003.

6. Sky News, 9 January 2007.

7. BBC Radio 4, *Today*, 30 May 2007.

8. Jones (2004a), p. 3.

9. Garnett (2006), p. 66.

10. Dairy Crest's Clover spread and St Ivel Utterly Butterly contain 29% and 9% reconstituted buttermilk respectively (2 February 2009).

11. Interview, 4 and 11 September 2008; email, 5 September 2008.

12. Bluefin tuna catches (illegal and legal) are currently estimated to be 50–60,000 t; scientists argue the quota should be 15,000 t, and that even if an immediate ban were introduced, bluefin tuna populations in the north-east Atlantic and Mediterranean will probably collapse. The quotas currently handed out are around 30,000 t and the International Commission for the Conservation of Atlantic Tunas (ICCAT) – nicknamed the 'International Conspiracy to Catch All Tunas' – called in 2008 for an insufficient reduction of just 5,000 t (*Economist* (2008a), p. 98).

13. Trivedi (2008), p. 32.
14. K10 would not tell me where it sourced its fish from.
15. Messages at the head office Customer Services to no avail; last attempt 5 September 2008, by telephone to the Fleet Street branch.
16. Environmental Protection Act (1990) Section 34, *Waste Management: The Duty Of Care: A Code Of Practice*, 2.1 (e) and 2.7. Cf also Regulation (EC) No 852/2004 On The Hygiene Of Foodstuffs, Ch. VI.3.

2. Supermarkets

1. Bartholomew (2002).
2. Kantor *et al.* (1997).
3. MAFF (Japan) (2007).
4. Per capita (1.6m t in UK = 26.7 kg/person/yr; 2.6m t in US = 8.3 kg/person/yr).
5. Jones (2004a) reports losses in US supermarket sector of 0.76%, attributing efficiency to price mark-downs and food donations. The only UK supermarket to have offered figures on a percentage basis is Marks & Spencer, which estimated that 4% of food on shelves did not get sold, though some of the remainder is donated or sold to staff.
6. WRAP's estimate is based on surveys of the nine largest grocery chains in the UK (Mark Barthel, personal communication, 20 March 2009). The Environment Agency's (EA) Commercial and Industrial Waste surveys (1998–9 and 2002–3) found similar levels of waste. The 2002–3 survey found 1.6m t of waste 'food' from the retail sector, though it cannot be discerned how much of this was once edible. However, this figure refers to England alone. To this should be added the retail food waste figures for Scotland (38,383 t) and Wales (15,000 t). The 2003–4 survey for Northern Ireland did not break down food waste by sector. So the total for the retail food waste for the UK according to the 2002–3 C&I survey is above 1.7m t. In addition to this, there was 4.9m t of mixed general waste from all kinds of retail outlet, some

of which will have been food. The figures for Wales and Scotland are much smaller than would be expected in comparison to England's figures based on the relative size of their populations.

7. Jones (2004a).

8. AEA (2007), pp. 37, 99. AEA used the EA Commercial & Industrial Waste survey 1998–9, which was conducted using completely different waste categories from the 2002–3 survey (UK categories rather than the EU ones used in 2002–3). Based on AEA's surveys in which companies were asked to estimate what percentage of their mixed wastes were food, AEA estimated that in total retailers wasted 455,000 t of food. Evolve (2007), pp. ii, 2–3.

9. Evolve (2007), p. 30, estimated that retailers wasted 500,000 t of food each year in the UK.

10. The source was the general manager of a London-based convenience food chain. Telephone interview, 11 September 2008.

11. Jones (2004b).

12. Eric Evans, personal communication, 2 March 2009.

13. Interview, Julian Walker-Palin, Asda, 10 October 2008. Reiterated by several other interviewees in the retail sector.

14. Interview, Chris Haskins, 25 July 2008.

15. David Lion, head of technical services, Ginsters, who used to manage an account with M&S, interview, 27 August 2008.

16. £60/t ÷ 5,000 for a 200 g product.

17. Defra (2007a), p. 27.

18. Interview, Rowland Hill, corporate social responsibility/sustainability manager, Marks & Spencer, 11 November 2008.

19. Garnett (2006), p. 65. Some of this will be sold to staff at a discount or redistributed through charitable schemes rather than being disposed of, but it may not include food waste further up the supply chain.

20. Interview, Alison Austin, Corporate and Social Responsibility, 13 November 2008. 60,000 t was disclosed by her colleague Lawrence Christensen: Pagnamenta (2008).

21. When I say 'store waste', this is the term used by the companies, and it may therefore not include waste in their distribution chains.

Co-op (2008); Morrisons (2008); John Lewis Partnership (2008); Asda (2008) and Walker-Palin (n.d.); Sainsbury plc (2008a), p. 38; Tesco (2008), p. 15.

22. Evans (2005), repr. Evolve (2007), p. 30. Eric Evans said that the categories covering raw meat (as opposed to cooked meat, which can be sent to landfill) were 'Meat and Fish' (15.1%) and 'Counter Meat' (3.8%). He was not entirely sure whether a further 12.1% ('Grocery Meat') would now be landfilled or not. Sainsbury's indicated that in 2004–5 fresh fruit and vegetable waste comprised only 25.5% of all food-related waste net of packaging (Garnett (2006), p. 62), so the proportion of waste containing raw meat could be even higher. Cooked meat, by law, is allowed to be disposed of with general waste for direct landfill disposal.

23. The 260 UK Asda Superstores (30,000+ sq ft) would, according to Walker-Palin's estimate, generate around 13,520 t/yr. By extrapolation on a floor area basis, each of the 29 Supercentres (70,000+ sq ft) would generate an additional 2.3 t/wk, equivalent to a total of 3,519 t/yr. This suggests that the 289 Asda Superstore and Supercentre stores, combined, generate an estimated 17,039 t of food waste per year. There are a further 47 small stores and 17 'Asda Living' stores.

24. Garner (2008).

25. See e.g. AEA (2007), p. 37.

26. Of the total grocery market worth of £133.3bn in 2007, £10.4bn are non-groceries such as electrical goods, £12.5bn tobacco and £20.8bn non-food groceries: IGD (2008).

27. Cf. AEA (2007), pp. 48–9.

28. Kroger's rate per £ clearly depends on exchange rate: Kroger annual sales in 2006 (Watson (2007)) $66.1bn (= £35.9bn at an average 2006 exchange rate of $1.84 to £1.00); Kroger donations in 2006 13,610 metric t = 0.379 t/£m. Sainsbury's store sales 2007 excl. VAT £16.9bn (Sainsbury plc (n.d.)); Sainsbury's donations in 2006/7 6,680 tonnes = 0.396 tonnes/£m.

29. Vidal (2005).

30. Sainsbury plc (2008a), p. 46.

31. E.g. interview, Julian Walker-Palin, 10 October 2008.

32. Marks & Spencer (2008).

33. John Lewis Partnership (2008) pledges that Waitrose will 'Recycle 75% of all Waitrose waste by year-end 2012'. Co-op (2008) pledges to 'Ensure that less than 50% of total waste arisings are landfilled by 2013' across its businesses.

34. Katie Wright, Waitrose Corporate Social Responsibility department, personal communication, 9 January 2009; Jess Hughes, Waitrose senior press officer, personal communication, 14 January 2009.

35. Waitrose (2008), p. 38.

36. Interview, Katie Wright, 9 January 2009.

37. Morrisons (2008).

38. Eric Evans, personal communication, 2 March 2009. Mark Barthel, personal communication.

39. Morrisons (2008). In 2006, Morrisons was at the bottom of the National Consumer Council *Greening Supermarkets* league table.

40. Dominic Burch, PR manager, Asda, personal communication, 1 December 2006.

41. Interview, Julian Walker-Palin, Asda, 10 October 2008. Interview, Maria Kortbech-Olesen, Fareshare, 4 March 2009.

42. Asda (2006).

43. Wal-Mart in Arkansas is trialling pyrolysis. Interview, Julian Walker-Palin, 10 October 2008.

44. Tesco (2008); Renton (2007).

45. In response to enquiries on the amounts of waste generated, the results were as follows. *Target Corp* (US): Brandy DuBreuil, Kaplow PR, contacted by email 5 March 2009. Michaela.Gleason@target.com replied referring to Target Corp (2008), where a recycling rate of 'roughly 70 percent' is claimed; and precise records of store waste are kept by the company, but not released. *Kroger Co.* (US): Meghan Glynn, contacted by email and telephone 5 and 6 March 2009; reply referred to Kroger (2008) and explained that Kroger does not publish specific data on food waste or amounts of waste disposed to landfill, though see p. 7 on rendering

and redistribution. *Supervalu* (US): contacted by email 5 March
2009 and by email and telephone 5 and 6 March 2009. Susie Bell
replied with a referral to Supervalu (n.d.), which has some figures
on recycling but does not report data on disposal overall or landfilled waste. *Wal-Mart* (US): email on 17 March 2009 referred to
publicity material on food donations: neither Wal-Mart (n.d.) nor
Wal-Mart (2008a) publishes tonnages on landfill disposal. The latter
notes an intention to 'create zero waste'. *Safeway* (US): Safeway
(n.d.) makes no specific mention of waste. *Carrefour* (France): email
17 March 2009 refusing to divulge waste data with a link to the
Fondation Internationale Carrefour, which, like Carrefour (2008),
contains no data on total waste arisings but advertises food donation
activity. *Aldi* (Germany/UK): email to Aldi stores 14 January 2009;
telephones to their PR service, Weber Shadwick, on 19 and
22 January 2009 – and several times thereafter – but these calls
were not returned. *Iceland* (UK): after several phonecalls to Iceland's press office around 12 January 2009, contact was established
with Tania Berry. Pauline Begley replied by email on 30 January
2009 but did not provide figures on waste landfilled or food waste.
A follow-up email of 30 January 2009 to Iceland requesting specific
information on amounts of food waste and waste sent to landfill
was not answered. *Lidl* (UK): contact was established with Emily
Sherman at Lidl PR around 12 January 2009. PR manager, Mr
Ivone, telephoned on 14 February 2009 but would not provide
information on amounts sent to landfill or on food waste. He
claimed that Lidl diverted/recycled 70% of its waste from landfill.
He also stated that Lidl were currently finalizing a food donation
contract with Fareshare.

46. TNS Retail Forward (2007).
47. Kroger (2008).
48. $110m according to Safeway (2009) and Anon. (2006).
49. Olson (2006).
50. Jeffrey M. Janes, Plaintiff-Appellee, v. Wal-Mart Stores Inc., Dba
    Sam's Club, Defendant-Appellant, and Gary Dawes, Defendant.
    No. 00-55611. 02 C.D.O.S. 1046. United States Court of Appeals

for the Ninth Circuit. D.C. No. CV-99-08777-GAF, http://
www.law.com/regionals/ca/opinions/feb/0055611.shtml; Monsterfodder (2006).
51. Wal-Mart (2008b); Wal-Mart (2008c).
52. Supervalu (n.d.).
53. Target Corp (n.d.).
54. Whole Foods Market (n.d.).
55. Food donation league table of three US grocers

	2006 sales ($m)*	Food donated (metric t)†	**Food donated per $m in kg**
Kroger Co.	60,553	13,610	**225**
Safeway	38,416	33,244‡	**865**
Wal-Mart Grocery Sales	91,988	at least 1,350(?)	**15**
	2007 sales: 134,000	2008 pledge: 31,750	**237**

* Source: Food Marketing Institute (2007)
† Source: company websites and other sources (see above). Kroger figure calculated
according to 2006 food redistribution; subsequently updated in Target Corp (2008)
with a slightly lower figure for 2007.
‡ Estimated tonnage based on value of $110m. This may not be comparable to other
companies owing to Safeway's inclusion of donations from customers and 'vendor
partners'.

3. Manufacturers

1. Bataille (1985), p. 121.
2. The Environment Agency Commercial and Industrial waste survey of 2002–03, published in 2005, sampled just 579 food and
drink manufacturers, out of a total of nearly 8,000 in the sector.
Of the 7.2m t of waste from the food and drink manufacturing
sector, 4.1m t was identified as 'food' waste, though this would
not include a great deal of food disposed of as 'mixed waste'. This
EA figure refers solely to England. WRAP publishes the figure
of 4.1m t on its website but attributes it to the whole of the
UK: WRAP (2009c). The waste arisings for Scotland, Wales and

Northern Ireland are not comparable, but in the Wales survey
there were found to be 271,000 t of 'animal and plant waste'; for
Scotland food-processing waste came to 210,000 t, with a further
100,000 t 'from agriculture, horticulture, aquaculture, forestry,
hunting and fishing'. Northern Ireland's data refers to 2004–5 and
identified 52,652 t of 'biodegradable kitchen and canteen waste'.
From commercial and industrial sectors as a whole, the total
amount of organic waste mainly consisting of food is 6.9m t,
not including waste from most agricultural processes (none in
England). The food service sector, including hotels, caterers and
restaurants, is accounted separately. (Environment Agency (England & Wales) (1999); Environment Agency (England & Wales)
(2005a); Environment Agency (England & Wales) (2005b);
Environment Agency (England & Wales) (2008); Environmental
Protection Agency, Scottish (n.d.); Environmental Protection
Agency (Northern Ireland) (n.d.).)

3. AEA (2007), p. 34.
4. From a total output of 45.2m t of 'dry' products, not including
   water in beverages.
5. The amended Waste Framework Directive (WFD) adopted in
   1991 introduced a new EU-wide definition of waste which was
   brought into force in the UK in May 1994. The interpretation of
   the WFD's definition of waste is now the subject of a substantial
   body of case law by the European Court of Justice (ECJ) (Defra
   (2009)), and in 2007 the European Commission published guidance (COM (2007a)) suggesting that food by-products such as
   animal feed should no longer be considered waste (Defra (2007a),
   p. 7; AEA (2007), p. 37).
6. C-Tech Innovation Ltd (2004), p. 2; cf. AEA (2007), p. 96.
7. Production and value of the European food industry by country
   based on Eurostat (2009). Awarenet (2004), Ch. 1, Annex 6.
   Awarenet waste figures are calculated on the difference between
   the weight of the raw material and the weight of the major
   product, which means that for some processes, such as potato
   starch production, much of the 'waste' is in fact water extracted

from the raw material in processing: Mahro and Timm (2007). Conversely, wastes often have a higher dry-matter content than the raw material, such as in juicing.

8. Timothy Jones, personal communication, 2008.
9. PICME (2006), p. 7.
10. See comments on Defra (2007a) below.
11. Cf. Stuart (2006a).
12. Ahlberg (2002); González Siso. (1996).
13. Kipler and Ornelas (eds.) (2000), II.1348.
14. Just-Food (2004); Nuthall (2005); Sogi *et al.* (2002).
15. Marks & Spencer (2009), Pledge 31 (my italics).
16. Interview, Maureen Raphael, occupational health manager, Hain Celestial Group, 21 August 2008. For more on waste and over-production in food manufacturing, see C-Tech Innovation Ltd (2004); Evolve (2007); Darlington and Rahimifard (2007); Darlington and Rahimifard (2006a); Darlington and Rahimifard (2006b); Morley and Bartlett (2008); AEA (2007); Cabinet Office (2008); Sustainable Development Commission (2008); PICME (2006); Food Processing Faraday Partnership (2008); Heller and Keoleian (2003); Knight and Davis (2007); Kantor *et al.* (1997); Defra (2007a), p. 16 and annex 2, p. 32; Taylor and Fearne (2006); Imperial College London (2007).
17. See e.g. Imperial College London (2007), p. 16.
18. Defra (2007a), pp. 24–5.
19. Interview, Shahin Rahimifard, 18 July 2008.
20. The Competition Commission called for a 'fair dealing' provision in the GSCOP as 'an important safeguard against the transfer of excessive risk and unexpected cost from grocery retailers to their suppliers . . . this overarching "fair dealing" provision provided a means of constraining behaviour while allowing commercial flexibility where it was appropriate': Competition Commission (2008), 11.315, p. 238. These excessive risks and unexpected costs are currently transferred from retailers to their suppliers, which therefore indicates that current practice is 'unfair'.
21. Defra (2007a), p. 32.

22. Defra (2007a), p. 33.
23. Jones (2006).
24. Imperial College London (2007). Some retailers and food manufacturers are switching to CO_2-based refrigeration systems which have lower greenhouse gas emission implications. On refrigeration see Garnett (2007a); Garnett (2008), pp. 44–5; Weidema *et al.* (2008), p. 65.
25. Imperial College London (2007), p. 9.
26. PICME (2006), p. 9.
27. Imperial College London (2007).
28. C-Tech Innovation Ltd (2004).
29. For firms with sales over £500m, branded averaged 13.6% compared to 4.4% for own-label: Evolve (2007), p. 12.
30. See e.g. Imperial College London (2007), pp. 13–14.
31. According to Mark Bartlett in August 2008, the Ginsters factory had an average daily production of around 115 t of final product (= 805 t/wk). The neighbouring Tamar factory produced 250 t/wk final product. Between both factories they sent around 72.5 t/wk food waste to the Holsworthy AD plant, 40% from the Ginster's factory and 60% from Tamar, so approximately 29 t/wk from Ginsters and 43.5 t/wk from Tamar. According to these figures, the mass of Ginsters' food waste represents about 3.6% of the mass of its final products. The mass of Tamar food waste represents about 17.4% of the mass of its final products. In an updated email from Mark Bartlett in March 2009 the figures differ significantly: the Ginsters factory had an average daily production of around 75 t of final product or 525 t/wk. The neighbouring Tamar factory produced 54 t/day or 376 t/wk final product. Between both factories they send 74 t/wk food waste to the Holsworthy AD plant, 30 t from the Ginsters factory and 30 t from Tamar, with the remaining 14 t consisting of organic material sieved out of the liquid effluent from both factories (= approx. 37 t each per week). According to these figures, the mass of Ginsters' food waste represents about 7% of the mass of its final products. The mass of Tamar food waste represents about 10% of the mass of its final

products. Site visit and interview with Mark Bartlett, environmental manager, Ginsters, 27 August 2008, and personal communications 26 and 27 March 2009. Manufacturer-owned brands tend to have longer shelf-life goods, whereas supermarket own-label products tend to be the more perishable items, but this explains only part of the discrepancy in levels of waste.

32. Evolve (2007), p. 12.
33. Interview, Chris Haskins, 25 July 2008.
34. Competition Commission (2008), 9.43, p. 165. See also Imperial College London (2007), p. 11, which points out that Tesco makes electronic point of sale (EPOS) data available to its suppliers, and Asda used software called Retail Link and CPFR (Collaborative Planning, Forecasting and Replenishment) to help suppliers predict demand, but that it is beyond the capacity of most small companies to process and analyse such data into useful forecasts. They also concluded that the data provided by retailers is often wholly inaccurate, sometimes by a surprising infinity per cent (p. 15).
35. Ogden (2005) found that fewer than one in four of food manufacturers responding to the survey had a good awareness of external support organizations; cf. Food Processing Faraday Partnership (2008), p. 18.
36. Interview, Shahin Rahimifard, 18 July 2008.
37. Henningsson *et al.* (2004); cf. Henningsson *et al.* (2001).
38. Defra (2007a), p. 14. The results are from the Envirowise Supply Chain Partnership Forum, www.envirowise.gov.uk/retail. Another project saved 80 companies £800,000: Food Processing Faraday Partnership (2008), p. 9.
39. PICME (2006), p. 10.
40. Evolve (2007), pp. 11–12.
41. Imperial College London (2007), pp. 9–10.
42. Interview, Shahin Rahimifard, 18 July 2008.
43. Andrew Parry and Mark Barthel, personal communication.
44. Defra (2007a), pp. 16, 25.
45. Competition Commission (2008), 9.47, p. 166; 9.84, p. 173. Of

the 380 concerns raised by suppliers with the Competition Commission, nearly half related to the transfer of excessive risks or unexpected costs from grocery retailers to suppliers, and one third related to requirements for retrospective payments or other adjustments to previously agreed supply arrangements. However, the Competition Commission were unable to raise these individually with the supermarket chains because the suppliers were too scared that it would damage their relations with the supermarkets. The Competition Commission concluded that these practices transferred 'excessive risks' onto the supplier, but the government has still not firmly requested that the practice stop (9.63, p. 169). See also Remedies 11.305–307, p. 236. Cf. Defra (forthcoming), project number FO0210 being conducted by the School of Management, Cranfield University; Cranfield University (2008).

46. SCOP was for any grocery with more than 8% market share, and the big four supermarkets became signatories to it.

47. Friends of the Earth (2003), which applied to farmer-suppliers, but the Competition Commission report found the same phenomenon among manufacturers; cf. Tescopoly (n.d.).

48. Competition Commission (2008), 9.47, p. 166; cf. 9.45, 9.46, p. 165 and n.

49. Interview, Rory Taylor, Competition Commission, 9 January 2009; and email 27 January 2009.

50. Competition Commission (2008), pp. 166, 173, 236; cf. Tescopoly (2008).

51. Interview, Rory Taylor, Competition Commission, 9 January 2009; Thompson (2009). Friends of the Earth argued that 'some practices are unfair whether agreed in advance or in retrospect including the supplier contributing to [cover the cost of] wastage which is clearly the result of an error made by the retailer and not due to any error on the part of the supplier.' Friends of the Earth (2008), p. 4.

52. Friends of the Earth (2003).

4. Selling the Sell–By Mythology

1. Interview, Chris Haskins, 25 July 2008.
2. The fact that food is perishable does not necessarily mean that it is hazardous to human health but the current EU food-labelling regulations do not make this distinction. At present the UK Food Standards Agency (FSA) is negotiating amendments to the proposed EU Food Information Regulations that would mean that only foods that are a potential danger to human health would require a *use-by* date. The original 1979 EC Directive 79/112, implemented in the UK in the early '80s, required a *use-by* or *best-before*, as does the new Food Labelling Directive (2000/13/ EEC) (see Article 10). In Great Britain, this is implemented under the 1996 UK Food Labelling Regulations (SI 1996/1499).
3. Safe thresholds for bacteria before they cause poisoning, established from studies involving human volunteers, are set by the European Food Safety Agency.
4. Stipulated in the EC's Poultry Meat Marketing Standards Regulations.
5. Stipulated in the EC's Egg Marketing Standards Regulations.
6. FSA (UK) (2003), my italics.
7. Interview, David Lion, 27 August 2008; interview, Chris Haskins, numerous occasions.
8. Defra (2007a), p. 16; a 2008 BBC poll suggested that 22% of the 1,530 consumers surveyed always threw their food away once the sell-by date had expired. Fruit, vegetables, bread and cheese were the items most likely to be eaten after their sell-by dates. Respondents were more cautious about eating out-of-date meat, poultry and fish – although a small percentage still did: PA News (2008). WRAP found that 'I in 5 say they won't take a chance with food close to its "best-before" date, even if it looks fine': WRAP (2007), p. 9. An FSA survey in 2007 found that just 55% understood the use-by date correctly while 32% incorrectly thought that the food would be past its best but not necessarily unsafe. 51% understood the best-before date correctly, but 36% incorrectly

believed that the food should not be eaten past the date: FSA (UK) (2008a).

9. The EC IMPRO project suggested that one of the most effective ways of reducing environmental damage across the EU would be to provide 'more appropriate consumer information . . . to prevent food being discarded because of misconceptions about freshness, colour, texture, and food safety issues' and to review 'too tight requirements on "preferably consumed before" dates, perverse measuring standards, and demands for what may be labelled "fresh" ': Weidema *et al.* (2008), pp. 9, 66.

10. Garnett (2008), p. 44.

11. WRAP (2008a), p. 4.

12. Mark Barthel, personal communication, 20 March 2009.

13. The advertising campaign information came from Dr David Jukes, senior lecturer in food regulation, Dept of Food Biosciences, University of Reading.

14. USDA (2007). The Food and Drug Administration Food Code is a food safety guideline for retail food operations and institutions. In a guide to the 2005 update to the Code (Food and Drug Administration (US) (2005)), the FDA gave guidance on date marking on which many states have based their state level regulation. Numerous websites, including the USDA's and that of the Food Marketing Institute, publish charts with recommended storage periods for different foods, and the USDA runs a Meat and Poultry hotline which handles calls from the public.

15. Severson (2001).

16. Senate Committee On Agriculture And Water Resources (1999).

17. WRAP says that this may have changed recently following an FSA consultation on how to include sustainability in their policy-making. WRAP also works closely with Defra and the FSA on the issue of food waste (Mark Barthel and Andrew Parry, personal communication).

5. Watching Your Wasteline

1. Opie and Opie (1951).
2. Avoidable food waste only. Adults living alone waste 36% more food than unrelated adults living in shared households: WRAP (2008a), p. 174.
3. Winnett (2008).
4. Poulter (2008).
5. Interview, Tim Lang, 22 July 2008.
6. Imperial War Museum, 'Don't Waste Bread', (1914–1918) IWM PST 6545. Cf. Osborne (1916), p. 5, col. B.
7. WRAP (2008a); this percentage was two and a half times greater than allowed for in MAFF (UK) (1982).
8. Engström and Carlsson-Kanyama (2004); cf. Getlinger *et al.* (1996); Barton *et al.* (2000); Shanklin and Ferris (1995); Dilly and Shanklin (1998).
9. Tim Hermann, Umweltbundesamt, email, 4 September 2008: in 2006, 3.8m t collected from domestic 'bio-bins' from half the population = 100 kg/participant/yr; but Hermann thought that just 15% of it was 'kitchen waste' = 15 kg/person. One website reports that 80% of German households are provided with bio-bins, with about 60% participation rate (http://www.compostnetwork.info/ad-workshop/presentations/05_kehrespdf) = 76 kg/person/yr. A further 48 kg per capita biodegradable garden and park wastes were collected in 2005: Statistisches Bundesamt Deutschland (2005).
10. Hogg *et al.* (2007a), p. 85. In 2006, Swedish Waste Management, Avfall Sverige, collected food waste totalling 135,000 t ÷ population of 9,113,257 = 14.8 kg/person/yr with a further category of 52 kg of 'biologically treated waste', but it is not clear what proportion of food waste is actually accounted for: Avfall Sverige (2007). In the Netherlands in 2007, 80 kg of organic waste was collected per capita: Netherlands Statline database (2008). If the proportion of this that is food is similar to that of the similar Italian waste collection scheme (61.4%), then this would equate to food waste of around 49 kg/person/yr.

11. The Australia Institute (2005), p. 12 and n. Based on a survey of 1,600 households in 2004, Australians admitted to wasting money on food at a rate more than three times greater than any other source of wasted goods and services, totalling AU$5.3bn/yr, though research by the Australian Food and Grocery Council (2003) showed that the total was more like AU$7.8bn.

12. Lundqvist, Fraiture and Molden (2008); 'Families with small children throw away about 25% of the food they have bought and carried home and that total losses and wastage in the food chain are close to 50%', KSLA (2007); Ennart (2007). Other studies suggest lower wastage: Naturvårdsverket (2007); cf. Weidema *et al.* (2008), p. 127; Sonesson *et al.* (2005). Groth and Fagt (1997) found discrepancies of 23% between supply and consumption of meat and cheese in Denmark. Torr (2008) reports 25% consumer food waste in Bahrain.

13. Timothy Jones, personal communication, and Jones (2004a).

14. Jones (2004a), Jones (2004b), Jones (2006) and personal communication: food waste of 467 lb/household/yr; average 2.2 inhabitants/household (less than national average of 2.6) = 96 kg/person/yr. UK equivalent of 'avoidable' food waste is 70 kg/person/yr.

15. WRAP (2008a), pp. 12–17, 211ff.

16. Cf. Agutter (1796), p. 7: 'The rites of hospitality and society will always require a due attention to our friends; yet in times of scarcity there can be no excuse for providing a needless abundance.'

17. Le Bailly (2007); Kitcho (2003); also the WRAP website www.lovefoodhatewaste.com, leftoverchef.com and kitchenscraps.com.

18. UK families with children waste 56% more food than all-adult families (avoidable food waste only, by weight), but *per capita* they actually waste less than average households: WRAP (2008a), pp. 32, 173–4.

19. Interview, Timothy Jones, October 2008 (estimate raised since rise in food prices; cf. Jones (2006)).

20. Councils in the UK pay for waste disposal partly through council tax.

21. As Tara Garnett has argued thoughtfully in Garnett (2008), p. 47, how people decide to spend this saved money affects the overall environmental benefit of reducing food waste. One might decide to use the extra money to buy organic food instead or invest in loft insulation, or perhaps use it to fly somewhere on holiday. Equally, one would need to account for how the land no longer used to grow the food would be used – whether building a golf course, or planting trees. In the end though, increased consumption of resource-intensive products (including food) is driving environmental degradation. It follows that reducing this will have the opposite effect, though this may be difficult to measure.

22. £10.2bn ÷ 60.9m people = £167.48/person = Pakistani Rs 19,261 (June 2008 exchange rate) = enough to buy 963 kg flour (@ Rs 20/kg, though price had previously been Rs 13/kg) = 3.4m kcal = food for 4.7 people for a year (@ 2,000 kcal/person/day) × 60.9m (UK population) = enough food for 284m people, 156% of Pakistan's population.

23. Hiza and Bente (2007); Blair and Sobal (2006); Pollan (2008), p. 74.

24. FAO (2003a); Smil (2001), pp. 250–51; Smil (2004).

25. Goklany (1998).

26. Cabinet Office (UK) (2008), p. vi.

27. US Dept of Agriculture (USDA) Economic Research Service (ERS) food expenditure tables for 2007. US Bureau Labor Statistics (BLS), 'Consumer expenditure survey 2004–2006', found that food bought to eat at home and in restaurants amounted to 13% of consumer expenditure.

28. Ivanic and Martin (2008), p. 1.

29. Brown University Faculty (1990).

30. The concept and method of calculating 'depth of hunger' are developed in FAO (2000). Updated figures on depth of hunger in the developing world for 2007 were provided in personal emails by Ricardo Sibrián, senior statistician, Statistics Division, FAO, Rome. My thanks to him for his generous assistance. Estimates of the number of hungry were increased by a further 40m in FAO

(2008b), but all calculations on depth of hunger here are based on the 2007 figures.

31. Braun (2007); World Bank (2009); Mitchell (2008); FAO (2008b), pp. 9–10; BBC (2008); WFP (2009).

32. Mitchell (2008); Reuters (2008).

33. FAO (2008a).

34. FAO (2008b), pp. 4–9: increase in malnourished people of at least 75m in 2007 to reach 923m, and in 2008 the figure rose again by an estimated 40m. This, according to FAO, is 'mainly as a result of high food prices' owing to the fact that 'growth in demand for food commodities continues to outstrip growth in their supply.' 100m into deeper poverty estimated by World Bank (2008a); on child mortality see World Bank (2008b). USDA, using a different hunger threshold of 2,100 kcal/person/day (in contrast to FAO's variable threshold of 1,600–2,000 kcal depending on age and gender distribution), estimated an increase of 133m people hungry in 70 countries analysed: Rosen *et al.* (2008). UNEP (2009) estimated additional 110m in poverty and 44m undernourished.

35. Dawe (2008); FAO (2008b), p. 10.

36. Avoidable and possibly avoidable waste of bakery and grain products in UK households. According to FAO, to lift one of the world's malnourished people out of hunger requires on average 250 kcal/day: see below for FAO's calculation of this depth of hunger and my use of it in this book.

37. Kantor *et al.* (1997) estimated losses of grain products to be 14,594m lb (6.6m t) = 17.7 trillion kcal, enough for a daily supplement of 250 kcal/day for 194m people.

38. See appendix.

39. The calorific content of the waste recorded in Kantor *et al.* (1997) would have been enough for a 250 kcal/day supplement for 1bn people. Dr Jones estimates that true waste levels are more than 20% higher than this. When indirect food waste through grains fed to livestock to produce meat and dairy products is included, the calorific value comes to 179 trillion kcal/yr, or enough to supplement the deficient diets of 2bn people. See appendix. In

fact, according to figures calculated by the USDA's ERS in 2004 (see ERS (2008b)), the food wasted by retailers, consumers and food services (not including grain used to produce wasted meat and dairy) is even higher than estimated here, amounting to enough for 1.4bn people (3,900 kcal/person/day (food supplies) – 2,717 kcal/person/day (after waste is allowed for) × 300m people × 365 days = 130 trillion kcal = enough for 1.4bn people to be given an extra 250 kcal/day).

40. There has been insufficient research in Europe to allow for anything more than a very approximate estimate of how much food is being wasted. WRAP estimates 11.3–13.3m t of food wasted by retailers, caterers, manufacturers, farmers and others in the industry in the UK (see ch. 12). Multiply this quantity according to the population of Europe (700m) relative to that of the UK (60.9m). Food wasted by an average UK householder (avoidable and possibly avoidable) contains on average enough calories to have lifted 15.5 malnourished people out of hunger for one year (not including indirect losses through grains used to produce wasted livestock products). If we were to say that commercial and industrial food waste may have just *half* the nutritional content of food wasted in households, total food wasted in the UK would be enough for 178.6m people, or much more if indirect losses in wasted meat and dairy were included. Supposing the whole of Europe were throwing away food as enthusiastically, it would be enough for 2.1bn people. If European households, retailers and food services were discarding as much food as in the US, and if indirect food waste through using grains to produce wasted meat and dairy products were included, the total would be more like 5bn people; added to the US total = 7bn.

41. Australia Institute (2005), p. 13. The survey covered the waste of several kinds of purchase, but most of the items considered were food products.

6. Losing Ground: Some Environmental Impacts of Waste

1. Tryon (1700), p. 280; cf. Stuart (2006a).
2. Bender (1994); Smil (2001); Kantor *et al.* (1997); Engström and Carlsson-Kanyama (2004).
3. Phillips (2008c).
4. Rice (2008).
5. Borger (2008b).
6. FAO (2008b), p. 2.
7. Farm input data from Williams *et al.* (2006). 1 gigajoule (GJ) = 239,000 kcal. FAO estimates that on average in 2007, according to the edible portion of food commodities (i.e. after husks have been removed), 1 t of cereals equates to 3.4m kcal.
8. 61,300 t tomatoes wasted in UK households, using 8m GJ primary energy input = enough to fuel the production of 3.2m t of wheat, at 3m kcal/t, which would provide 105m of the world's under-nourished with the 250 kcal they require (according to FAO) to lift them out of undernourishment. In addition to household waste, one study showed that tomatoes imported from Spain were being wasted in the supply chain at an average of 7.7% (adding 14,700 extra tonnes from just that single supply chain): Imperial College London (2007), p. 22.
9. Figures on land, energy and water requirements for respective crops taken from Williams *et al.* (2006).
10. Lundqvist, Fraiture and Molden (2008), p. 30, made a similar calculation for water losses represented in the waste of food on US irrigated farms; he wrote 30% waste, but his calculation is actually for 33.3%.
11. Heller and Keoleian (2003).
12. WRAP used the Swedish assessment, Uhlin (1997); see also Garnett (2008), pp. 20–21; Garnett estimates that food is responsible for 19% of UK emissions, not including emissions from land use changes. Garnett estimates that around half of this is from agriculture. There is no definitive study on total emissions from food consumption. IPCC *Climate Change 2007* report estimates

that agriculture is responsible for 10–12% total emissions, but this does not take into account emissions from land use changes.

13. Based on an analysis of data from Defra's Family Food & Expenditure Survey.

14. Based on Manchester Business School (2006).

15. Garnett (2008), pp. 3, 36.

16. EC (2006).

17. It could be argued that reducing food waste would not significantly impact upon, say, the number of shopping trips made by car, the amount of energy expended washing up or even cooking. But there are scenarios where avoiding food waste could indeed do this. For example, deciding to eat a meal of leftovers, instead of throwing them away and cooking a new meal, eliminates an entire cooking and washing scenario, and could delay the next shopping trip by one day. Until studies have examined these scenarios, it is impossible to say what proportion of environmental impacts can be improved through food waste reduction.

18. Bellarby *et al.* (2008); Garnett (2008), pp. 20–21. Duxbury (1994); Horrigan *et al.* (2002) found that in 2002, the food production system accounted for 17% of all fossil fuel use in the US.

19. Steinfeld *et al.* (2006); cf. Eshel and Martin (2006).

20. FAO (2003a), global DEC 2,332 kcal/person/day from vegetable foods and 477 kcal/person/day from animal products (including fish).

21. Not including land use changes (EC (2006)): the literature review found that animal-based food was responsible for 19–38% of emissions in the food category.

22. de Haan *et al.* (2001), p. 18; Faminov and Vosti (1998).

23. Chomitz (2006); Forster *et al.* (2007), p. 139.

24. MCT (2004); Greenpeace International (2006), p. 5; Bugge (2004); Osava (2004).

25. Cowling *et al.* (2004); Cox *et al.* (2006); Monbiot (2007); Hopkin (2007).

26. Jones *et al.* (2009); Adam (2009).

27. Monbiot (2007), pp. 9–10, 15; UNEP (2009).

28. Kaimowitz *et al.* (n.d.), p. 3.
29. Gelder *et al.* (2008).
30. *Economist* (2008b), p. 7; Steinfeld *et al.* (2006), pp. 65–6; Garnett (2008), p. 21; Greenpeace International (2006); Gold (2004); Vidal (2006); Barrionuevo (2008); Phillips (2008a); Phillips (2008b); Lawrence (2009); USDA (2004).
31. Dalgaard *et al.* (2007). See below for Dalgaard's discussion of this in Dalgaard *et al.* (2008); cf. Garnett (2007b).
32. Garnett (2008), p. 3.
33. This was changes to grasslands, not forests: Sevenster and Jong (2008), p. 1.
34. Dalgaard *et al.* (2008), and see further calculations below.
35. Millennium Ecosystem Assessment (2005) lists land use change as the leading cause of biodiversity loss.
36. Tekelenburg *et al.* (2003); Thomas *et al.* (2004); Nijdam *et al.* (2005), p. 153.
37. UNEP (2009), p. 7.
38. For a quantified estimate of the indirect impacts of demand for agricultural land, see RFA (2008).
39. Land use requirements for commodities taken from Williams *et al.* (2006); waste data from WRAP (2008a) and Kantor *et al.* (1997). See appendix.

PART II: Squandered Harvests

7. Farming: Potatoes Have Eyes

1. *King James Version.*
2. Pe'ah (Jerusalem Talmud) 2:5; Hullin, 134b; Gittin, 59b; Maimonides, Mishneh Torah, 1:14; Wikipedia, 'tzedakah'.
3. Pailliet (1837), Code Pénal, Book IV, II.1470n.
4. ibid., Book IV, II.1470; [Bonaparte, ed.] (1810), Livre IV, Ch. II, Section 1, Article 471, 10°; Varda (2000).
5. The Code Pénal remained in place until it was replaced by the Nouveau Code Pénal on 1 March 1994. The new penal code can

be read at http://www.legifrance.gouv.fr; cf. Dalloz (ed.) (2001),
p. 2382, Art. R. 26.

6. Ministère de la Justice, Luxembourg (2001), Titre X, Art. 553, 2;
(Algeria) (n.d.), Livre IV, Titre I, Chapitre I, Section 5, Art.
464, 2°; Système d'informations juridiques, institutionnelles et
politiques (Côte-d'Ivoire) (1969), Ch. 1.11.

7. Another restriction was that farmers did not usually allow gleaning
in fields of barley or beans, which were considered animal feed
and thus could encourage the poor to keep pigs and chickens that
they could not otherwise maintain without stealing grain. Gleaners
were expected to collect wheat for their own consumption.

8. Morgan and Nuti (1982), pp. 155–9; Burn *et al.* (1831), II.1063.

9. Polian (2003), p. 87; Gardner (2006); cf. Wikipedia, 'Gleaning'.

10. Leake (2005); Friends of the Earth (2002); Friends of the Earth
(2005). Henningsson *et al.* (2004) suggested that 40–50% of raw
vegetables and salad are discarded at the stage of processing.

11. Visit to M. H. Poskitt Carrots, 19 December 2007.

12. M. H. Poskitt Ltd QC Sheet, 10 December 2008, T. Hey First
Load. Comparable figures were obtained by a study from the
Institute of Grocery Distribution, where 40% of an organic
grower's carrots were rejected by the supermarket: 19% for being
the wrong size or shape; 5% damaged in harvesting. These levels
were estimated to be similar for conventional growers. Garnett
(2006), p. 63.

13. Michael Mann, EC Spokesman for Agriculture and Rural Development, personal communication, 11 February 2009.

14. This was one of the complaints made by Farmers Link to the
Competition Commission (2008), p. 22; cf. also Friends of the
Earth (2005), which found that 57% of apple growers surveyed
said that they have to apply more pesticides to meet the cosmetic
standards of the supermarkets.

15. Defra carrot specification as supplied to Guy Poskitt, application
of EC No. 730/1999.

16. EC No. 85/2004.

17. Brown (2008).

18. ibid.
19. Apricots, artichokes, asparagus, aubergines, avocados, beans, Brussels sprouts, carrots, cauliflowers, cherries, courgettes, cucumbers, cultivated mushrooms, garlic, hazelnuts in shell, headed cabbage, leeks, melons, onions, peas, plums, ribbed celery, spinach, walnuts in shell, water melons and witloof/chicory.
20. EC (2008c).
21. Michael Mann, personal communication, 13 February 2009; Copa and Cogeca (2008).
22. Graeme Skinner, St Nicholas Court Farm, 23 September 2008.
23. Sainsbury plc (2008b).
24. Michael Mann, personal communication, 13 February 2009. This view was shared by several other informants, including Philip Hudson, chief horticultural advisor, NFU, and Nick Twell, ex-member of the British Potato Council Committee. There are a few exceptions; Waitrose, for example, claimed they were debarred from selling onions they wished to stock.
25. Interview with Chris Brown, Asda, 25 January 2008.
26. Interview with Simon Fisher, NFU, 17 December 2007; interview with Robert Baird, Greenvale AP, 12 February 2008.
27. Interview with Nick Twell, 7 October 2008. C-Tech Innovation Ltd (2004) found that around 624,000 t or 12% of the mass of fruit and vegetables entering British food and drink manufacturers end up as waste. 14% vegetables, 44% of fruit and 40% of potatoes in the UK are processed. British consumers waste 40% of the edible mass of vegetables they purchase; half of that consists of whole vegetables and other avoidable wastage, and the other half consists only of those peelings and so forth which are nutritious and could potentially have been eaten (i.e. not including inedible peels like broad bean husks).
28. WRAP (2008a), p. 4; an article in the *Observer* made a similar calculation on the basis that 60% of salad was thrown away by consumers, but it is 60% by cost, and 40% by weight: Renton (2009).
29. Doug Warner, personal communication, 3 March 2009; Garnett (2006), p. 63.

30. Friends of the Earth (2002).

31. Friends of the Earth (2002).

32. Timothy Jones, personal communications, 2008.

33. Food Chain Centre (2006).

34. Garnett (2006), pp. 108–9.

35. Food Chain Centre (2006).

36. Interview, Richard Hirst, Hirst Farms grower and chairman of the Horticulture Board, NFU, 4 August 2008.

37. Interviews, Simon Fisher, NFU, 17 December 2007; Robert Baird, 12 February 2008.

38. Interview Robert Baird, 12 February 2008; also confirmed by Phil Bradshaw, supply chain manager at the British Potato Council, August 2008; Graeme Skinner and Nick Twell, BPC.

39. Interviews, Robert Baird, 12 February 2008; Graeme Skinner, 23 September 2008; Chris Brown, Asda, 25 January 2008.

40. A month after I went to see Robert Baird, key figures at Greenvale AP were arrested for allegedly paying bribes to a Sainsbury's procurement manager to ensure that the supermarket continued to buy Greenvale produce. In the wake of the scandal, Sainsbury's halted deliveries from Greenvale and thus had to source its spuds from elsewhere. It turned to Chris Haskins, retired chairman of Northern Foods, who is also a potato grower. It now wanted to buy from him the very potatoes which it had formerly rejected for being sub-standard. Once again, reject spuds became perfectly acceptable.

41. Potato waste in 2005 was 1.1m t; in 2006, 0.8m t. Between 1988 and 2006 wastage averaged 0.9m t/yr: Potato Council (Britain) (2008).

42. Farmers invest £60–70 in seed, labour and machinery for each tonne they produce (interview, Phil Bradshaw, August 2008).

43. Avoidable waste figures from WRAP (2008a), pp. 63–4, 126; 'possibly avoidable' waste figures from Andrew Parry, personal communication.

44. Interview, Phil Bradshaw, August 2008.

45. Interview, Phil Britton, 29 January 2008.

46. Robert Baird, Greenvale AP, reported 35% pick-offs, 12% of which goes to stock feed in a normal year, and much more in bad years like 2007 (interview, Robert Baird, 12 February 2008). However, Paul Leiss, working in quality control at Greenvale said the packing and grading house receives 20–25 loads a day, each weighing 26–27 t, which adds up to 4,174 tonnes a week. They sell on only 2,000 t of packed potatoes to the retailers each week, which would imply that the pick-off rate may have been as high as 50%. A different supplier reported that 15–20% of potatoes would fail to make it to market (Richard Hirst, 4 August 2008).

47. Interview, Graeme Skinner, 23 September 2008.

48. ibid.

49. Turff (2008), pp. 10–11; interview, Nick Twell, 7 October 2008.

50. Interview, Philip Hudson, 30 July 2008.

51. Interview, Nick Twell, 7 October 2008.

52. Interview and farm visit, Richard Hirst, 4 August 2008; personal communication, 27 March 2009.

53. Zhang *et al.* (2005).

54. Locke (1690).

55. Michael Mann, spokesman for Agriculture and Rural Development, personal communication, 25 February 2009. Intervention (i.e. where the EU buys products at a guaranteed price) still exists for sugar, wheat (max. 3m t/yr), butter (max. 30,000 t/yr) and skimmed milk powder (max. 109,000 t/yr). Beyond these limits, it is theoretically possible to buy more but the price would be set by tender.

56. Michael Mann, 'Exports Refunds Expenditure – Financial Year 1998 to 2008', personal email, 25 February 2009.

8. Fish: The Scale of Waste

1. For the confusion between the terms 'discards' and 'by-catch' see Kelleher (2005), 2.2.1, and FAO (2009), p. 74. A recent European Commission (EC) paper defines discards as commercial species retained by a fishing gear that have been brought on board a

fishing vessel and are thrown back into the sea, effectively ignoring non-commercial species.

2. UNEP (2009), p. 28; Bettoli and Scholten (2006).
3. 'The Northeast Atlantic and Northwest Pacific jointly account for 40% of estimated discards, attributable to high discards in many EU fisheries and in some Japanese fisheries.' Kelleher (2005), 3.1.1. COM (2007b); EC STECF (2006). This is not an entirely new phenomenon; in 1872 one observer complained of the British fishing industry's 'wholesale destruction of codling': Anon. (1872).
4. Siddique (2007).
5. Kelleher (2005), 3.2.1.
6. WWF (Germany) (2008a).
7. EC (2007a).
8. WWF (Germany) (2008a).
9. EC (2007b); EC STECF (2006), p. 15.
10. OSPAR Commission (2008a), p. 3; OSPAR Commission (2008b), pp. 115–17; COM (2007b); cf. Press Association (2004).
11. COM (2007b), p. 3.
12. Vidal (2008).
13. Kelleher (2005), 3.1.2, 3.2.1; Natural Environment Research Council (2003), 6; COM (2007b); EC STECF (2006).
14. *Economist* (2009a), pp. 9, 11.
15. Fleming (2008).
16. See WWF's website http://panda.org/what_we_do/knowledge_centres/marine/our_solutions/sustainable_fishing/sustainable_seafood/seafood_guides/; Marine Stewardship Council, http://www.msc.org/where-to-buy; and Marine Conservation Society, http://www.fishonline.org/buying_eating/purchasing_guide.php.
17. The packet label does not specify the fish species, but in 2008 it was listed on the Tesco website as Alaskan pollock.
18. MSC (2007).
19. *Economist* (2009a), p. 13; James Simpson, MSC, personal communication, 5 January 2009. The North Pacific Fisheries Management Council, responsible for managing the fishery, is trying to reduce the by-catch rate, particularly with regard to Chinook

salmon, caught at higher rates during 2007, which caused concern over the potential impact on remote Alaskan native communities which rely heavily on the salmon.

20. Meikle (2008), p. 12.
21. Russ and Alcala (2004).
22. Roberts and Hawkins (2000); Sarsby (2001); Clover (2004), pp. 215–29.
23. There are a small number of no-fish blocks in EU waters.
24. Aldred and Benjamin (2007).
25. EC (2008d).
26. Scottish Government (2008).
27. Evironmental News Service (2008); Karoline Schacht, personal communication, 2 March 2009. The initiative referred to started in 2007 with a communication of the COM announcing a long-awaited 'Policy to reduce unwanted by-catches and eliminate discards in European fisheries', which was followed by several consultation processes and finally the NON-paper in April 2008, a necessary element within the EU legislative process. In the middle of the discussion and just a few weeks before the regulation was expected to be published, the COM decided to withdraw the initiative. The subsequent EU/Norway negotiations meant that some of the proposed measures have been adopted for the North Sea, and also the commitment to adopt by-catch measures in the future.
28. WWF (2008b); WWF (2008c); cf. Siddique (2007); Gianni (2002).
29. COM (2007b).
30. Alverson *et al.* (1994).
31. Kelleher (2005).
32. Paul (1994).
33. Gilman *et al.* (2007); Dobrzynski *et al.* (2002).
34. FAO (2009), p. 76. Ironically, this document argues that it is essential to define the meaning of 'by-catch', but it then shifts in one sentence, without making a definition of its own terms, from a statement on discards to one on by-catch (which can include retained non-target species, for example).

35. UNEP (2009), p. 23.
36. Clover (2004), p. 64.
37. Bettoli and Scholten (2006).
38. WWF (2008a).
39. *Economist* (2009a), pp. 15–17.
40. Clover (2004), p. 180; *Economist* (2009a).
41. Clover (2004), pp. 57, 73–4.
42. Valdemarsen *et al.* (2007).
43. Kelleher (2005), Table 3.
44. FAO (2009), p. 76; WWF (2006).
45. *Economist* (2009a), pp. 14–15.
46. Clover (2004), pp. 175, 180–81.
47. Bettoli and Scholten (2006).
48. Kelleher (2005), 3.2.1.
49. ibid., 3.1.1 and Annex A.
50. Ababouch (2003).
51. UNEP (2009), p. 30; the reference cited (DieiOuadi, 2007) is in fact a forthcoming work (June 2009) on spoilage but not discards (Yvette DieiOuadi, personal communication, 17 March 2009).
52. FAO (2009), pp. 43–5.
53. ibid., pp. 3, 43–5.
54. Jowit (2008); *Economist* (2009a), p. 11; cf. Blanco *et al.* (2007).

9. Meat: Offal isn't Awful

1. Keckchen (2004).
2. UNEP (2009), p. 15.
3. Steinfeld *et al.* (2006), p. 12; FAO (2008c).
4. UNEP (2009), p. 25.
5. Gold (2004), p. 22.
6. Steinfeld *et al.* (2006).
7. By calorific value: FAO (2003a), all animal feed (including oil-cakes): all animal food minus caught fish (68% of all fish (FAO 2009)), global conversion ratio from animal feed into animal food is 3.3:1.

8. Steinfeld *et al.* (2006); Worldwatch Institute (2006), pp. 25–6; Tudge (2004).

9. FAO (2006); UNEP (2009).

10. UNEP (2009), p. 27. This calculation should be treated with caution. UNEP estimated that by 2050 around 1.45bn t (which it unfortunately mistakenly wrote as 1.45m t) of cereals will be used as animal feed. It assumes that it takes 3 kg of cereal to produce 1 kg of meat and that the calorific value of cereals (3,000 kcal/kg) is around twice that of meat (1,500 kcal/kg), yielding a conversion ratio of 6:1. This may be representative of some industrial farming models, but not global agriculture as a whole. Many of the cereals fed to livestock are not used for meat production but for egg and dairy production which have higher conversion ratios, and average calorific value of meat calculated by taking global production of each major meat type (pork, poultry, lamb/goat and beef), multiplying the tonnage of each type of meat by its calorific value (FSA (2008b)) and dividing by total tonnage of those meats produced = 2,246 kcal/kg.

11. FAO (2003b).

12. Keyzer *et al.* (2005).

13. Lundqvist, Fraiture and Molden (2008), pp. 9–10; Peden *et al.* (2007), p. 488; Nierenberg (2005).

14. Falkenmark and Rockström (2004); Lundqvist (2008); Chapagain and Hoekstra (2008); Halweil and Nierenberg (2008), pp. 61–74; Steinfeld *et al.* (2006).

15. Lakshmi (2008).

16. Peden *et al.* (2007), p. 495, predicts growth rate in milk consumption at 3.8%/yr in China, leading to consumption of 16 kg/person/yr by 2020, though FAO (2003a) already has Chinese consumption at 16.6 kg/person/yr. An annual 3.8% increase to 2020 would lead to a consumption level of 31 kg/person/yr; cf. Fraiture *et al.* (2007); Lundqvist, Fraiture and Molden (2008), p. 9.

17. Includes the calorific value of the dairy cow's carcase at the end its life: Harris (1986), pp. 51–3.

18. Peden *et al.* (2007), pp. 485–514; Lundqvist, Fraiture and Molden (2008), pp. 9–10.

19. By calorific value (FAO (2003a)): total animal feed inputs including oilcakes (net of exports and imports): total output of animal food minus caught fish.

20. Waste of red meat at consumer level estimated at 30.3% and 7% at retail level: ERS (2008b).

21. By calorific value: US diet surveys suggest that people's average intake = 2,000–2,600 kcal/person/day (see below). FAO (2003a): US livestock are given 4,369 kcal/person/day (incl. oilcakes), and if imported meat is included with a feed conversion ratio US/EU average, indirect feed inputs are around 5,787 kcal/person/day.

22. Sahib Haq, WFP, noted that this was a significant factor in the food crisis in Pakistan.

23. This did not even include damage due to deforestation and other land use changes abroad. Among the most effective ways of reducing damage to the environment, it concluded, is for consumers of meat and dairy 'to reduce food losses (wastage)': Weidema *et al.* (2008), pp. 3, 5, 6, 9, 63–5, 86, 97, 103–4, 128; cf. also Carlsson-Kanyama (1998).

24. Kantor *et al.* (1997).

25. WRAP (2008b); East Malling Research (EMR) is leading a new project supported by WRAP that aims to help consumers reduce their fresh fruit and vegetable food waste. On the potentially negative impact of increasing the use of refrigerators, see Garnett (2008), pp. 44–5; Garnett (2007a).

26. Engström and Carlsson-Kanyama (2004).

27. FAO (2003a).

28. Spencer (2008).

29. It has been estimated that humans directly consume only 68% of a chicken, 62% of a pig, 54% of a bovine animal and 52% of a sheep or goat: Nordberg and Edström (2003); C-Tech Innovation Ltd (2004), p. 9. Cf. Competition Commission (1985).

30. Arvanitoyannis and Ladas (2008); cf. Singer (1979).

10. Moth and Mould: Waste in a Land of Hunger

1. Interview, 11 May 2008.
2. According to the WFP survey to the end of March 2008, numbers rose 28% from 60m to 77m compared to the previous year.
3. IRIN (2008a).
4. Global grain harvest in 2007 was 2.1bn t, more than any previous year and more than 2006 by nearly 5%: FAO (2008c).
5. Abder-Rahman *et al.* (2000); cf. FAO (1981), Recommendation 11: 'Every country should be cautioned against the use of hazardous protective agricultural chemicals until their effects are fully understood and farmers are educated in their use'; Baloch *et al.* (1994), §4.2.
6. A study conducted in 1984–5 found that on average 5.6% of Pakistan's grain was lost after spending less than half a year in storage, and in some cases, over longer periods, losses in storage alone reached 15%: Baloch *et al.* (1994).
7. It may not be true that a mini-silo would be the most cost-effective solution for Hayat, but there certainly would be improvements he could make. In different circumstances, a farmer's decision to sell harvest immediately can be a 'wise' decision based on cash-flow versus investment. On the ongoing debate between the different scenarios – whether farmers are 'forced to sell', or whether they sell because they regard storage as too costly, risky or unprofitable, see e.g. Proctor (ed.) (1994), ch. 1. There also needs to be a balance struck, so that strategic storage is satisfied without excessive 'hoarding'.
8. Baloch *et al.* (1994).
9. ibid.
10. Interview, Zahir Shah, Shangla Development Society, Besham, Kohistan District, NWFP, Pakistan, 6 August 2008.
11. cf. Proctor (ed.) (1994), Annex 1.
12. *The Nation* (2008).
13. Interview, Sahib Haq at WFP, Islamabad, 9 May 2008.
14. According to Khan (2008) wheat and rice were wasted at levels

from 15% to over 17% respectively, including that lost at the consumer level.

15. ibid.

16. Interview, Sahib Haq, WFP, Islamabad, 9 May 2008; citing from FAO (2003a).

17. Interview, Sahib Haq, 9 May 2008.

18. In fact some Pakistani companies do export orange peels, e.g. Sh. Mohammad Ibrahim & Co. (SMICO); cf. Ajila *et al.* (2007).

19. United Nations University Press (1979); cf. Grolleaud (2001), ch. 1.

20. Smil (2004).

21. FAO (1981).

22. The resolution was passed in the 7th Special Session of the UN General Assembly in 1975.

23. OECD (2008); DFID (2004); UNEP (2009), p. 12.

24. Kader and Rolle (2004), p. 2.

25. FAO (1981); Brown University Faculty (1990); Kader (2005); Lundqvist, Fraiture and Molden (2008), p. 31.

26. Smil (2004).

27. Society for Bettering the Condition and Increasing the Comforts of the Poor (1802), III.12.

28. Smil (2004); Liang *et al.* (1993).

29. World Resources Institute (1998).

30. Calverley (1996); Grolleaud (2001), §3.2.1.

31. Smil (2004).

32. Ricardo Sibrián, senior statistician at the UN FAO, reported that 24.3m t of grain (plus an extra 10% to allow for losses) would satisfy the depth of hunger for all 907m of the developing world's undernourished people in 2007 (personal communication).

33. Bender (1994).

34. Proctor (ed.) (1994); Smil (2001), p. 187.

35. Armitage (2004).

36. Fehr *et al.* (2000).

37. Institute of Post Harvest Technology (Sri Lanka) (2002), p. 5; cf. Senewiratne (2006).

38. Cf. FAO (1981), Recommendation 5.
39. Institute of Post Harvest Technology (Sri Lanka) (2002); Sene-
    wiratne (2006).
40. Khushk and Memon (2006).
41. Daniyal Mueenuddin, personal communication, 1 March 2009.
42. CII-McKinsey & Co. (1997), cited in World Bank (2007).
43. *Economic Times* (2008).
44. *Economic Times* (2008).
45. World Bank (2007).
46. Rabo India Finance (2005), p. 32. This costing was made before
    the recent food price rises: cf. Lok Sabha (2005), pp. 21, 47 (Rs
    515bn); Lok Sabha (2007), §1.7, 3.2; cf. Rediff (2007).
47. Expensive failures have left what one UN report refers to as
    'rusting monuments to inappropriate development assistance'
    (Proctor (ed.) (1994)); cf. Friendship and Compton (1991). The
    Pakistan Agricultural Research Council, for example, found that
    the storage loss reductions promised by a number of proposals to
    replace bag with bulk handling were grossly exaggerated, and that
    it would be more cost-effective to improve on existing, traditional
    methods of handling and storage: Coulter (1991); cited in Proctor
    (ed.) (1994); Baloch *et al.* (1994).
48. Coulter (n.d.), pp. 19–20.
49. World Resources Institute (1998); Grolleaud (2001); cf. Proctor
    (ed.) (1994), Annex 1, on cost-effectiveness.
50. Bokanga (1996).
51. World Resources Institute (1998); IRRI (2007); Bas (2008).
52. United Nations Trust Fund for Human Security (2007). In
    addition to waste of food itself, there are enormous inefficiencies
    in farm inputs in the developing world. Rivers are drained and
    water tables depleted to irrigate crops, only to allow up to 60%
    of the water to go to waste because of sub-optimal irrigation
    systems (interview, Sahib Haq, Islamabad, 9 May 2008). Vaclav
    Smil estimates that in the production of Asian rice 69–75% of
    nitrogen fertilizer escapes into the environment rather than being
    taken up by plants, while even developed-world cereal cultivation

can lose 25–50%. The world over, he estimates that at least 50m t of nitrogen, or 40% of total usage, are lost in this way every year, producing a financial deficit of $50bn, equivalent to one tenth of the value of the world's entire agricultural trade, not including the costs of the ensuing environmental damage (Smil 2004).

53. Baloch *et al.* (1994).

54. Interview, M. Shafi Niaz, Islamabad, 10 May 2008; cf. Niaz (2008).

55. Interview, Ali Tauqir Sheikh, LEAD, Islamabad, Pakistan, 12 May 2008.

56. Sowell (2007), p. 58.

57. Proctor (ed.) (1994), ch. 1.

58. 10 May 2008.

59. Cf. Dawe (2008); Ivanic and Martin (2008), p. 5.

60. Interview, 11 May 2008.

61. Current opinion varies widely on how much 'suitable' land is available for agricultural expansion. The Gallagher report identified between 790 and 1,215m ha (around one fifth of present agricultural land) (RFA (2008)); other estimates are as low as 50–400m ha (MNP (2008)).

11. The Evolutionary Origins of Surplus

1. *King James Version.*

2. Diamond (1992), pp. 342–7; Martin and Klein (eds.) (1984), pp. 345–53; cf. Dunnell and Greenlace (1999).

3. The role of climate change in megafaunal extinctions instead of, or as well as, human hunting is still a fiercely debated topic; cf. Martin (1973); Mosimann and Martin (1975); Martin and Klein (eds.) (1984); Diamond (1992), pp. 342–7; Diamond (1999), pp. 42–4, 47, 175; Martin (2005); Hopkin (2005a); Hopkin (2005b).

4. Diamond (2006); Diamond (1992), pp. 333–5; Kipler and Ornelas (2000), II.1124–5.

5. Stuart (2006b).

6. Diamond (1999), p. 111; Kipler and Ornelas (eds.) (2000), II.1124–5. A team of archaeologists co-directed by Ian Kuijt working in

the Jordan Valley have found the earliest climate-controlled granaries in permanent dwellings, dated as 11,500 years old. Preservation technologies for other foodstuffs came later: fermentation from 6000 BC, dairy churns from 4500 BC and salting from possibly as early as the fifth millennium BC. Sterile canning, or 'bottling', was only perfected in 1795 by the Frenchman Nicolas Appert to supply Napoleon's armies.

7. United States, Office of War Information Poster No. 58, Division of Public Inquiries (US Government Printing Office, Washington, DC, 1943), http://digital.library.unt.edu/permalink/meta-dc-603. See also Stone (1800), pp. 65–7.

8. Kipler and Ornelas (eds.) (2000), II.1416.

9. Harris (1975), pp. 20–21.

10. FAO (2008b), pp. 9–10; Proctor (ed.) (1994).

11. FAO does not disseminate data on minimum dietary energy requirements (MDER) for developed countries, but for the purposes of my work FAO statistician Cinzia Cerri kindly revealed that FAO estimates a minimum value of MDER for western Europe, from 2000 to 2007, between 1,900 and 1,910 kcal/person/day, while the maximum, over the same period, is around 2,000 kcal/person/day. In North America the minimum is 1,950 and the maximum 1,980; in Oceania (developed) the minimum is 1,940 and the maximum 1,950 (personal communication, 18 December 2008). Cf. Smil (2001), p. 236.

12. Smil (2004) suggests 130% of needed mean or no more than 2,600 kcal/person/day; Bender (1994) suggested 130% of requirements; Bender and Smith (1997) and Lundqvist, Fraiture and Molden (2008), p. 18, recommend supplies of 2,700 kcal/person/day; others suggest 3,000 kcal (Bruinsma (ed.) (2003)). On food entitlements, Sen (1987); Sen (1981).

13. EU supply from FAO (2003a); US supply from Hiza and Bente (2007). Minimum requirement in the US of 1,950 is exactly half of the current estimated supply of 3,900 kcal/person/day.

14. Harris (1975), p. 32.

15. Smil (2004).

16. Bracken (1997), pp. 167–8.

17. Harris (1975), pp. 111–27.

18. Bataille (1985), p. 121; Douglas (1990); Jonaitis (ed.) (1991), p. 162; Bracken (1997); footage of a 'reconstructed' potlatch scene can be seen in Curtis (1914).

19. Harris (1975), p. 127. Cf. Mauss (1990).

20. Kipler and Ornelas (eds.) (2000), II.1424.

21. Doyle (2007).

22. Veblen (1970), p. 64; cf. Frow (2003).

23. Jeffery and Harnack (2007).

24. Wright (2008); Shell (2003); Geiger *et al.*

25. Baylis (2008).

26. [Malthus] (1798), chs. 3, 7, 10, p. 187; Malthus (1826), II.25–7.

27. Agutter (1796), pp. 3–8, 20–22.

28. Society for Bettering the Condition and Increasing the Comforts of the Poor (1802), III.66–7. Tom Holland informed me that in early medieval Europe it was common practice for parochial bishops to sacrifice their oxen for the benefit of the poor in times of dearth. In the bad harvest year of 1972 in Russia chicks were killed because it became unaffordable to feed them.

29. Harris (1986), pp. 47–66; Harris (1975), pp. 12–32.

30. Diamond (2006), pp. 292, 440, 542. Diamond assumes that Tikopia chiefs presided over the pig extermination, but it seems plausible that this was a rebellious grassroots movement, only retrospectively adopted by the pork-eating elite.

12. Adding It All Up and Asking . . . 'What if?'

1. Peacock (1817), II.147–50; cf. Roy (1976), p. 29, which proposes that food waste may even be considered desirable since it represents a slack which can to some extent be taken up in times of necessity.

2. WRAP (2009c); cf. Evolve (2007), pp. 31–2. WRAP estimated the figure on catering waste based on information from Horizon among other sources. The EA waste survey for 2002–3 calculated waste of all types from the catering sector to be around 3.3m t.

For the UK as a whole, the figure is likely to be 3.5–4m t. For England, only 153,000 t was registered as segregated food waste, but the vast majority (2.2m t) was mixed general waste, an unknown quantity of which will have been food. Some studies have shown that only around 23% of waste from this sector typically consists of food. In total, therefore, food wasted in catering institutions in the UK may be under 1m t.

Lillywhite and Rahn (2005) estimated food waste by taking the fraction of municipal waste believed to be food (10–21%) and adding this to the Environment Agency surveys on commercial and industrial food waste. They came up with a total of 9m t in the UK. This did not include food wasted on farms. Roy (1976) documented waste between farmgate/dockside and the household, but much of the data relied on estimates from the 1960s. Cf. also Singer (1979), Osner (1982) and Mellanby (1975).

3. Research on behalf of Food Chain Centre: Ambler-Edwards *et al.* (2009), p. 27.
4. Kantor *et al.* (1997). The ERS says that, as of December 2008, it 'has not been updated' (Jean Buzby, senior economist (ERS), personal communication, 14 November 2008).
5. Kantor *et al.* (1997).
6. Neither internal nor external documentation to explain the estimated rate of losses was kept.
7. Buzby and Perry (2006); Wells and Buzby (2007); ERS (2008b); cf. Putnam (1999).
8. Jean Buzby, personal communication, 10 March 2009.
9. Personal communication, 3 March 2009.
10. Blair and Sobal (2006). Some of the discrepancy between actual losses and estimated losses can be attributed to the fact that people trim off and throw away fat cuts of meat, which means they are throwing away the most calorie-dense portion and this may be under-represented on the food loss estimates. Cf. also Dowler and Seo (1985); Miller (1979).
11. Jones's study, which focused on calorific values that cannot be translated directly into mass, found that the households wasted

14% of food purchased; extrapolated on to a national scale would mean wastage of 29m t, which would mean purchases of roughly 209m t; USDA in 1994 estimated that the entire national food supply was 161m t (Kantor *et al.* (1997)).

12. Jones (2004a). Figures converted into metric tonnes.

13. ibid. and personal communications, 2008.

14. Jones (2004a).

15. UNEP (2009), p. 32; Jones (2004a). The estimate was $90–100bn, but owing to food price increases in 2008, Dr Jones increased this, personal communication. Cf. also Kader and Rolle (2004), p. 1.

16. Also called 'conversion factor': not all loss at this stage is waste; it also includes weight loss due to evaporation. Contract ends 20 September 2009, though at the time of writing a decision has not been made whether or not the data will be published.

17. Buzby *et al.* (2009). Jean Buzby, ERS, personal communication, 4 December 2008 and 10 March 2009.

18. For each food in the ERS Food Availability Data System, using the Nielsen Company's Homescan data and comparing National Health and Nutrition Examination Survey (NHANES) data and other supplemental data sources. Report due to ERS on 31 March 2009, though publication has not been decided on at the time of writing. Jean Buzby, personal communication, 11 March 2009.

19. Andrew Parry, personal communication, 10 March 2009.

20. EPA (US) (2008), p. 35. All figures converted from US tons (2,000 lb) into metric tonnes (1,000 kg). Organic fractions of MSW from all nations can be found at UN Statistics Division (2007). Low-income countries (GDP less than US$5,000 per capita), for example India and African countries, produce 150–250 kg municipal solid waste (MSW) per capita each year, 50–80% of which is food or putrescible material. High-income countries (GDP greater than $20,000 per capita) such as the US and western European countries, produce 350–750 kg MSW per capita per year, 20–40% of which is food or putrescible material: Lacoste *et al.* (eds.) (2007), pp. 4–6. Only some of the biodegradable waste is food waste. Cf. Adhikari *et al.* (2006).

21. 76% is the proportion of UK household waste that was 'avoidable' or 'possibly avoidable'. Both Jones's and Kantor's studies of food waste accounted only for 'edible or once edible food', which does not perfectly correspond with either of WRAP's categories.

22. Total organic waste (including garden waste) sent for centralized composting in 2007 was 21.7m t.

23. Other outputs from livestock, such as traction and manure, appear not to be accounted for.

24. Smil (2001), pp. 236–7. Smil was following on the work of Bender (1994).

25. Smil (2001), pp. 209–10.

26. Smil (2004); cf. also Dowler and Seo (1985).

27. Smil (2001), p. 210.

28. WFP in 2008 calling for around 25m t of grain to alleviate the hunger of the world's malnourished.

29. Sometimes known as the crude energy gap: see Baines and Hollingsworth (1961); Serra-Majem *et al.* (2003).

30. Vaclav Smil, personal communication, 29 February 2008.

31. US energy supply taken from Hiza and Bente (2007). Minimum dietary energy requirement of 1,950 kcal/person/day is estimated by FAO.

32. See appendix.

33. See appendix and Bender (1994). Some empirical studies of food losses have found a linear relationship between affluence and levels of food wastage, with richer countries and wealthier households wasting a greater proportion of their food than poor ones: e.g. Fung and Rathje (1982), Chun *et al.* (1986); Jones (2004a). However, other studies have found no correlation between income and food waste (e.g. Wenlock *et al.* (1980); Dowler (1977)). Others found that food waste and household income were related in a non-linear fashion: Van De Reit (1985). WRAP (2008a), pp. 190–93, found that there was an *inverse* relationship between income and food waste, with the least well-off households wasting a larger quantity of food. However, WRAP showed that this correlation disappeared when waste per capita was accounted for – i.e. that

the difference was accounted for by poorer households having on average more people in them, and thus wasting a larger quantity of food. However, the WRAP study did not account for food wasted out of the home. Higher-income families may consume a greater proportion of their food out of the home, and will therefore have a correspondingly smaller amount of food waste in their household rubbish bins. If this were taken into account, high-income families may be found to waste a higher proportion of the food they consume in the home. See Sibrián *et al.* (2008).

34. Bender used the figure of 130% of MDER as his standard, suggesting that there were rich nations who achieved this level. However, I suspect that this may have been on the assumption that all rich nations' MDER was 2,000 kcal/person/day, in which case Japan would indeed appear to have only 130% of MDER. However, developed Asian nations such as Japan have populations with average heights significantly lower than those of Western nations, among other things, and as the FAO points out, their MDER is lower at 1,810 to 1,890 kcal/person/day (Cinzia Cerri, FAO, personal communication, 18 December 2008), in which case, Japan's energy supply is greater than 130% of MDER.

35. Bender (1994) estimated that 7.4% of global food consumption would be saved by reducing institutional and household losses to the extent that national food supplies in any country were no more than 130% of requirements, equivalent to best practice *in high-income countries*. If one takes into account that some of the food wasted was meat, which required cereals to produce it, the potential savings rise to 12.5% of global food consumption. Reducing post-harvest losses to 4%, as achieved by the best-performing high-income countries, Bender conservatively estimated could save 1.65% of global food supply in the case of cereals and 5.45% in the case of tubers. Total 19.6% of total food supply. (In fact, Bender's estimates of the proportion of meat wasted may have been an overestimate since he assumed that it was wasted in the same proportion as all other foods, whereas studies in the US and UK indicate that a lower proportion of meat than other foods

is wasted in the home. However, it is often fat – with the highest calorific content – which is wasted at household levels.)

36. Vaclav Smil, personal communication, 3 December 2008.

37. See appendix for further details and method of calculation.

38. 5.7 trillion kcal/day divided by FAO estimated average MDER for global population = 1,840 kcal/person/day. Depth of hunger of malnourished in 2007 = 250 kcal/person/day.

39. Global Land Cover (2000) database: 1,500m ha cropland. In 1993 estimates were 1,450m ha cropland and 3,360m ha pastureland: Goklany (1998). Total liberated land = 948.5m ha.

40. Williams *et al.* (2006); yield of bread wheat on good-quality arable land.

41. This could be implemented through biomass boilers and district heating systems, or gasification (Elsayed *et al.* (2003)) and injection into existing gas mains (National Grid (2009)).

42. RFA (2008).

43. Engström and Carlsson-Kanyama (2004) says that if Salix was used to produce energy on 1.5m ha it could produce about 260,000 TJ (0.173 TJ/ha), enough to heat 2.8m houses = 0.173 TL/ha. 948.5 Mha × 0.173 = 164.4m TJ.

44. 164.4m TJ = 45.7 trillion kWh multiplied by carbon savings of using renewable fuel instead of conventional heating (0.206 $kgCO_2e$/kWh) = total carbon saving of 9.408bn t CO_2equivalent (henceforth CO_2e) which is 35% of the 27bn tCO_2e ($GtCO_2e$) global emissions (not including land use changes) in 2005: Forster *et al.* (2007), p. 139. For a corroborating calculation, see Braschkat *et al.* (2003): this estimates carbon savings of around 10.2 tCO_2e/ha poplar coppice or miscanthus, which, multiplied by the 948.5 Mha liberated agricultural land minus that needed to feed the world's malnourished would come to a saving of 9.67bn tCO_2e.

In line with the publications above, I have used gross greenhouse gas emissions savings for willow coppice production (not taking into account emissions from cultivation and processing). This is fair because in calculating their potential carbon saving (0.206 $kgCO_2e$/kWh) I have not taken into account the comparable upstream emissions for fossil fuel extraction, transportation, refining etc. Upstream emissions for willow coppice are in any case very small: see Elsayed *et al.* (2003), especially p. 15, Appendix G, Appendix S1 comment b., Appendix S5, and Table 1; Heller *et al.* (2003).

45. 19.6% of 30 = 5.88. 5.88 + 34.53 = 40.41. 5.88 + 43.95 = 49.83. For global emissions from food production, see pages 91–5 above.

46. RFA (2008), p. 27, points out that carbon savings can be made by using pastureland for biofuel-yielding crops such as palm oil and sugar cane, and for woody plants such as willow.

47. Waggoner and Ausubel (2001).

48. Silver *et al.* (2000) calculated carbon sequestration potential on abandoned tropical deforested land to be 7.5 tC/ha/yr (1 kg of carbon (C) combines with 2.6667 kg of oxygen to form 3.6667 kg of CO_2, so that's 27.5 tCO_2/ha/yr) for the first 20 years × 948.5 Mha = 26bn tCO_2/yr for the first 20 years = 97% of global GHG emissions in 2005 (not including land use changes). A global afforestation programme would include non-tropical land. Nilsson and Schopfhauser, (1995) results suggest that if 948.5 Mha were afforested, the potential carbon fixing would be 2.9 GtC/yr for the first 100 years or 4.1 GtC/yr (15 $GtCO_2$/yr) 60 years on. Ovandoa and Caparrós (2009) reviewed recent studies on sequestration potential of European afforestation, with carbon fixing of up to 4 tC/ha/yr. 948.5 Mha × 4 tC/yr = 3.8 GtC/yr (13.9 $GtCO_2$/yr) = over 50% of the 27 $GtCO_2$e/yr GHGs emitted.

PART III: Where There's Muck There's Brass

13. Reduce: Food is for Eating

1. Taylor (2006).

2. Webster (2008), blog responses.

3. Interview, Mark Barthel, 26 August 2008. WRAP should have new estimates for waste in the supply chain by the end of 2009.

4. WRAP (2009d).

5. Engström and Carlsson-Kanyama (2004).

6. Cathcart and Murray (1939), p. 45.

7. The 1976 studies in the UK were commissioned by the Ministry of Agriculture, Fisheries and Food (MAFF) and the results published in Dowler (1977); Wenlock and Buss (1977); Wenlock *et al.* (1980). The waste in this survey did not include such items as outer leaves of vegetables, potato peelings, cores, skins etc., but meat bones and chicken carcasses were collected and any remaining tissues were scraped off for assessment. Studies in the US include Adelson *et al.* (1961); Harrison *et al.* (1975); Zaehringer and Early (1976). Waste in British schools in the 1960s and '70s was found to be 8–10% (Singer (1979)).

8. Cf. e.g. Singer (1979), p. 185.

9. Poppendieck (1986); CNN (2008).

10. Defence of the Realm Regulation. Cf. *The Times* (1917).

11. Ophüls (1969).

12. (US) Office of War Information poster, no. 77. 1943. 23 × 16.

13. Second World War poster designed by James Fitton in 1943. From 1936 the Nazis staged a massive campaign, *Kampf dem Verderb* ('Fight the Waste'), following reports that 10% of Germany's food went bad: Berghoff (2001), p. 180.

14. WRAP (2008a), pp. 174–5.

15. Goodchild and Thompson (2008).

16. Quoted in Ristow (2008).

17. Ashton (2006); Associated Press (2008).

18. Timothy Jones, personal communications, 2008.

19. WRAP (2009d).

20. Diamond (2006), p. 428.

21. One particularly popular one near Sheffield, called the Company Shop (also known as Marrens), requires a membership card to get in, only available to NHS and emergency services staff.

22. Carbon Trust (n.d.). Survey conducted among 1,159 consumers from across the UK by GfK NOP on 20–22 and 27–9 October 2006.

23. Imperial College London (2007), p. 10.

24. ibid. One supermarket employee claimed, on the BBC 4 programme *You and Yours*, 17 February 2009, that the resultant loss of control from people on the shop floor actually increased waste.

25. Defra (2007a), p. 22; LeGood and Clarke (2006).

26. Andrew Parry and Mark Barthel, personal communication.

27. Interview, 22 August 2009.

28. Interview, Mark Barthel, 26 August 2008, and personal communication. This cost was estimated as part of the FISS process and importantly only relates to existing regulated sites (some 340 sites subject to IPPC) and not the whole of the food and drink industry, for which the cost would be considerably higher. Cf. Defra (2007a), pp. 9–10.

29. WRAP (2009b); WRAP (2008c).

30. Interview, Tim Lang, 22 July 2008; Sustainable Development Commission (2008).

31. Cabinet Office (UK) (2008), Action Point 5.9 (my italics).

32. Mark Barthel, personal communication, 20 March 2009.

33. WRAP (2009a).

34. Initially, the challenge was to reduce the food manufacturing industry's food and packaging waste by 15–20% by 2010: Defra (2006a), p. 44. Six industry-led 'Champions' Groups' were set up to report on the feasibility of achieving this and diluted this already unambitious target. The result of these voluntary discussions was published in Defra (2007a), 'Recommendation 23: The Group considers that a realistic target for the food manufacturing industry to adopt is to reduce its food and packaging wastes by 3% a year over 5 years from a 2006 baseline.' The Environment Agency (EA) National Waste Production surveys cannot be used to establish a 2006 baseline because (1) the definition of waste has changed since these were conducted, (2) they do not give sufficient detail on the breakdown of waste and (3) they are inaccurate owing to poor sample sizes. An EA analysis found that the relative standard of error in the 1998–9 survey data was <12%. The relative standard error in the 2002–3 EA survey data was 9.6%. As the need is to detect changes of 3% a year, this makes both sets of EA survey

data unsuitable. The 2007 FISS CGW report expressed a hope that the IPPC data for the UK's 340 largest food and drink manufacturers, representing around a third of the industry, 'might serve as a surrogate waste baseline for the sector, subject to confirmation of how representative this information is of the industry as a whole. It would be helpful if amendments to food and drink permits could be made that would allow food waste arisings to be distinguished in the returns from other wastes.' This aspiration is yet to be implemented, and therefore sufficient data is still lacking for either establishing a baseline or monitoring progress – particularly for the majority of manufacturers not covered by IPPC. See Defra (2007a), pp. 9–13. All studies on waste have called for more reporting. For example, the government-commissioned report, AEA (2007), criticized 'the almost total lack, in some areas, of baseline data' and recommended 'as a matter of priority, that baseline studies be initiated to obtain essential data on current resource consumption/waste production'. The lack of data 'undermines the purpose of setting such waste reduction targets'.

35. Defra (2007a), pp. 25–6; AEA (2007), pp. 44–5.
36. FDF (2008a).
37. FDF (2008b); Morley and Bartlett (2008). The survey found that overall, from the total 835,000 t of food and packaging waste produced by those food companies that responded to the survey, 686,000 t (82%) were recycled or recovered in some way with just under 138,000 t of waste sent to landfill in 2006 (16.5% of total tonnage). An additional 512,000 t of potential waste was avoided through the use of by-products, in e.g. animal feed. WRAP suggests that its current work with Cranfield University and the logistics company DHL should provide more robust data than anyone has had before because DHL already keep records of some kinds of waste within the food industry as their contracts with companies oblige them to record their efficiency and success (Mark Barthel, personal communication, 20 March 2009).
38. Veolia Environment tipping fees at Pitsea landfill near London approx. £25/t in February 2009.

39. The figures and opinion were given by Gus Atri, Northern Foods, and were echoed by numerous others in the industry.

40. In the UK an estimated 15m t of Biodegradable Municipal Solid Waste (BMSW) was landfilled in 1995, so the targets are to reduce BMSW landfilled to 11.25m t/yr by 2010; 7.5m t by 2013; and 5.25m t by 2020. The government left it until 2003 before developing its Waste Implemention Programme and little was achieved before 2006, when Defra finally set up the more pro-active Waste Infrastructure Delivery Programme.

41. This is the Landfill Allowance Trading Scheme (LATS).

42. National Audit Office (2009).

43. EC (2005a) and EC (2005b). Cf. Fehr *et al.* (2002).

44. Environmental Protection Agency (Northern Ireland) (2007).

45. EPA (US) (2008), p. 7; Hogg *et al.* (2007b), p. ii. The EU throws away 2bn t of waste a year (including 700m t agricultural waste); 67% of the 1.3bn t of municipal waste is either burnt or sent to landfill (EC (2009)). According to WRAP, of the 6.7m t of household food waste in the UK, 5.9m t currently goes to landfill with the remainder being poured down the sink, fed to pets, composted at home or collected and treated separately by councils.

46. http://customs.hmrc.gov.uk/.

14. Redistribute: The Gleaners

1. Teresa (1996), p. 51.

2. Interview, Maria Kortbech-Olesen, Fareshare, 4 March 2009.

3. EC (2008b), p. 30, suggested that 'logistical assistance could be offered to make [supermarkets] more willing to participate in food recuperation schemes'. However, it claimed – without any discussion – that 'Supermarkets cannot be obliged to donate produce to charities.'

4. Interview, Maria Kortbech-Olesen, 4 March 2009.

5. The Climate Change Act (2008) makes it the duty of the Secretary of State to ensure that the net UK carbon account for the year 2050 is at least 80% lower than the 1990 baseline.

6. Evolve (2007), p. ii.
7. Hawkes and Webster (2000), pp. 10–11. Sustain accounted for just 3,200 t of food redistributed in the 1990s; adding to that Fareshare's massive expansion to 4,000 t/yr, their total would come to around 6,700 t/yr.
8. Feeding America comprises a network of more than 200 individual food banks and food rescue organizations serving all 50 states, the District of Columbia and Puerto Rico. Feeding America secures and distributes more than 900,000 metric t (2bn lb) of donated food and grocery products annually. It is not known what proportion of this is from genuine 'surpluses' and how much comprises 'charitable' donations, and the distinction between these is fuzzy in any case. 'Food Pantries' – a non-profit organization – provides a directory of food banks and soup kitchens across America. Notable examples of local initiatives include the New York City Food Bank and Washington's Food Lifeline organization. In 2000 one publication estimated the total food redistributed to be 660,000 t: Hawkes and Webster (2000), p. vii.
9. Hawkes and Webster (2000).
10. Interview, Timothy Jones, 9 October 2008.
11. Evolve (2007), p. ii.
12. Hawkes and Webster (2000), pp. 16–17.
13. From 1972, EC regulation 1035/72 had allowed distribution to charities, persons receiving public assistance, schools, prisons, hospitals and old people's homes. The priorities were changed by regulation 2200/96. Hawkes and Webster (2000), pp. 10, 16; *The Food Magazine* (1997); Acheson (1998).
14. The Bill Emerson Good Samaritan Food Donation Act, 1 October 1996. The Food and Nutrition Act of October 2008 aimed 'to help to achieve a fuller and more effective use of food abundances; to provide for improved levels of nutrition among low-income households through a cooperative Federal–State program of food assistance'. The Act sets outs and revises the authorization process for participating food stores and wholesalers. From 1 October 2008 the Supplemental Nutrition Assistance Program (SNAP) became

the new name for the US government's federal Food Stamp Program, which operates independently and alongside the voluntary NGO food redistribution sector. Unlike SNAP, there are no statutory requirements regarding eligibility of individuals for food donated through food banks. However, USDA does issue published Income Guidelines on eligibility and issues a standard Eligibility Form.

15. Sec. 170(e)(3) of the Internal Revenue Code. Cf. Hawkes and Webster (2000), p. 13.

16. Hawkes and Webster (2000), p. 16.

17. http://www.secondharvest.org/learn_about_hunger/katrina_rita_study.html.

18. BBC Radio 4, *You and Yours*, 22 December 2003; interview, Tim Lang, 22 July 2008.

19. In Hawkes and Webster (2000), p. vi, the food organization Sustain, which Lang headed for many years, wrote that the expansion of food redistribution networks 'is unwelcome in that it indicates that Britain has an increasing inability to provide all its citizens with adequate mechanisms, financial or social, to obtain food in culturally acceptable ways'. However, it does not follow that an increase in food redistribution necessarily indicates an increased inability to provide for citizens; it could, rather, be due to a rise in corporate and public awareness of a pre-existing problem and an increased willingness to use food more sustainably (and to avoid rising disposal costs). Furthermore, it is a circular argument (and counter-factual) to suggest that food redistribution is culturally unacceptable. Similar objections have been raised for over a century: see e.g. Loch (1885).

20. Interview, Maria Kortbech-Olesen, 4 March 2009.

15. Recycle: Compost and Gas

1. Clement of Alexandria (2008), p. xiv.

2. Kasting (2006).

3. Kangmin and Ho (2005).

4. Offering a global warming potential saving of up to 174% as compared to diesel: RFA (2008), p. 24.

5. Cabinet Office (2008), p. 91; Defra (2007b), and numerous government and other publications; cf. e.g. Callaghana *et al.* (2002); Al Seadi and Holm-Nielsen (eds.) (2003).

6. Grid injection is used in Germany, France and Austria. In the US, landfill gas from the Staten Island landfill is purified and injected into the New York Grid. Methane collected from landfill sites currently provides around 1% of the UK's gas needs, but the Renewable Obligation scheme which provides a premium for renewable electricity but not renewable gas or heat means that this biogas, along with that produced in anaerobic digestion (AD) plants, is used to generate electricity. National Grid (2009), p. 4, argues that gas mains injection would be a more efficient use of biogas and claims that in the long-term food waste could provide 1% of gas demand in the UK, while all sources of biogas, including some energy crops, could provide 18% of demand.

7. There is ongoing research into what proportion of methane from landfills actually escapes rather than being collected and either flared or used for power generation. In the US, 340 of 2,975 landfills recover biogas; in the EU biogas collection is now mandatory for new landfills and older landfills are being retro-fitted: Lacoste *et al.* (eds.) (2007), p. 19. There is also the question of how much organic matter in landfills actually fails to break down at all, and whether some of it is effectively fossilized, remaining as buried carbon for centuries or even longer. See e.g. Themelis and Ulloa (2007): theoretical and experimental studies indicate that complete anaerobic digestion of municipal solid waste (MSW) generates about 200 m³ of methane per dry tonne of contained biomass; the reported rate of generation of methane in industrial AD reactors ranges from 40 to 80 m³/t of organic wastes. Several US landfills report capturing as much as 100 m³ of methane per (US) ton of MSW landfilled in a given year. These findings led to a conservative estimate of methane generation of about 50 m³ of methane per ton of MSW landfilled. Therefore, for the estimated global

landfilling of 1.5bn t annually, the corresponding rate of methane generation at landfills is 75bn m³. Less than 10% of this methane was being captured and utilized.

8. Hogg *et al.* (2007b) estimates that sending 3.2m t (60% of 5.4m t) of household food waste to AD rather than landfill could save 1.6–3.6m tCO₂e of which electricity generation was responsible for 0.18–0.29m t.

9. Mata-Alvarez *et al.* (2000).

10. Interview, Mark Barthel, 26 August 2008. Madsen and Kromann (2004) and Danish Ministry of the Environment (2004) suggest that incineration can be better for the environment than centralized AD schemes, especially if the AD collection uses plastic bags rather then reusable bins.

11. Bruun *et al.* (2006) found that the application of processed biodegradable MSW on to agricultural land resulted in increased GHG and ammonia emissions; that heavy metals and organic compounds from the compost can potentially migrate to drinking water and accumulate in food crops; and that leached nitrogen can cause eutrophication. See also *Economist* (2009b).

12. Weiland (2000).

13. WRAP (2008d).

14. Ruddock (2007); cf. Cabinet Office (2008), Action 5.9, p. 94.

15. Almost all of this (63,464 t) was in Staffordshire. By comparison, just 34,607 t of unmixed food waste were sent for composting. 93,975 t were sent for incineration, pyrolysis or gasification with energy recovery (though it is not specified whether these processes had net energy output). 216,345 t of liquid food wastes and sludges were spread on agricultural land. A further 162,633 t were sent through unspecified 'recycling' processes. Just 17,569 t were sent to landfill. (Much greater volumes of mixed food and packaging waste (109,686 t) went to landfill.) It is not at all clear whether these figures are representative of the food industry: Morley and Bartlett (2008), p. 10. Food Processing Faraday Partnership (2008), p. 7, found that of the 466,702 t of waste produced by food- and drink-processing companies per year in the East Midlands region,

167,318 t are segregated and recycled and 144,843 t of Category 3 Animal By-Products waste are treated in approved processes (such as composting). This leaves 111,900 t of mixed waste that are sent to landfill sites every year, in which it was calculated there is only enough capacity for the next eight years.

16. Silsoe Research Institute (2004), §4, 'Results of the Life Cycle Assessment'. Cf. Sandars *et al.* (2006), I.203–207.

17. Interview, Jake Prior, The Summerleaze Group, July 2008.

18. According to Matteson and Jenkins (2007), recovering food waste for AD to generate energy that competes with the market price of conventionally produced power in the US ($0.053/kWh) would require charging food waste producers a tipping fee of nearly $46/t.

19. Lundie and Peters (2005) found that for a household producing 182 kg/yr food waste, the emissions from various disposal methods were as follows (emissions given in $kgCO_2e$): food waste processor (13); home composting in aerobic conditions (0.3); home composting with anaerobic conditions (273); centralized composting (52); landfilling (82). Hogg *et al.* (2007a); Hogg *et al.* (2007b); Fisher *et al.* (2006); Mendes *et al.* (2003); Hirai *et al.* (2001).

20. Ogino *et al.* (2007) found that, compared to incineration, feeding food waste to pigs could be five times better from the point of view of greenhouse gas emissions and could save 99.9% of water consumption. Ogino – with whom I spoke at length during a visit to Japan – compared his results to those of Lundie and Peters (2005) and found that feeding food waste to pigs in centralized collection systems produced lower direct emissions than all waste management options except aerobic home composting. Mendes *et al.* (2003) and Hirai *et al.* (2001) reported that GHG emissions associated with biogasification and composting ranged from less than 50 kg to 210 $kgCO_2e$/ton of waste. Ogino points out that feeding food waste to pigs under best-practice methods beats all these options when the avoided emissions of producing commercial pig feed are taken into account. Furthermore, most of the emissions associated with feeding food waste to pigs in Ogino's study came from heat-treating the food waste. In Europe, where

feeding animal by-products to livestock is banned, there is no need to heat-treat those former foodstuffs that are free from animal by-products, such as bread and vegetables.

Lee *et al.* (2007) compares environmental impacts of treating food waste by animal feed manufacturing, incineration, composting and landfilling (with and without power generation from landfill gas). Once again, 'In the global warming category, feed manufacturing showed the least impact.' This study included biogenic carbon in its final assessment of the global warming potential of each treatment method. When biogenic carbon emissions are stripped out, incineration with power generation performed better than animal feeding. However, the study considered the one method of turning food waste into animal feed that Ogino *et al.* (2007) had found performed worse than incineration – drying it out, which necessarily demands a large input of fossil fuel energy. Ogino showed that feeding pigs heat-treated liquid food waste produced one quarter of the emissions of drying the food waste. Braschkat *et al.* (2003), p. 13, compares the environmental impacts of biogas, combustion and animal feed for cultivated rape seed, not food waste. See also Russ and Meyer-Pittroff (2004).

21. Interview, Michael Chesshire, Greenfinch, 31 March 2008. The food waste has a dry-matter content of 27.5%.

22. Gross output is 300 kWh/t food waste, and around 15% of this is needed to power the plant, leaving a net output of 255 kWh/t. Michael Chesshire indicates that a more efficient larger engine than the one in the Greenfinch plant could provide a gross output of 370 kWh/t food waste. The Greenfinch AD process is 'mesophilic' – using microbes that operate at lower temperatures. Other plants use a thermophilic process, which can yield higher volumes of biogas but require more energy to provide heating; Greenfinch found that it was not suitable for the digestion of general kitchen waste.

23. 1 kWh renewable energy supplied to the UK National Grid saves 0.43 kgCO$_2$e by avoiding new conventional energy generation: Defra (2008b), pp. 7–8. 0.43 × 255 kWh = 109.65 kgCO$_2$e/t of

food waste. Many thanks to Jo Howes from E4Tech for confirming the validity of this approach. Thanks also to Robyn Kimber and Natalia Zglobisz of Imperial College, London: their calculations assumed a methane yield of just 70.8 m³/t, but a carbon saving of 0.53 kgCO$_2$/kWh (the grid rolling average which is usually used when calculating avoided emissions from using less electricity now rather than the effect of changes to generation), and came to an almost identical figure of 109.4 kgCO$_2$e/t food waste. A government-funded study in Britain calculated that the renewable energy generated by sending 3.2m t of food waste for AD could save 0.18–0.29m tCO$_2$e = 56–91 kgCO$_2$e/t food waste sent to AD: Hogg *et al.* (2007b), pp. 57–8. Their lower figure for potential savings can be attributed to their assumption of a lower methane yield, and lower generator efficiency and their assumption that power from AD would replace gas-fired power with carbon emissions of 0.382 kgCO$_2$/kWh. Cf. Cabinet Office (2008), p. 91.

24. Greenfinch capacity = 100 t/wk, but in 2008 intake was 70 t/wk = 3,650 t/yr. Chesshire estimates annual yield of 600,000 kWh of hot water ÷ 3,650 t = 164 kWh/t. Emissions factor for heat from natural gas is 0.206 kgCO$_2$e/kWh: Defra (2008c), annex 1. 164 kWh × 0.206 kgCO$_2$e/kWh = 33.78 kgCO$_2$e/t of food waste. Plus 109.65 =143.43 kgCO$_2$e/t. This carbon saving is before you start deducting the emissions associated with transporting the food waste or building the plant. It also leaves out the avoided emission of methane from landfill sites and the potential savings by replacing nitrogen fertilizers. However, this is the savings figure required for the purpose of comparing AD with alternative methods of treating food waste, such as feeding it to pigs, which require similar collection and distribution inputs, and which also prevent the food waste from going to landfill (though pig-feeding would require less in respect of building treatment facilities).

25. Conventional pig feed = 730 kgCO$_2$e/t: Dalgaard *et al.* 2007. Dry-matter content of conventional feed is 85%, so is equivalent to 3.09 t of good-quality food waste swill with similar nutritional profile with a dry-matter content of 27.5%. 730 kgCO$_2$e/t ÷

$3.09 = 236.2$ kgCO$_2$e/t. These are approximate figures because each type of food waste has different nutritional qualities and methane yields. But, for example, bread slices which are currently sent in massive quantities to AD would have an even greater carbon saving of 516 kgCO$_2$e/t if fed to pigs. 221 kcal per 100 g sliced bread is about 55% of the calorific content of feed barley; Dalgaard *et al.* (2007) attributes feed barley with 934 kgCO$_2$e/t; so feeding waste bread to pigs has carbon saving of $934 \times 0.55 = 516$ kgCO$_2$e/t. Ogino *et al.* (2007) standardized feed nutrient content in his study, which backs up these findings.

Dalgaard *et al.* (2007) assumes that demand for pig feed stimulates demand for soy and barley with a proportional distribution in pig feed of 85% locally grown barley and 15% Argentine soymeal. Barley and soymeal cause the emission of 934 kgCO$_2$e/t and 694 kgCO$_2$e/t respectively. This yields a total of 730 kgCO$_2$e/t pig feed with a 85:15 mix. Dalgaard *et al.* (2008) explains that emissions from producing and importing soymeal is higher, but Dalgaard's consequential Life Cycle Analysis (LCA) approach assumes that producing and crushing soybeans, using the meal for animal feed, and then selling the extracted vegetable oil will *reduce* the demand for palm oil, so the emissions from producing palm oil are *deducted*. Another method of calculating emissions associated with soymeal production is to take the total emissions from growing soybeans and divide them according to the relative *value* of the meal and the oil, and this yields a similar result of 726 g CO$_2$e/kg soymeal. None of these figures include emissions from deforestation or other land use changes.

26. 236.2 is 163% of 144.9.

27. Rightly so, since that is not an additional benefit from feeding food waste to pigs rather than conventional feed; because you could use manure from conventionally fed pigs too.

28. Dalgaard *et al.* (2008) calculates impact of land use based on the assumption of *average* soy yields in Argentina, rather than *marginal* yields – i.e. the yield achieved by those farmers who are actually responding to increased demand by extending the agricultural

frontier into virgin land such as forests. Since this land is sometimes relatively poor, yields can be below average, therefore more forest has to be cut per tonne of crop, which would raise the emissions per tonne still further. The most controversial variable, however, is the question of depreciation, i.e. the period over which you spread the emissions from turning forest into cultivated land. Dalgaard assumes the arbitrary, though reasonable, depreciation period of 20 years (a standard now adopted in UK policy and in PAS 2050), and thus gives a total of 5,700 $kgCO_2/t$ soymeal. It could be argued that depreciation periods should be longer – even as long as deforested land remains cultivable. (Adopting this approach would mean that annual harvests of European cereals, for example, would have to take a portion of the burden of prehistoric deforestation.) It would also be arguable that emissions from land use changes should only be attributed to soymeal in the proportion that soy grown on virgin land comprises all production (more than 40% of increased soy production in Argentina has come from encroachment into virgin land including savannahs and forests: Pengue (2006), p. 20). However, as confirmed in discussion with consequential Life Cycle Analysis specialists, for a study like this, it is fair to say that the extra tonne of soymeal causes the extra deforestation – all of it, now. In other words, the depreciation period is one year and the carbon emissions from a tonne of soymeal according to Dalgaard's calculations would be 100,700 $kgCO_2e/t$ soymeal. Dalgaard *et al.* (2008) calculated this level of emissions on the basis of there being an average of 94 tons of carbon per hectare in *above ground biomass* for all tropical forests (according to Houghton (2005)). However, as shown in Schmidt (2007), pp. 215–16, when losses of carbon from soils, below ground biomass and other sources from tropical American forests specifically are all included, according to figures in IPCC (2003), ch. 3, p. 43, and ch. 3, p. 157, the loss of carbon incurred when replacing tropical American forest with soy cultivation is 205,000 kgC/ha, equivalent to 752,000 $kgCO_2/ha$, or (using yields assumed in Dalgaard *et al.* (2008)) 218,800 kgCO$_2$/t soymeal. In addition to carbon emissions, Schmidt calculates losses of 7,500 kgN/ha, of which N$_2$O emissions are 147 kgN$_2$O/ha. GWP of N$_2$O over 100 years is 310 times that of CO$_2$: Forster *et al.* (2007), p. 212. 147 × 310 = 45,570 kgCO$_2$e/ha = 13,300 kgCO$_2$e/t soymeal. Total emissions from land use change = 13,300 kgCO$_2$e + 218,800 kgCO$_2$e = 232,100 kgCO$_2$e/t soymeal. Added to the emissions from cultivation, transport etc. of soymeal (from Dalgaard *et al.* (2008)) = 232,800 kgCO$_2$e/t soymeal.

29. This is an approximate value based on the assumption that soymeal (dry-matter content 85%) or equivalent food wastes with similar nutritional content (e.g. animal by-products) would yield approximately three times more electricity in an AD plant than the mixed food waste digested by the Greenfinch plant (dry-matter content 27.5%). This would mean that the emissions savings of putting this food waste into AD would be 448 kgCO$_2$e/t. Using this food waste for animal feed instead would save from 11,600 to 232,800 kgCO$_2$e/t, which is 26–224 times greater than the 448 kgCO$_2$e/t yielded by sending it for AD.

30. Silsoe Research Institute (2004), §4, 'Results of the Life Cycle Assessment'. Pig manure also causes problems, but feeding food waste to pigs replaces commercial feed, without a corresponding increase in faecal output; cf. Braschkat *et al.* (2003), p. 13. Lee *et al.* (2007) found that using food waste for feed manufacture performed 17 times better than composting for acidification and three times better for ecotoxicity; it performed worse for eutrophication, but this is largely due to the waste-water management associated with the technique of drying the food waste to form pig feed, rather than feeding it wet to the pigs: see Ogino *et al.* (2007).

31. Ogino *et al.* (2007).

32. Retail price of 11.5p/kWh of renewable electricity (Good Energy Standard prices effective from 15 July 2007), and 5p/kWh of gas heating from EDF energy (8 November 2008, 7p/kWh up to the first 2,680 kWh, and thereafter 3.05p).

33. Feed conversion ratio of food waste to edible portions of pork of 15:1; retail price of pork at £5/kg.

34. BBC Radio 4 (2005).

35. Mixed food waste in the Greenfinch plant potentially saves 143 $kgCO_2e/t$; tomatoes may yield around half of the methane of mixed food waste: estimate based on results in Zhang *et al.* (2007); Miller and Clesceri (2003); Mahro and Timm (2007). Half of 143 $kgCO_2e = 71.5 \ kgCO_2e$. By comparison, the global warming potential of producing 1 t of tomatoes is 9,400 $kgCO_2e$: Williams *et al.* (2006) – not including the emissions from transport, refrigeration or packaging further down the supply chain, which would probably approximately double the emissions. 9,400 divided by 71.5 = 131.

36. Jess Hughes, Waitrose senior press officer, personal communication, 14 January 2009; Spear (2006); and http://www.metro.co.uk/news/climatewatch/article.html?in_article_id=76228&in_page_id=59.

37. Hogg *et al.* (2007b), pp. 57–8, estimates potential of 477–761 GWh/yr from household food waste, equivalent to the electricity used by 103,000–164,000 households (median 133,500) of a total 24.5m UK households using 72 TWh/yr. This estimate does not take into account energy requirements for building AD facilities or for collecting food waste separately. For total energy consumption in UK, see Monbiot (2007), p. 133.

38. Marie-Louise Terbeek, environment manager, McDonald's, 20 March 2009; Griffiths (2008); Eccleston (2008) quotes Helen Humphrey, vice-president National Operations: 'We are always looking for ways in which we can move closer to our goal of sending zero waste to landfill by 2010.'

39. Interview with Alison Austin, Sainsbury's, 13 November 2008; cf. Pagnamenta (2008).

40. Shipping is not included under Kyoto Protocol; however, the UK Climate Change Act (2008), 30.3, stipulates that international shipping and aviation should be included unless the Secretary of State explains why not by the end of 2012.

16. Omnivorous Brethren: Pigs and Us

1. Nelson (1998), pp. 2–3, 18–21, 103, 110; Nemeth (1995); Sauer (1969).
2. Nemeth (1995).
3. Markham (1676), p. 100; Thirsk (1985), p. 445; Stuart (1995), pp. 5–6.
4. Mortimer (1707), p. 184; Mastoris and Malcolmson (2001), pp. 34–5; cf. Laurence (1726), p. 147.
5. Mastoris and Malcolmson (2001), pp. 37–9.
6. ibid., pp. 39–40.
7. ibid., p. 37.
8. Harris (1986), pp. 67–76; Harris (1975), pp. 40–44; Harris (1978), pp. 131–2.
9. Shaler (2007), p. 142.
10. Giedion (1948), pp. 213–16.
11. Washington (1980), p. 670; Carrier (1998), p. 139.
12. *The Times* (1918); Bathurst (1921); Mastoris and Malcolmson (2001), p. 123.
13. Mastoris and Malcolmson (2001), pp. 123–6.
14. BBC Radio 4 (2008a).
15. BBC WW2 archive (2005); cf. *The Times* (1946).
16. Allen (1998).
17. Proctor (1999), pp. 120–21, 131–2; Hitler (2000), pp. 230–31.
18. Kipler and Ornelas (eds.) (2000), II.1318.
19. Steinfeld *et al.* (2006), p. 12 and *passim*; Garnett (2008), p. 12.
20. Westendorf (ed.) (2000), p. 7.
21. Interview and visit to John Rigby's farm, 6 and 14 February 2008.
22. According to the USDA, in 1995 there were over 2,200 licensed garbage feeders in the United States and nearly 3,000 in Puerto Rico. The 1980 Swine Health Protection Act requires that food waste should be boiled (212 °F or 100 °C) for 30 minutes: Westendorf and Meyer (2003).
23. Interview and visit to John Rigby's farm, 6 and 14 February 2008.
24. *Economist* (2007), p. 77.

25. Defra (2007d), Annex 1.

26. Westendorf and Meyer (2003); Westendorf (ed.) (2000).

27. Jones *et al.* (2004); Myer *et al.* (2000).

28. Kwak *et al.* (2006).

29. Hansard (2004), 4.13pm, Column 62WH.

30. ibid. The other problem with the law was that it required farmers to boil swill at 100 °C for one hour, which requires a great deal of energy and actually denatures many of the proteins and micronutrients and is unnecessary on hygiene grounds; recent research shows that food waste treated at 65 °C for 20 minutes is sufficiently sanitary based on microbiological analyses: Sancho *et al.* (2004).

31. Hansard (2004), 4.30pm, Column 68WH.

32. Dr Martin Blissit, advisor for notifiable disease, telephone interview, 12 November 2007.

33. Awarenet (2004), ch. 1, ch. 3 and Annex 6: meat, fish and dairy 82,392,218 t; fruit and vegetable 139,711,957 t.

34. Assumed conversion ratio of 15:1 and retail price of pork of £5/kg.

35. Parliamentary and Health Service Ombudsman (2007), p. 19, Item 50.

36. Interview, Maureen Raphael, Hain Celestial Group, 21 August 2008.

37. See ch. 15 for assumptions. 232,800 $kgCO_2e/t$ soymeal × 3m = 698.4m tCO_2e. This calculation does not include avoided methane emissions from not sending the food waste to landfill.

38. Animal By-Products Order (1999). Followed in 2001 by the British Swill-Feeding Ban in response to the foot-and-mouth outbreak. This was followed by EU Regulation 1774/2002 October 2002. The UK Animal By-Products Regulations 2003/1482 is the enforcement of EU Regulation 1774/2002. This came into force in England on 1 July 2003. Transitional measures allowed former foodstuffs of animal origin to be disposed of to landfill until 31 December 2005.

39. Though cf. National Renderers Association (US) (2006).

40. Sancho *et al.* (2004).
41. Steinfeld *et al.* (2006), pp. 49–50. Cf. UNEP (2009).
42. UNEP (2009), p. 19.
43. Hansard (2004), 4.13 pm, Column 62WH.
44. Interview, Tim Lang, 22 July 2008.
45. Interview, Chris Haskins, 25 July 2008.
46. The separation plans are known in Britain as a HACCP (Hazard Analysis and Critical Control Point) procedure.
47. The veterinary advisor for the Scottish government, Sheila Voas, said on the telephone and confirmed in an email that 'material generated from a high street bakery is CATERING WASTE and cannot be fed to pigs . . . The difference between catering waste and former foodstuff is that catering waste is supplied to the consumer, ready to eat from a retail outlet.' Published government guidelines, however, specifically state that waste from retail outlets selling ready-to-eat food – such as supermarkets, 'bakers and bread shops' – can indeed be fed to livestock even if it contains an element of animal by-products such as melted fat or lard. Others in Animal Health strongly discourage farmers from using any former foodstuffs even when it is legal – which, according to Defra officials, is well beyond their remit. Telephone interview with Sheila Voas, 30 January 2008 (12.40 pm); email from Paul Oliver sent on behalf of Sheila Voas, 5 February 2008; interview with Neil Leach, Animal By-Products & Feed Policy Team, Defra HQ, 30 January 2008 (16.30–17.00). If the outlet handles meat, there must be HACCP (Hazard Analysis and Critical Control Points) procedures in place to ensure that it cannot come into contact with the material intended to be fed to livestock. A member of staff must be trained and must ensure separation at every stage, and make a written note of the procedure. Some Animal Health officers are oblivious as to the nature of HACCP (imagining it to be a regulating institution, for example); and there remains confusion in government guidelines as to whether an outlet is *obliged* or *advised* to notify the Local Authority of its separation procedures: Defra (2006b), p. 18; Defra (2008a). The actual wording of the

regulation (EC) No. 197/2006 is as follows: '(4) Certain former foodstuffs, such as bread, pasta, pastry and similar products, pose little risk to public or animal health providing they have not been in contact with raw material of animal origin such as raw meat, raw fishery products, raw eggs and raw milk. In such cases, the competent authority should be permitted to allow the former foodstuffs to be used as feed material if the authority is satisfied that such practice does not pose a risk to public or animal health.'

48. http://sugarmtnfarm.com/blog/2007/01/feeding-bread-to-pigs. html (accessed 19 November 2007).

49. Interview, Maureen Raphael, 21 August 2008; interview, Andy Powell, Hain Celestial, 20 March 2009.

50. Wlcek and Zollitsch (2005).

51. The same study, covering the Yorkshire and Humber region in the UK, found that around one third of the 8,900 t of biodegradable waste sent by food manufacturers to landfill each year was confectionery, pastry and bread by-products that could be reutilized as animal feed: Ogden (2005); cf. Morley and Bartlett (2008). C-Tech Innovation Ltd (2004) estimated that of the 1.9m t of segregated biodegradable waste discarded (not recycled) by UK food and drink manufacturers, 0.8m t were meat and meat products and 1.1m t were comprised of cereals, fruit, vegetables, dairy and other non-meat products. These totals do not include the majority of food that ended up in the mixed general waste category. However, this study was based on figures collected before the animal by-products regulations which will have diverted more organic waste into the waste stream. Based on Environment Agency figures, AEA (2007), p. 36, estimated that only 8% of biodegradable industrial waste from the food industry in England went to landfill. Ogden (2005) found that 31% of food waste was landfilled. Garnett (2006) estimated that a third of the 522,000 t of biodegradable waste produced by the processed fruit and vegetable sector in the UK is landfilled. A recent one-year study by the National Industrial Symbiosis Programme in Britain has found that only 7% of industrial food waste was landfilled while the majority was fed to

livestock; however, as Eric Evans has pointed out, their study selected two very large unrepresentative manufacturers: a brewer and a potato-processing plant, both of which belong to sub-sectors that have traditionally sent their food waste to livestock; Eric Evans, Bio-Recycle, personal communication, 2 March 2009.

52. Interview, Julian Walker-Palin, 10 October 2008. The government-funded study AEA (2007), p. 41, also calls for greater levels of segregation for animal feed. Segregated former foodstuffs of non-animal origin may also have to be approved by the Feed Materials Assurance Scheme (FEMAS) requirements to assure the safety of feed materials.

53. Cited in Mastoris and Malcolmson (2001), pp. 123–6.

17. Islands of Hope: Japan, Taiwan and South Korea

1. Maathai (2005); Maathai (2008).
2. UK per capita figure calculated here from WRAP figures for total business and household waste minus agricultural waste to allow for fair comparison with Japanese figures which do not include agriculture.
3. MAFF (Japan) (2007); Yakou-san at Agri-Gaia estimated 22m t.
4. MAFF (Japan) (2007).
5. Fitzpatrick (2005).
6. *China Daily* (2008).
7. These published figures relate to the monetary value of the food, rather than the nutritional value.
8. *China Daily* (2008): this article estimates that after waste, therefore, only 1,891 kcal/person/day are actually eaten.
9. *Japan Times* (2007a).
10. *Japan Times* (2007b).
11. *China Daily* (2008).
12. Lewis (2008).
13. Arita (2008).
14. Act on the Promotion to Recover and Utilize Recyclable Food Resources (2001). In 2002, Japan was only recycling 10% of its

total volume of waste food, but 45% of its commercial and industrial food waste. MAFF (Japan) (2008).

15. MAFF (Japan) (2007).

16. Different sectors within the food industry have different targets
depending on the relative ease of recycling the different kinds of
waste. Food manufacturers will be obliged to recycle 85% of their
waste and already recycle 81% (according to Mr Tanami's figures
for 2005). The wholesale sector have to recycle 70% and currently
recycle 21%. Restaurants and hotels will have to recycle 40% and
currently recycle 21%. Convenience stores and supermarkets will
have to recycle 45% and currently recycle 31%. The average of
66% is calculated by taking an average from the 4.9m t of food
waste from the manufacturing sector at a recycling rate of 85% and
the 6.3m t from the other sectors at an average recycling rate of
52%.

17. Household food waste collection does happen on a smaller scale
in certain districts – about 2% of Japan's domestic food waste is
recycled.

18. Interview, 10 June 2008. Dr Kawashima works alongside Akifumi
Ogino, author of the brilliant paper Ogino *et al.* (2007).

19. *Pig International* (2007).

20. Cf. MAFF (Japan) (2007). That is 22% of the total commercial
food waste produced in Japan.

21. Interview, 3 June 2008.

22. For every 1% of crude protein in feed, there is ¥1,000/t, and
Tanami's dried feed is 23% protein and thus should achieve
¥23,000/t.

23. It costs only ¥12.5/kg for companies to send their garbage to be
incinerated in Tokyo, whereas they have to pay Alfo ¥23/kg to
have it recycled. As Shimadu-san at the Ministry pointed out to
me, other prefectures, such as Tama-city, set a higher price for
garbage collection, and this shifts much more waste into recycling.

24. Furthermore, frying food at such high temperatures denatures
some of its most valuable nutrients. Tanami-san argued that water-
soluble proteins are not badly affected because they are cooked in

oil, but Dr Kawashima was doubtful if this was the case and agreed that nutritional value would be impaired; cf. Kawashima (n.d.).

25. For every tonne of food waste that Alfo collects, it makes a gross taking of ¥25,200 (¥23,000 collection fee and ¥2,200 for the resulting 220 kg of animal feed after drying), but about ¥5,000 of that goes to cover the cost of electricity, gas and water. Capital costs were ¥1.3bn for the machinery, ¥754m for the building and ¥750m for the land.

26. Yakou-san finds that machinery can remove soft plastic satisfactorily, but not brittle hard plastic.

27. The Odakyu Food Ecology Centre cost ¥200m for machinery and adjustments to the building, but unlike Alfo it did not buy the building or land. However, the Odakyu plant currently employs ten people, against Alfo's three, and its capacity is less than a third of Alfo's.

28. Interview, 7 June 2008.

29. The rate varies from ¥20 to ¥25/kg for low-nutrition food like vegetables and ¥15 for rice.

30. Takahashi has written a forthcoming article, 'Generation Control of Food Waste', outlining this phenomenon.

31. Cf. Yang *et al.* (2006); Ruiz López *et al.* (2003).

32. Commercial feed is ¥50/kg dry matter, and he sells his for ¥25/kg, unless the farmer comes to collect it himself, in which case he will sell it for ¥7–8/kg to save on the high transport costs.

33. Interview, 3 June 2008.

18. Action Plan: A Path to Utrophia

1. Stevenson (1948), p. 719.
2. Adrien Assous, personal communication.
3. Jones (2004a).
4. Food Chain Centre (2006).

Afterword

1. UN Population Division (2007).
2. UNEP (2009).
3. In addition, the Brazilian government launched the Amazon Fund in August 2008 to collect money from foreign governments to pay for the protection of the Amazon. The Norwegian government pledged $1bn (£668m) for the fund in September 2008: Phillips (2008b).
4. Diamond (2006), pp. 311–28.
5. The fertility controls now actually permit more than one child in certain regions and some circumstances, such as if both parents of a couple were single children.
6. Diamond (2006), pp. 358–77.
7. ibid., pp. 360, 373.

Bibliography

Ababouch, L. (2003), 'Impact of fish safety and quality on food security', in *Report of the Expert Consultation on International Fish Trade and Food Security, Casablanca, Morocco, 27–30 January 2003*, FAO Fisheries Report, No. 708, Rome, FAO

Abder-Rahman, H. *et al.* (2000), 'Aluminum phosphide fatalities. New local experience', *Med. Sci. Law* 40:2, 164–8

Acheson, D. (1998), *Independent Inquiry into Inequalities in Health*, London, The Stationery Office

Ackerman, F. (1997), 'Environmental impacts of packaging in the U.S. and Mexico', *Phil & Tech* 2:2

Adam, D. (2009), 'Amazon could shrink by 85% due to climate change, say scientists', *Guardian*, 16 March 2009

Adelson, S. *et al.* (1961), 'Household records of foods used and discarded', *J. Amer. Dietetic Assoc.* 39, 578

Adelson, S. *et al.* (1963), 'Discard of edible food in households', *J. Home Econ.* 55, 633

Adhikari, B. *et al.* (2006), 'Predicted growth of world urban food waste and methane production', *Waste Management and Research* 24:5, 421

AEA (2007), *Resource Use Efficiency in Food Chains: Priorities for Water, Energy and Waste Opportunities*, report to Defra, Didcot, AEA Energy & Environment, http://sciencesearch.defra.gov.uk/Document.aspx?Document=WU0103_4830_FRA.pdf

Agutter, W. (1796), *The Sin of Wastefulness*, London, F. & C. Rivington

Ahlberg, P. (2002), 'Whey-hey! Humble dairy by-product makes good', 7 January 2002, www.just-food.com

Ajila, C. *et al.* (2007), 'Valuable components of raw and ripe peels from two Indian mango varieties', *Food Chemistry* 102, 1006–11

Al Seadi, T. and J. Holm-Nielsen (eds.) (2003), *The Future of Biogas in Europe II*, University of Southern Denmark, 2–4 October

Aldred, J. and A. Benjamin (2007), 'Campaigners condemn lack of commitment to full marine bill', *Guardian*, 6 November 2007

[Algeria (n.d.)], *Contraventions et leurs sanctions*, http://www.lexinter.net/DZ/contraventions_et_leurs_sanctions.htm

Allen, K. (1998), 'Sharing scarcity: bread rationing and the First World War in Berlin, 1914–1923', *J. Social History*, December 1998

Alverson, D. *et al.* (1994), *A Global Assessment of Fisheries By-Catch and Discards*, Fisheries Technical Paper No. 339, Rome, FAO

Ambler-Edwards, S. *et al.* (2009), *Food Futures: Rethinking UK Strategy*, Chatham House report, London, Royal Institute of International Affairs, http://www.chathamhouse.org.uk/files/13248_r0109food futures.pdf

Anon. (1872), 'Waste of food', Letters to the Editor, *The Times*, 2 September 1872, p. 9, Issue 27471; col. E

Anon. (2006), 'Safeway donating 1.8 million holiday meals to food banks', *Progressive Grocer*, 7 December 2006

Anon. (n.d.), 'The European waste catalogue', http://212.104.147.54/media/pdf/q/3/The_European_Waste_Catalogue.pdf (accessed 26 March 2009)

Arita, E. (2008), 'Hazoken delights', *Japan Times*, 20 April 2008

Armitage, T. (2004), 'Post-harvest loss costs East African milk market $90m', *Dairy Processing & Markets*, 27 October 2004

Arvanitoyannis, I. and D. Ladas (2008), 'Meat waste treatment methods and potential uses', *Int. J. Food Sci. and Technol.* 43, 543–59

Asda (2006), 'Asda launches "Zero waste to landfill" target', press release, 25 July 2006, http://www.asda-press.co.uk/pressrelease/35

Asda (2008), 'Asda unveils UK's first "zero waste to landfill" supermarket', press release, 26 October 2008, http://www.asda-press.co.uk/pressrelease/59 (accessed 16 January 2009)

Ashton, E. (2006), 'Eat up or pay up', *Guardian*, 21 April 2006

Associated Press (2008), 'NYC eatery imposes surcharge for unfinished food', www.foxnews.com, 11 December 2008

Australia Institute (2005), *Wasteful Consumption in Australia*, Canberra, https://www.tai.org.au/?q=node/8&offset=3

Avfall Sverige (2007) 'Swedish waste management', http://www.

avfallsverige.se/m4n?oid=english&_locale=1 (accessed 12 March 2009)

Awarenet (2004), *Handbook for the Prevention and Minimisation of Waste and Valorisation of By-Products in European Agro-Food Industries*, http:// eea.eionet.europa.eu (accessed 1 March 2009)

Baines, A. and D. Hollingsworth (1961), *Family Living Studies – A Symposium. A Survey of Food Consumption in Great Britain*, Geneva, ILO

Baloch, U. (n.d.), 'Wheat: Post-harvest operations', in *Compendium on Post-Harvest Operations*, Rome, FAO, http://www.fao.org/inpho/ content/compend/text/ch06-02.htm (accessed 26 March 2009)

Baloch, U. *et al.* (1994), 'Loss assessment and loss prevention in wheat storage – technology development and transfer in Pakistan', in E. Highley (ed.), *Stored Product Protection*, Wallingford, CABI Publishing

Barrionuevo, A. (2008), ' "Stagnation" made Brazil's environment chief resign', *The New York Times*, 16 May 2008

Bartholomew, D. (2002), *The Big Beat of Dave Bartholomew: 20 of His Milestone New Orleans Productions 1949–1960* (album), Crescent City Soul

Barton, A. *et al.* (2000), 'High food wastage and low nutritional intakes in hospital patients', *Clinical Nutrition* 19:6, 445

Bas, R. (2008), 'Enthusiasms and forebodings', *Manila Times*, 21 April 2008

Bataille, G. (1985), *Visions of Excess: Selected Writings, 1927–1939*, tr. Allan Stoekl, Manchester University Press

Bathurst, C. (1921), *Potatoes and Pigs with Milk, as the Basis of Britain's Food Supply*, 2nd edn, London, Hugh Rees

Baylis, K. (2008), 'Food tax reform – digesting the $6 billion mandate', presentation at Centre for Science in the Public Interest conference, *Championing Public Health Nutrition*, Ottawa, 22–23 October, http:// www.cspinet.org/canada/2008conference/program.html

BBC (2005), 'Inside out: supermarket landfills', 21 February 2005, http://www.bbc.co.uk/insideout/yorkslincs/series7/supermarket _landfills.shtml

BBC (2008), 'The cost of food: facts and figures', 16 October 2008, http://news.bbc.co.uk/2/hi/in_depth/7284196.stm

BBC Radio 4 (2005), Costing the Earth, 14 April 2005

BBC Radio 4 (2008a), The Food Programme, 6 January 2008

BBC Radio 4 (2008b), The Moral Maze, 26 March 2008

BBC WW2 archive (2005), People's WW2, http://www.bbc.co.uk/ww2peopleswar/stories/26/a4124026.shtml

Bellarby, J. et al. (2008), Cool Farming: Climate Impacts of Agriculture and Mitigation Potential, report produced by the University of Aberdeen for Greenpeace, Amsterdam

Bender, W. (1994), 'An end use analysis of global food requirements', Food Policy 19:4, 381–95

Bender, W. and M. Smith (1997), 'Population, food, and nutrition', Population Bull. 51:4, 2–47

Berghoff, H. (2001), 'Enticement and deprivation: the regulation of consumption in pre-war Nazi Germany', in M. Daunton and M. Hilton (eds.), The Politics of Consumption, Oxford, Berg Publishers, 165–84

Bettoli, P. and G. Scholten (2006), 'Bycatch rates and initial mortality of paddlefish in a commercial gillnet fishery', Fisheries Research 77:3, 343–7

Blair, D. and J. Sobal (2006), 'Luxus consumption: wasting food resources through overeating', Agriculture and Human Values 23, 63–74

Blanco, M. et al. (2007), 'Towards sustainable and efficient use of fishery resources: present and future trends', Trends in Food Sci. & Technol. 18:1, 29–36

BLS (2008), 'Consumer expenditure survey', Bureau of Labor Statistics (BLS) of US Dept of Labor, http://www.bls.gov/cex/

Boersma, L. and I. Murarka (1995), Waste Management and Utilization in Food Production and Processing, Task Force Report No. 124, Ames, Iowa, Council for Agricultural Science and Technology

Bokanga, M. (1996), 'Cassava: post-harvest operations', in Compendium on Post-Harvest Operations, Rome, FAO, http://www.fao.org/inpho/content/compend/text/ch12-04.htm#TopOfPage

[Bonaparte, N. (ed.)] (1810), *Code Pénal de 1810*, 1st edn, http://ledroit criminel.free.fr/la_legislation_criminelle/anciens_textes/code _penal_1810/code_penal_1810_4.htm

Borger, J. (2008a), 'UN chief calls for review of biofuels policy', *Guardian*, 5 April 2008

Borger, J. (2008b), 'Rich countries launch great land grab to safeguard food supply', *Guardian*, 22 November 2008

Bracken, C. (1997), *The Potlatch Papers: A Colonial Case History*, Urbana, Chicago University Press

Braschkat, J. *et al.* (2003), 'Biogas versus other biofuels: a comparative environmental assessment', in T. Al Seadi and J. Holm-Nielsen (eds.), *The Future of Biogas in Europe II*, University of Southern Denmark, 2–4 October 2003

Braun, J. (2007), *The World Food Situation*, Food Policy Report 18, Washington DC, IFPRI

Brown, C. (2008), 'Knobbly fruit and veg back on menu as EU plans to scrap uniformity laws', *Independent*, 16 June 2008

Brown University Faculty (1990), 'Overcoming hunger. Promising programmes and policies', *Food Policy* 15:4, 286–98

Bruinsma, J. (ed.) (2003), *World Agriculture: Towards 2015/2030. A FAO Perspective*, Rome/London, FAO/Earthscan Publications

Bruun, S. *et al.* (2006), 'Application of processed organic municipal solid waste on agricultural land – a scenario analysis', *Environmental Modeling and Assessment* 11, 251–65

Bugge, A. (2004), 'Amazon burning makes Brazil a top greenhouse gas polluter', 19 July 2004, http://www.mongabay.com/external/ brazil_top_Co2_polluter.htm

Burn, R. *et al.* (1831), *The Justice of the Peace and Parish Officer*, London, S. Sweet

Buzby, J. and J. Perry (2006), 'ERS food availability data under revision', *Amber Waves: Behind the Data* 4:3, Economic Research Service, US Department of Agriculture, http://www.ers.usda.gov/ AmberWaves/June06/Indicators/behinddata2.htm

Buzby, J. *et al.* (2009), 'Supermarket loss estimates for fresh fruit, vegetables, meat, poultry, and seafood and their use in the ERS loss-adjusted food availability data', *Economic Information Bulletin*, No. EIB-44, March 2009, Economic Research Service, US Department of Agriculture, http://www.ers.usda.gov/Publications/EIB44 /EIB44.pdf

Cabinet Office (UK) (2008), *Food Matters: Towards a Strategy for the 21st Century*, July 2008, The Strategy Unit, Cabinet Office

Callaghana, J. *et al.* (2002), 'Continuous co-digestion of cattle slurry with fruit and vegetable wastes and chicken manure', *Biomass and Bioenergy* 22:1, 71–7

Calverley, D. (1996), *A Study of Loss Assessment in Eleven Projects in Asia Concerned with Rice,* Programme PPA/PFL, Rome, FAO

Carbon Trust (n.d.), 'Climate change and equity valuations', www .carbontrust.co.uk

Carlsson-Kanyama, A. (1998), 'Climate change and dietary choices – how can emissions of greenhouse gases from food consumption be reduced?', *Food Policy* 23:3/4, 277–93

Carrefour (2008), *2007 Sustainability Report: Carrefour Group Building Responsible Relationships*, http://www.carrefour.com/docroot/ groupe/C4com/Commerce%20responsable/Publications/RDD% 202007%20GB.pdf

Carrier, L. (1998), *Illinois: Crossroads of a Continent*, Urbana, University of Illinois Press

Cathcart, E. and A. Murray (1939), 'A note on the percentage loss of calories as waste in ordinary mixed diets', *J. Hygiene* 39, 45–50

Chalmin, P. and C. Gaillochet (2009, forthcoming), *World Waste Survey 2009*, CyclOpe report, Economica

Chapagain, A. and A. Hoekstra (2008), 'The global component of freshwater demand and supply: an assessment of virtual water flows between nations as a result of trade in agricultural and industrial products', *Water International* 33:1, 19–32

China Daily (2008), 'Japanese dispose one-fourth of available food', *China Daily: Asianewsnet*, 3 July 2008, http://www.chinadaily.com .cn/world/2008-07/03/content_6817431.htm

Chomitz, K. (2006), *At Loggerheads: Agricultural Expansion, Poverty*

Reduction, and Environment in the Tropical Forests, World Bank Policy Research Report, Washington DC, World Bank

Chun, S. *et al.* (1986), *Food Handling Practices and Processing at Village and Household Levels in Republic of Korea*, Rural Nutrition Institute, Rural Development Administration for FAO

CIA (Central Intelligence Agency) (2009), *The 2008 World Factbook*, Washington DC, CIA, www.cia.gov

CII-McKinsey & Co. (1997), *Modernizing the Indian Food Chain: Food & Agriculture Integrated Development Action Plan*, New Delhi, Confederation of Indian Industry and McKinsey & Co.

Clement of Alexandria (2008), 'Who is the Rich Man That Shall be Saved? (*Qui Dives Salvetur*)', tr. W. Wilson, New Advent, http://www.newadvent.org/fathers/0207.htm

Clover, C. (2004), *The End of the Line: How Overfishing is Changing the World and What We Eat*, London, Ebury Press

CNN (2008), 'Great Depression holds lessons for surviving tough economy: Great Depression in Globe, Arizona', CNN.com, 13 October 2008, http://www.ireport.com/docs/DOC-109218

Cohen, M. (1977), *The Food Crisis in Prehistory: Overpopulation and the Origins of Agriculture*, New Haven, Conn., Yale University Press

COM (2007a), *Communication from the Commission to the Council and the European Parliament on the Interpretative Communication on Waste and By-Products*, Brussels, Commission of the European Communities, http://eur-lex.europa.eu/LexUriServ/LexUriServ.do?uri=COM:2007:0059:FIN:EN:DOC (accessed 26 March 2009)

COM (2007b), *Communication from the Commission to the Council and the European Parliament: A Policy to Reduce Unwanted By-Catches and Eliminate Discards in European Fisheries*, 136 final, Brussels, Commission of the European Communities

Competition Commission (1985), 'The market for the supply of animal waste in Great Britain', in Competition Commission, *Animal Waste: A Report on the Supply of Animal Waste in Great Britain*, Great Britain, HMSO, http://www.competition-commission.org.uk/rep_pub/reports/1985/fulltext/187c02.pdf (accessed 25 March 2009)

Competition Commission (2008), *The Supply of Groceries in the UK: Market Investigation*, Great Britain, Stationery Office Books, http:// www.competition-commission.org.uk/rep_pub/reports/2008/full text/538.pdf (accessed 25 March 2009)

Co-op (2008), *The Co-op Sustainability Report 2007/08. Altogether Different and Making a Difference*, http://www.co-operative.coop/ Corporate/PDFs/sustainability%20report%20200708_final.pdf (accessed 16 January 2009)

Copa and Cogeca (2008), 'EC ignoring interests of European fruit and veg sector in dismantling marketing standards – industry body', 13 November 2008, http://www.flex-news-food.com/pages/ 20442/European-Commission/Fruit/Vegetable/ec-ignoring -interests-european-fruit-veg-sector-dismantling-marketing -standards–industry-body.html

Coulter, J. (1991), 'The case for bulk storage and handling of wheat in Pakistan', in *Compendium on Post-Harvest Operations*, Rome, FAO, http://www.fao.org/inpho/content/compend/text/ch06-02.htm

Coulter, J. (n.d.), 'Making the transition to a market-based grain marketing system', submitted for publication, www.nri.org/docs/ grainmarket.pdf

Cowling, S. *et al.* (2004), 'Contrasting simulated past and future responses of the Amazonian forest to climate change', *Philosophical Trans. R. Soc.* 359, 539–47

Cox, P. *et al.* (2006), 'Conditions for sink-to-source transitions and runaway feedbacks from the land carbon cycle', in H. Schellnhuber *et al.* (eds.), *Avoiding Dangerous Climate Change*, Cambridge University Press

Cranfield University (2008), 'Waste in the food chain', unpublished scoping presentation, 2 July 2008 (due to report in June 2009)

Crooks, E. and F. Harvey (2008), 'Suspend biofuel rules, say MPs', *Financial Times*, 2 May 2008

C-Tech Innovation Ltd (2004), *United Kingdom Food and Drink Processing Mass Balance: A Biffaward Programme on Sustainable Resource Use*, http://www.ctechinnovation.com/publications.htm

Curtis, E. (1914), *In the Land of the War Canoes: A Drama of Kwakiutl*

Life in the North West (originally *In the Land of the Head-Hunters*), restored (1972) by B. Holm *et al.*, Canada and USA, University of Washington Press/Milestone Film and Video

Curtis, K. (1997), 'Urban poverty and the social consequences of privatised food assistance', *J. Urban Affairs* 19, 207–26

Dairy Crest (2008), 'Our brands & products', http://www.dairy crest.co.uk/our-brands–products.aspx (accessed 2 February 2009)

Dalgaard, R. *et al.* (2007), 'Danish pork production: an environmental assessment', *DJF Animal Science* 82.

Dalgaard, R. *et al.* (2008), 'LCA of soybean meal', *Int. J. Life Cycle Assessment* 13:3, 240–54

Dalloz (ed.) (2001), *Code Pénal*, Dalloz, Paris

Danish Ministry of the Environment (2004), 'Industrial food waste: from pig swill to biogas', *Danish Environment Newsletter*, 29 February 2004, http://www.mex.dk/uk/vis_nyhed_uk.asp?id=6441&ny hedsbrev_id=939

Darlington, R. and S. Rahimifard (2006a), 'A responsive demand management framework for the minimization of waste in convenience food manufacture', *Int. J. Computer Integrated Manufacturing* 19:8, 751–61

Darlington, R. and S. Rahimifard (2006b), 'Improving supply chain practices for minimizing waste in chilled ready-meal manufacture', *Food Manufact. Eff.* 1, 15–23

Darlington, R. and S. Rahimifard (2007), 'Hybrid two-stage planning for food industry overproduction waste minimization', *Int. J. Production Research* 45:18–19, 4273–88

Dawe, D. (2008), *Have Recent Increases in International Cereal Prices been Transmitted to Domestic Economies? The Experience in Seven Large Asian Countries*, ESA Working Paper No. 08-03, Rome, Agricultural Development Economic Division, FAO

de Haan, C. *et al.* (2001), *Livestock Development: Implications for Rural Poverty, the Environment and Global Food Security*, Washington DC, World Bank

Defra (2006a), *Food Industry Sustainability Strategy (FISS)*, London, Department for Environment, Food and Rural Affairs

Defra (2006b), 'Regulation (EC) No 1774/2002 Laying down health rules concerning animal by-products not intended for human consumption. Guidance notes (non statutory) on the disposal of animal by-products, including former foodstuffs of animal origin, from food outlets', http://www.defra.gov.uk/animalh/by-prods/pdf/ffguidance1774-2002.pdf (accessed 12 March 2009)

Defra (2007a), *Report of the Food Industry Sustainability Strategy Champions' Group on Waste (CGW)*, Department for Environment, Food and Rural Affairs, London, www.defra.gov.uk/farm/policy/sustain/fiss/pdf/report-waste-may2007.pdf (accessed 25 March 2009)

Defra (2007b), *UK Biomass Strategy*, London, Department for Environment, Food and Rural Affairs, http://www.defra.gov.uk/ENVIRONMENT/climatechange/uk/energy/renewablefuel/pdf/ukbiomassstrategy-0507.pdf (accessed 25 March 2009)

Defra (2007c), 'Annex 1: Implications of rising agricultural commodity prices', in 'Monthly farming and food brief, September 2007', https://statistics.defra.gov.uk/esg/publications/monthly%20brief/Annex%201%20Food%20and%20farming%20brief%20-%20impact%20of%20high%20commodity%20prices.pdf (accessed 25 March 2009)

Defra (2007d), 'Monthly farming and food brief, September 2007', https://statistics.defra.gov.uk/esg/publications/monthly%20brief/Sept07_FarmingFood_Brief.pdf (accessed 25 March 2009)

Defra (2008a), 'Animal by-products: former foodstuffs – questions and answers', http://www.defra.gov.uk/animalh/by-prods/wastefood/formerfoodstuffs-qa.htm#7 (accessed 12 March 2009)

Defra (2008b), 'Guidelines to Defra's greenhouse gas conversion factors for company reporting', http://www.defra.gov.uk/environment/business/envrp/pdf/ghg-cf-guidelines2008.pdf (accessed 25 March 2009)

Defra (2008c), 'Guidelines to Defra's greenhouse gas conversion factors for company reporting – Annexes', http://www.defra.gov.uk/environment/business/envrp/pdf/ghg-cf-guidelines-annexes2008.pdf (accessed 25 March 2009)

Defra (2009), 'What happens to waste', www.defra.gov.uk/environment/waste/topics/ (accessed 12 March 2009)

Defra (forthcoming), 'Evidence on the role of supplier–retailer trad ing relationships & practices on waste generation in the food chain', Defra project number FO0210, http://randd.defra.gov.uk/Default.aspx?Menu=Menu&Module=More&Location=None&ProjectID=15806&FromSearch=Y&Publisher=1&SearchText=FO0210&SortString=ProjectCode&SortOrder=Asc&Paging=10#Description (accessed 25 March 2009)

Defra (forthcoming), 'Greenhouse gas impacts of food retailing', Defra project number FO0405, Brunel University, http://randd.defra.gov.uk/Default.aspx?Menu=Menu&Module=More&Location=None&ProjectID=15805&FromSearch=Y&Publisher=1&SearchText=FO0405&SortString=ProjectCode&SortOrder=Asc&Paging=10#Description (accessed 25 March 2009)

DFID (2004), 'Official development assistance to agriculture', UK Department for International Development, http://www.isgmard.org.vn/Information%20Service/Report/General/oda%20in%20agriculture-DFID.pdf

Diamond, J. (1992), *The Third Chimpanzee: The Evolution and Future of the Human Race*, London, HarperCollins

Diamond, J. (1999), *Guns, Germs and Steel: The Fates of Human Societies*, New York, W. W. Norton

Diamond, J. (2006), *Collapse: How Societies Choose to Fail or Survive*, London, Penguin Books

Dilly, G. and C. Shanklin (1998), 'Characterization of waste in two military hospital foodservice operations', *J. Amer. Dietetic Assoc.* 98:9, Supplement 1, A09

Dobrzynski, T. *et al.* (2002), *Oceans at Risk: Wasted Catch and the Destruction of Ocean Life*, Washington DC, Oceana

Douglas, M. (1990), 'Introduction', in M. Mauss, *The Gift: The Form and Reason for Exchange in Archaic Societies*, London and New York, Routledge

Dowler, E. (1977), 'A pilot survey of domestic food wastage', *J. Human Nutrition* 31, 171–80

Dowler, E. and Y. Seo (1985), 'Assessment of energy intake: estimates of food supply v measurement of food consumption', *Food Policy* 10:3, 278–88

Doyle, L. (2007), 'US food aid is "wrecking" Africa, claims charity', *Independent*, 17 August 2007

Dunnell, R. and D. Greenlace (1999), 'Late Woodland period "waste" reduction in the Ohio river valley', *J. Anthropol. Archaeol.* 18:3, 376–95

Dunphy, J. (1995), 'Food banks fight against hunger and wastage in Europe', Accueil: Fédération Européene des Banques Alimentaires

Duxbury, J. (1994), 'The significance of agricultural sources of greenhouse gases', *Nutrient Cycling in Agroecosystems* 38:2, 151–63

EC (2000), 'Directive 2000/13/EC of the European Parliament and of the Council of 20 March 2000 on the approximation of the laws of the Member States relating to the labelling, presentation and advertising of foodstuffs', http://eur-lex.europa.eu/LexUriServ/LexUriServ.do?uri=CONSLEG:2000L0013:20070112:EN:PDF

EC (2005a), 'Report from the Commission to the Council and the European Parliament on the national strategies for the reduction of biodegradable waste going to landfills pursuant to Article 5(1) of Directive 1999/31/EC on the landfill of waste', 30 March 2005, http://eur-lex.europa.eu/LexUriServ/LexUriServ.do?uri=CELEX:52005DC0105:EN:NOT (accessed 26 March 2009)

EC (2005b), 'Commission staff working document; annex to the report from the Commission to the Council and the European Parliament on the national strategies for the reduction of biodegradable waste going to landfills pursuant to Article 5(1) of Directive 1999/31/EC on the landfill of waste {Com(2005) 105 Final}', 30 March 2005, http://ec.europa.eu/environment/waste/pdf/sec_2005_404.pdf (accessed 26 March 2009)

EC (2006), *Environmental Impact of Products (EIPRO) Analysis of the Life Cycle Environmental Impacts Related to the Final Consumption of the EU-25, Main Report*, Brussels, Joint Research Centre, European Commission

EC (2007a), 'Catching only what we need', 30 March 2007, http://

ec.europa.eu/news/agriculture/070330_1_en.htm (accessed 26 March 2009)

EC (2007b), 'What are the reasons for discarding at sea?', 28 March 2007, http://ec.europa.eu/fisheries/cfp/management_resources/conservation_measures/reasons_en.htm

EC (2008a), 'Proposal for a Council Regulation amending Regulation (EC) No. 1290/2005 on the financing of the common agricultural policy and Regulation (EC) No. 1234/2007 establishing a common organisation of agricultural markets and on specific provisions for certain agricultural products (Single CMO Regulation) as regards food distribution to the most deprived persons in the Community', 17 September 2008, http://eur-lex.europa.eu/LexUriServ/LexUriServ.do?uri=COM:2008:0563:FIN:EN:PDF (accessed 26 March 2009)

EC (2008b), 'Commission Staff Working Document accompanying the proposal for a Council Regulation amending Regulations (EC) No. 1290/2005 on the financing of the common agricultural policy and (EC) No. 1234/2007 establishing a common organisation of agricultural markets and on specific provisions for certain agricultural products (Single CMO Regulation) as regard food distribution to the most deprived persons in the Community: impact assessment {Com(2008) 563}', http://ec.europa.eu/agriculture/markets/freefood/fullimpact_en.pdf (accessed 26 March 2009)

EC (2008c), 'The return of the curvy cucumber: Commission to allow sale of "wonky" fruit and vegetables', IP/08/1694, 12 November 2008, Brussels, http://europa.eu/rapid/pressReleasesAction.do?reference=IP/08/1694&format=HTML&aged=0&language=EN&guiLanguage=en (accessed 26 March 2009)

EC (2008d), 'EU and Norway reaffirm their determination to reduce discards', 10 October 2008, http://ec.europa.eu/fisheries/press_corner/press_releases/2008/com08_70_en.htm (accessed 26 March 2009)

EC (2008e), 'Free food for Europe's poor', 17 September 2008, http://ec.europa.eu/agriculture/markets/freefood/index_en.htm (accessed 26 March 2009)

EC (2009), 'Waste', http://ec.europa.eu/environment/waste/index.
htm (accessed 26 March 2009)

EC STECF (2006), *Discarding by EU fleet*, report of the Scientific,
Technical and Economic Committee for Fisheries – Commission
Staff Working Paper, 9–12 October 2006, Brussels, European Com-
mission, http://ec.europa.eu/fisheries/legislation/reports_en.htm

Eccleston, P. (2008), 'McDonald's: waste recycling scheme a success',
Daily Telegraph, 7 April 2008

Eckholm, E. (1976), *Losing Ground: Environmental Stress and World Food
Prospects*, W. W. Norton, New York

Economic Times (India) (2008), '72% of India's fruit, vegetable goes to
waste', 12 May 2008

Economist (2007), 'Cheap no more', 8 December 2007

Economist (2008a), 'Managed to death', 1 November 2008

Economist (2008b), 6 December 2008

Economist (2009a), 'A special report on the sea', 3 January 2009

Economist (2009b), 'A special report on waste', 28 February 2009

Elsayed, M. *et al.* (2003), *Carbon and Energy Balances for a Range of
Biofuels Options*, report for UK Department of Trade and Industry
(DTI) Sustainable Energy Programmes, Resources Research Unit,
Sheffield Hallam University, http://www.berr.gov.uk/files/file
14925.pdf

Engström, R. and A. Carlsson-Kanyama (2004), 'Food losses in food
service institutions: examples from Sweden', *Food Policy* 29, 203–13

Ennart, H. (2007), 'Var fjärde matkasse slängs' ('Every fourth bag of
food is thrown away'), *Svenska Dagbladet*, 25 April 2007, http://
svd.se/nyheter/inrikes/artikel_221563.svd

Environment Agency (England & Wales) (2005a), 'National waste
production survey for commerce and industry: strategic waste
management assessment', http://www.defra.gov.uk/environment/
statistics/waste/wreuwastestats.htm or http://www.defra.gov.uk/
environment/statistics/waste/wrindustry.htm; see also AEA (2007)
and C-Tech Innovation Ltd (2004)

Environment Agency (England & Wales) (2005b), 'Waste production
survey for commerce and industry: Wales', http://www.environ

ment-agency.gov.uk/regions/wales/816243/1220048/1223323/
?version=1&lang=e

Environment Agency (England & Wales) (2008), 'Commercial &
industrial waste in Wales 2002–3', http://www.environment-
agency.gov.uk/research/library/publications/103522.aspx (accessed
25 March 2009)

Environmental News Service (2008), 'One million tons of North Sea
fish discarded every year', 5 November 2008, http://www.ens-news
wire.com/ens/nov2008/2008-11-05-03.asp

Environmental Protection Agency (Northern Ireland) (2007), *National
Waste Report 2007*, Wexford, Environmental Protection Agency,
http://www.epa.ie/downloads/pubs/waste/stats/epa_national
_waste_report_20071.pdf (accessed 26 March 2009)

Environmental Protection Agency (Northern Ireland) (n.d.), 'Waste
data surveys', http://www.ni-environment.gov.uk/waste/waste-
publications-2.htm (accessed 26 March 2009)

Environmental Protection Agency, Scottish (n.d.), 'Commercial and
industrial waste reporting', http://www.sepa.org.uk/waste/waste
_data/commercial_industrial_waste/c_i_waste_reporting.aspx
(accessed 25 March 2009)

EPA (US) (2008), *Municipal Solid Waste in the United States: 2007 Facts
and Figures, EPA530-R-06-011*, Washington DC, Environmental
Protection Agency, http://www.epa.gov/osw/nonhaz/municipal/
pubs/msw07-rpt.pdf

ERS (2008a), 'Food expenditure tables', Economic Research Service,
US Department of Agriculture, http://www.ers.usda.gov/briefing/
CPIFoodAndExpenditures/Data/

ERS (2008b), 'Loss-adjusted food availability (also known as food
guide pyramid servings data), updated 15 March 2008, Economic
Research Service, US Department of Agriculture, http://www.er-
s.usda.gov/data/foodconsumption/ (accessed 1 March 2009)

Eshel, G. and P. Martin (2006), 'Diet, energy, and global warming',
Earth Interactions 10:9, 1–17

Eurostat (2009), 'Prodcom: statistics on the production of manufactured goods', http://epp.eurostat.ec.europa.eu/portal/page?_pageid

=2594,63266845&_dad=portal&_schema=PORTAL (accessed 25 March 2009)

Evans, E. (2005), *Food Waste Opportunities in the North East*, Bio Recycling Solutions Ltd

Evolve (2007), *Study on the Economic Benefits of Waste Minimisation in the Food Sector: Final Report, January 2007*, ed. M. Betts *et al.*, Integrated Skills Limited in co-operation with Fareshare, London, http://www.integrated-skills.com/ISL/Files/Evolve%20Food%20 Waste%20Minimisation%20Final%20Report%208-1-07.pdf

Fairlie, S. (2008), 'Can Britain feed itself?', *The Land* 4, 18–26

Falkenmark, M. and J. Rockström (2004), *Balancing Water for Humans and Nature: The New Approach in Ecohydrology*, London, Earthscan Publications

Faminov, D. and S. Vosti (1998), 'Livestock–deforestation links: policy issues in the Western Brazilian Amazon', in A. Nell (ed.), *Proc. Int. Conf. Livestock and the Environment*, 16–20 June 1997, International Agricultural Centre, Ede/Wageningen, The Netherlands, http://www.fao.org/wairdocs/LEAD/X6130E/X6130E00 .htm#TOC

FAO (1973), 'Food balance sheets', FAOSTAT (Food & Agriculture Organization Statistics) database, Rome, FAO, http://faostat.fao .org/ (accessed 12 March 2009)

FAO (1981), *Food Loss Prevention in Perishable Crops*, FAO Agricultural Services Bulletin No.43, Rome, FAO

FAO (1989), *Prevention of Post-Harvest Food Losses: Fruits, Vegetables and Root Crops*, Rome, FAO

FAO (2000), *The State of Food Insecurity in the World*, Rome, FAO

FAO (2001), *Food Balance Sheets: A Handbook*, Rome, FAO

FAO (2003a), 'Food balance sheets', FAOSTAT (Food & Agriculture Organization Statistics) database, Rome, FAO, http://faostat.fao .org/site/368/default.aspx (accessed 12 March 2009)

FAO (2003b), *World Agriculture: Towards 2015/2030*, Rome, FAO, ftp://ftp.fao.org/docrep/fao/004/y3557e/y3557e.pdf

FAO (2005), *The State of Food and Agriculture 2005*, Rome, FAO, http://www.fao.org/docrep/008/a0050e/a0050e00.htm (accessed 26 March 2009)

FAO (2006), *World Agriculture: Towards 2030/2050*, Rome, FAO, http://www.fao.org/es/ESD/AT2050web.pdf

FAO (2008a), *The State of Food and Agriculture 2008*, Rome, FAO, http://www.fao.org/docrep/011/i0100e/i0100e00.htm (accessed 26 March 2009)

FAO (2008b), *The State of Food Insecurity in the World 2008*, Rome, FAO

FAO (2008c), *Crop Prospects and Food Situation*, Rome, FAO, http://www.fao.org/docrep/010/ai465e/ai465e01.htm (accessed 26 March 2009)

FAO (2008d), *Food Security Statistics*, updated 7 August 2008, Rome, FAO, http://www.fao.org/faostat/foodsecurity/ (accessed 26 March 2009)

FAO (2009), *The State of World Fisheries and Aquaculture 2008*, Rome, FAO

FAO (n.d.), *Compendium on Post-Harvest Operations*, Rome, FAO, http://www.fao.org/inpho/content/compend/text/ch06-02.htm (accessed 26 March 2009)

FDF (2008a), 'Food and drink manufacturers cut carbon emissions by 17%', press release, 27 November 2008, London, Food and Drink Federation, http://www.fdf.org.uk/news.aspx?article=4181&news indexpage=2 (accessed 5 March 2009)

FDF (2008b), *Our Five-Fold Environmental Ambition: Progress Report 2008*, London, Food and Drink Federation, https://www.fdf.org.uk/publicgeneral/environment_progress_report_finalversion.pdf

Feeley, K. *et al.* (2007), 'Decelerating growth in tropical forest trees', *Ecol. Lett.* 10, 461–9

Fehr, M. *et al.* (2000), 'A practical solution to the problem of household waste management in Brazil', *Resources, Conservation and Recycling* 30:3, 245–57

Fehr, M. *et al.* (2002), 'The basis of a policy for minimizing and recycling food waste', *Environmental Sci. & Policy* 5, 247–53

Fisher, F. (1963), 'A theoretical analysis of the impact of food surplus disposal on agricultural production in recipient countries', *J. Farm Economics* 45:4, 863–75

Fisher, K. *et al.* (2006), *Carbon Balances and Energy: Impacts of the Management of UK Wastes*, Defra R&D Project WRT 237, Final Report, ERM and Golder Associates

Fitzpatrick, M. (2005), 'What a waste – Japan faces up to food waste mountain', www.just-food.com, 2 June 2005

Fleming, N. (2008), 'Man's effect on world's oceans revealed', *Daily Telegraph*, 15 February 2008

Food and Drug Administration (US) (2005), 'The 2005 FDA Food Code questions and answers', http://vm.cfsan.fda.gov/~dms/fc05-qa.html

Food Chain Centre (2006), *Cutting Costs: Adding Value in Organics*, Watford, Institute of Grocery Distribution

Food Magazine (1997), 'Fruit for schoolchildren fed to animals', *The Food Magazine* 39

Food Marketing Institute (2007), 'Top U.S. supermarket & grocery chains (by 2006 grocery sales)', http://www.fmi.org/docs/facts_figs/faq/top_retailers.pdf#search="top%20u.s.%20supermarket%20%20%20grocery%20chains"

Food Processing Faraday Partnership (2008), 'Cutting out waste in food and drink: a resource efficiency strategy for the East Midlands food industry', March 2008, http://www.emda.org.uk/upload documents/CuttingoutWasteinFoodandDrinkExecutiveSummary. pdf

Forster, P. *et al.* (2007), 'Changes in atmospheric constituents and in radiative forcing', in S. Solomon *et al.* (eds.), *Climate Change 2007: The Physical Science Basis. Contribution of Working Group I to the Fourth Assessment Report of the Intergovernmental Panel on Climate Change*, Cambridge University Press

Fraiture, C. *et al.* (2007), 'Looking ahead to 2050: scenarios of alternative investment approaches', in D. Molden (ed.), *Water for Food, Water for Life: A Comprehensive Assessment of Water Management in Agriculture*, Colombo, Sri Lanka/London, IWMI/Earthscan Publications

Friends of the Earth (2002), 'Supermarkets and Great British fruit', media briefing, October 2002, http://www.foe.co.uk/resource/ briefings/supermarket_british_fruit.pdf (accessed 26 March 2009)

Friends of the Earth (2003), 'Farmers and the supermarket code of practice', press briefing, 17 March 2003, http://www.foe.co.uk/ resource/briefings/farmers_supermarket_code.pdf (accessed 26 March 2009)

Friends of the Earth (2005), 'Checking out the environment? Environmental impacts of supermarkets', briefing, June 2005, http://www. foe.co.uk/resource/briefings/checking_out_the_environme.pdf (accessed 26 March 2009)

Friends of the Earth (2008), 'Consultation response. Competition Commission: Provisional decision on remedies relating to supply chain practices', briefing, March 2008, http://www.foe.co.uk/ resource/consultation_responses/supply_chain_remedies.pdf (accessed 26 March 2009)

Friendship, C. and J. Compton (1991), 'Bag or bulk? A decision-making checklist', *Natural Resources Inst. Bull.* 45

Frow, J. (2003), 'Invidious distinction: waste, difference, and classy stuff', in G. Hawkins and S. Muecke (eds.), *Culture and Waste: The Creation and Destruction of Value*, Oxford, Rowman & Littlefield

FSA (UK) (2003), 'Use by date guidance notes', 27 February 2003, Food Standards Agency, http://www.food.gov.uk/foodindustry/ guidancenotes/labelregsguidance/usebydateguid (accessed 10 March 2009)

FSA (UK) (2008a), *Consumer Attitudes to Food Standards: Wave 8 UK Report Final*, London, Food Standards Agency, http://www.food. gov.uk/multimedia/pdfs/cas2007ukreport.pdf (accessed 10 March 2009)

FSA (UK) (2008b), *McCance and Widdowson's the Composition of Foods Integrated Dataset*, London, Food Standards Agency, http://www

.food.gov.uk/science/dietarysurveys/dietsurveys/ (accessed July 2008)

Fung, E. and W. Rathje (1982), 'How we waste $31 billion in food a year', in J. Hayes (ed.), *The 1982 Yearbook of Agriculture*, Washington DC, US Government Printing Office

Gadre, S. van and M. Woodburn (1987), 'Food discard practices of householders', *J. Amer. Dietetic Assoc.* 87:3, 322–9

Galer-Unti, R. (1995), *Hunger and Food Assistance in the United States*, New York, Garland Publishing Inc.

Gallo, A. (1980), 'Consumer food waste in the United States', *National Food Review* Fall, 13–16

Gardner, R. (2006), 'Repression cycles in the USSR', http://www.indiana.edu/~workshop/colloquia/papers/gardner_paper2.pdf

Garner, E. (2008), 'December market report 2008', TNS WorldPanel, http://www.tnsglobal.com/_assets/files/TNS_Market_Research_Market_Share_Nov08.htm (accessed 26 March 2009)

Garnett, T. (2006), *Fruit and Vegetables & UK Greenhouse Gas Emissions: Exploring the Relationship*, working paper produced as part of the work of the Food Climate Research Network, University of Surrey, Centre for Environmental Strategy

Garnett, T. (2007a), *Food Refrigeration: What is the Contribution to Greenhouse Gas Emissions and How Might Emissions be Reduced?*, working paper produced as part of the work of the Food Climate Research Network, University of Surrey, Centre for Environmental Strategy

Garnett, T. (2007b), 'Animal feed, livestock and greenhouse gas emissions: what are the issues?', paper presented to the Society of Animal Feed Technologists, Coventry, 25 January 2007

Garnett, T. (2008), *Cooking Up a Storm: Food, Greenhouse Gas Emissions and Our Changing Climate*, Food Climate Research Network, University of Surrey, Centre for Environmental Strategy

Geiger, B. *et al.* (2008), 'Evidence for defective mesolimbic dopamine exocytosis in obesity-prone rats', *The FASEB Journal* 22, 2740–46

Gelder, W. van *et al.* (2008), 'Soy consumption for feed and fuel in the European Union', research paper prepared for Milieudefensie (Friends of the Earth Netherlands)

Getlinger, M. *et al.* (1996), 'Food waste is reduced when elementary-school children have recess before lunch', *J. Amer. Dietetic Assoc.* 96:9, 906–8

Gianni, M. (2002), 'The entry into force of the 1995 UN Fish Stocks Agreement: an NGO perspective', *Reporter*, January/February 2002, No. 2, Internet Guide to International Fisheries Law, http://www.intfish.net/igifl/archive/ops/olp/papers/8.htm

Giedion, S. (1948), *Mechanization Takes Command: A Contribution to Anonymous History*, New York, Oxford University Press

Gillies, M. (1978), *Animal Feeds from Waste Materials*, Food Technology Review, Noyes Data Corporation

Gilman, E. *et al.* (2007), *Shark Depredation and Unwanted By-Catch in Pelagic Longline Fisheries: Industry Practices and Attitudes, and Shark Avoidance Strategies*, Honolulu, Western Pacific Regional Fishery Management Council

Goklany, I. (1998), 'Saving habitat and conserving biodiversity on a crowded planet', *BioScience* 48:11, 941–53

Gold, M. (2004), *The Global Benefits of Eating Less Meat*, Godalming, Compassion in World Farming Trust

Goletti, F. (2003), 'Current status and future challenges for the post-harvest sector in developing countries', *Acta Hort.* 628, 41–8

Goletti, F. and C. Wolff (1999), 'The impact of postharvest research', MSSD Discussion Paper No. 29, Washington DC, International Food Policy Research Institute

González Siso, M. (1996), 'The biotechnological utilization of cheese whey: a review', *Bioresource Technol.* 57, 1–11

Goodchild, C. and A. Thompson (2008), *Keeping Poultry and Rabbits on Scraps* (1941), London, Penguin

Greenpeace International (2006), *Eating Up the Amazon*, Amsterdam, Greenpeace

Griffiths, S. (2008), 'McDonald's hails success of waste-to-energy trial', 14 April 2008, http://www.computing.co.uk/business-green/news/2214193/mcdonald-hails-success-waste

Grolleaud, M. (2001), *Overview of the Phenomenon of Losses during the Post-Harvest System*, Rome, FAO

Groth, M. and S. Fagt (1997), *Danskernes kostvaner 1995. Teknisk rapport 1. Undersøgelsens tilrettelæggelse, gennemførelse og datakvalitet*, Søborg, Instituttet for Levnedsmiddelkemi og Ernæring, Levnedsmiddelstyrelsen

Halweil, B. and D. Nierenberg (2008), 'Meat and seafood: the global diet's most costly ingredients', in *State of the World: Innovations for a Sustainable Economy*, Washington DC, Worldwatch Institute

Hammond, R. (1954), *1911 – Food and Agriculture in Britain, 1939–45: Aspects of Wartime Control*, Stanford, Calif., Stanford University Press

Hanley, M. (1991), 'After the harvest', *World Development* 4:1, 25–7

Hansard (House of Commons Daily Debates) (2004), Westminster Hall Debate, 16 March 2004, http://www.publications.parliament.uk/pa/cm200304/cmhansrd/vo040316/halltext/40316h05.htm #40316h05_spnew0

Harris, M. (1975), *Cows, Pigs, Wars and Witches: The Riddles of Culture*, London, Vintage

Harris, M. (1978), *Cannibals and Kings: Origins of Cultures*, London, Collins

Harris, M. (1986), *Good to Eat: Riddles of Food and Culture*, London, Allen & Unwin

Harrison, C. *et al.* (1975), 'Food waste behaviour in an urban population', *J. Nutrition Education* 17, 13

Hawkes, C. and J. Webster (2000), *Too Much and Too Little: Debates on Surplus Food Redistribution*, London, Sustain

Hawkins, G. and S. Muecke (2003), *Culture and Waste: The Creation and Destruction of Value*, Oxford, Rowman & Littlefield

Heller, M. and G. Keoleian (2003), 'Assessing the sustainability of the US food system: a life cycle perspective', *Agricultural Systems* 76, 1007–41

Heller, M. *et al.* (2003), 'Life cycle assessment of a willow bioenergy cropping system', *Biomass and Bioenergy* 25:2, 147–65

Henningsson, S. *et al.* (2001), 'Minimizing material flows and utility use to increase profitability in the food and drink industry', *Trends in Food Sci. & Technol.* 12:2, 75–82

Henningsson, S. *et al.* (2004), 'The value of resource efficiency in the food industry: a waste minimisation project in East Anglia, UK', *J. Cleaner Production* 12:5, 505–12

Herzka, A. and R. Booth (eds.) (1981), *Food Industry Wastes: Disposal and Recovery*, London/Englewood, NJ, Applied Science Publishers

Hinchcliffe, D. (ed.) (2005), *The Government's Public Health White Paper (Cm 6374)*, London, The Stationery Office

Hirai, Y. *et al.* (2001), 'Life cycle assessment on food waste management and recycling' (in Japanese), *J. Japanese Soc. Waste Management Experts* 12, 219–28

Hitler, Adolf (H. Picker *et al.* (eds.)) (2000), *Hitler's Table Talk 1941–1944*, tr. Cameron and Stevens, London, Phoenix Press

Hiza, H. and L. Bente (2007), *Nutrient Content of the U.S. Food Supply, 1909–2004: A Summary Report*, Home Economics Research Report no. 57, Washington DC, Center for Nutrition Policy and Promotion, US Department of Agriculture

Hogg, D. *et al.* (2007a), *Managing Biowastes from Households in the UK: Applying Life-Cycle Thinking in the Framework of Cost-Benefit Analysis: A Final Report for WRAP*, Bristol, Eunomia Research & Consulting Ltd

Hogg, D. *et al.* (2007b), *Dealing with Food Waste in the UK: Report for WRAP*, Bristol, Eunomia Research & Consulting Ltd

Hopkin, M. (2005a), 'Did climate shift kill giant Australian animals?', *Nature News*, 30 May 2005, http://www.nature.com/news/2005/050530/full/news050523-11.html (accessed 26 March 2009)

Hopkin, M. (2005b), 'Fire-starters blamed for Australian extinctions', *Nature News*, 7 July 2005, http://www.nature.com/news/2005/050707/full/news050704-10.html (accessed 26 March 2009)

Hopkin, M. (2007), 'Rising temperatures "will stunt rainforest growth"', *Nature News*, 10 August 2007, http://www.nature.com/news/2007/070810/full/news070806-13.html (accessed 26 March 2009)

Horrigan, L. *et al.* (2002), 'How sustainable agriculture can address the environmental and human health harms of industrial agriculture', *Environmental Health Persp.* 110, 445–56

Houghton, R. (2005), 'Aboveground forest biomass and the global carbon balance', *Global Change Biology* 11, 945–58

IAASTD (2008), *Summary for Decision Makers of the Global Report*, Washington DC, International Assessment of Agricultural Knowledge, Science and Technology for Development, http://www.agassessment.org/docs/IAASTD_GLOBAL_SDM_JAN_2008.pdf

IGD (2007), *Beyond Packaging: Food Waste in the Home*, Watford, IGD, www.igd.com

IGD (2008), *UK Grocery Outlook 2007*, Watford, IGD, www.igd.com

Imperial College London (2007), *Sustainable Waste Management in the Chilled Food Sector*, Final Report, Defra Research Project FT0348, London, Centre for Environmental Policy, Applied Economics and Business Management Research Section, http://randd.defra.gov.uk/Default.aspx?Menu=Menu&Module=More&Location=None&Completed=0&ProjectID=12996

Imperial War Museum (1940), *How to Start a Pig Club*, Department of Printed Books, K 7308, Leaflet No.1, May 1940, London

Institute of Post Harvest Technology (Sri Lanka) (2002), *Institute of Post Harvest Technology Annual Report*, Colombo

IPCC (2003), *Good Practice Guidance for Land Use, Land-Use Change and Forestry* (2003), Intergovernmental Panel on Climate Change (IPCC), IPCC National Greenhouse Gas Inventories Programme, Japan, Institute for Global Environmental Strategies (IGES) for the IPCC, http://www.ipcc-nggip.iges.or.jp/public/gpglulucf/gpglulucf_contents.html

IRIN (2008a), 'Pakistan: fears of worsening child malnutrition', 10 June 2008

IRIN (2008b), 'Global: bad ethanol, good ethanol', 4 June 2008

IRIN (2008c), 'Global: why everything costs more. A rough guide to why food prices keep going up', 8 July 2008

IRRI (2007), 'Postproduction management', Irrigated Rice Research Consortium, www.irri.org/irrc/postharvest/index.asp

Ivanic, M. and W. Martin (2008), *Implications of Higher Global Food Prices for Poverty in Low-Income Countries*, Policy Research Working

Paper 4594, Washington DC, World Bank Development Research Group Trade Team, http://www-wds.worldbank.org/external/default/WDSContentServer/IW3P/IB/2008/04/16/000158349 _20080416103709/Rendered/INDEX/wps4594.txt

Iwamoto, Y. (2007), 'Interview: Japanese producer leader calls for cost savings', *Pig International*, 1 January 2007, http://www.pig-inter national.com/ViewArticle.aspx?id=25300

Japan Times (2007a), 'Heat's on the food industry' (editorial), 3 November 2007, http://search.japantimes.co.jp/cgi-bin/ed 20071103a2.html (accessed 26 March 2009)

Japan Times (2007b), 'McDonald's outlets hit over dated food', 28 November 2007, http://search.japantimes.co.jp/cgi-bin/nn 20071128a4.html (accessed 26 March 2009)

Jeffery, R. and L. Harnack (2007), 'Evidence implicating eating as a primary driver for the obesity epidemic', *Diabetes* 56, 2673–6

John Lewis Partnership (2008), 'Waste and recycling', http://www.johnlewispartnership.co.uk/Display.aspx?MasterId=eadb0559-e0b 5-4ed6-b9e0-f547279ad193&NavigationId=665 (accessed 16 January 2009)

Jonaitis, A. (ed.) (1991), *Chiefly Feasts: The Enduring Kwakiutl Potlatch*, Seattle/London, University of Washington Press

Jones, C. *et al.*, (2009), 'Committed ecosystem change due to climate change', in *Climate Change: Global Risks, Challenges and Decisions*, IOP Conf. Series: *Earth and Environmental Science* 6, Institute of Physics

Jones, J. *et al.* (2004), 'Case study: comparison of swine diets containing a food waste product made with wheat middlings and corn or a corn/soybean diet', *Professional Animal Scientist*, October 2004

Jones, T. (2004a), 'Using contemporary archaeology and applied anthropology to understand food loss in the American food system', unpublished paper, http://www.communitycompost.org/info/usa food.pdf (accessed 26 March 2009)

Jones, T. (2004b), 'What a waste!', ABC *The Science Show*, 4 December 2004, http://www.abc.net.au/rn/science/ss/stories/s1256017.htm (accessed 26 March 2009)

Jones, T. (2006), 'Addressing food wastage in the US', ABC *The*

Science Show, 8 April 2006, http://www.abc.net.au/rn/scienceshow/
stories/2006/1608131.htm (accessed 26 March 2009)

Jowit, J. (2008), 'Krill fishing threatens the Antarctic', *Observer*,
23 March 2008

Just-Food (2004), 'Tomato waste could be used as food additive, says
EU project', 13 February 2004, www.just-food.com

Kader, A. (2005), 'Increasing food availability by reducing post-harvest
losses of fresh produce', in F. Mencarelli, and P. Tonutti (eds.),
Proc. 5th International Postharvest Symposium, Acta Horticulturae, 682

Kader, A. and R. Rolle (2004), *The Role of Post-Harvest Management in
Assuring the Quality and Safety of Horticultural Produce*, Rome, FAO,
http://www.fao.org/docrep/007/y5431e/y5431e00.HTM

Kaimowitz, D. *et al.* (n.d.), *Hamburger Connection Fuels Amazon Destruction: Cattle Ranching and Deforestation in Brazil's Amazon*, Bogor
Barat, Indonesia, Center For International Forestry Research, http://
www.cifor.cgiar.org/publications/pdf_files/media/Amazon.pdf#
search=%22Hamburger%20Connection%20Fuels%20Amazon%20
Destruction%22

Kamensek, M. (2008), 'Left Over Chef', www.leftoverchef.com

Kangmin, L. and M.-W. Ho (2005), 'Biogas China', http://www.i-sis
.org.uk/DreamFarm.php

Kantor, L. *et al.* (1997), 'Estimating and addressing America's food
losses', *Food Review* 20:1, 2–12

Kasting, J. (2006), 'Ups and downs of ancient oxygen', *Nature* 443,
643–5

Kawashima, T. (n.d.), 'The use of food waste as a protein source for
animal feed – current status and technological development in Japan',
http://www.fao.org/docrep/007/y5019e/y5019e0i.htm

Keckchen, L. (2004), 'Hard Cases', teleplay, *The Wire*, series 2, episode
4, HBO Original Programming

Kehres, B. (2008), 'Long-term perspectives for separate collection and
recycling of biowastes', in *The Future for Anaerobic Digestion of Organic
Waste in Europe*, Weimar, ECN/ORBIT e.V. Workshop 2008,
http://www.compostnetwork.info/ad-workshop/presentations/05
_kehrespdf

Kelleher, K. (2005), *Discards in the World's Marine Fisheries. An Update*, FAO Fisheries Technical Paper 470, Rome, FAO, http://www.fao.org/docrep/008/y5936e/y5936e00.htm

Keyzer, M. A. *et al.* (2005), 'Diet shifts towards meat and the effects on cereal use: can we feed the animals in 2030?', *Ecological Economics* 55:2, 187–202

Khan, I. (2008), 'How to achieve food security', *Dawn*, 28 April–4 May 2008

Khushk, A. and A. Memon (2006), 'Making harvesting of mangoes productive', *Dawn*, 8 May 2006, http://www.dawn.com/2006/05/08/ebr5.htm

Kipler, K. and K. Ornelas (eds.) (2000), *The Cambridge World History of Food*, Cambridge University Press

Kitcho, C. (2003), *The Use It Up Cookbook: Creative Recipes for the Frugal Cook*, Naperville, Ill., Cumberland House Publishing

Knight, A. and C. Davis (2006), 'What a waste!', Surplus Fresh Foods Research Project, SCRATCH, November 2006, updated May 2007, http://www.veoliatrust.org/docs/Surplus_Food_Research.pdf

Kroger (2008), 'Doing our part: 2008 sustainability report', http://www.kroger.com/company_information/community/Pages/feeding_america.aspx (accessed 27 February 2009)

KSLA (Royal Swedish Academy of Agriculture and Forestry) (2007), *Den beresta maten – matens kvalitet i ett globalt perspektiv*, KSLAs Tidskrift 10

Kwak, W. *et al.* (2006), 'Effect of feeding food waste-broiler litter and bakery by-product mixture to pigs', *Bioresource Technology* 97, 243–9

Lacoste, E. *et al.* (eds.) (2007), *From Waste to Resource. 2006 World Waste Strategy*, Paris, Economica

Lakshmi, R. (2008), 'Bush comment on food crisis brings anger, ridicule in India', *Washington Post*, 8 May 2008

Laurence, J. (1726), *A New System of Agriculture*, London, T. Woodward

Lawrence, F. (2009), 'Amazon rainforest: upping the anti', *Guardian*, 4 March 2009

Le Bailly, P. (2007), *Second Time Around: Ideas and Recipes for Leftovers*, Oxford, Trafford Publishing

Leake, J. (2005), 'Picky stores force farmers to dump veg', *Sunday Times*, 17 July 2005

Leather, S. (1996), *The Making of Modern Malnutrition*, London, The Caroline Walker Trust

Ledward, D. (1983), *Upgrading Waste for Feeds and Food*, Oxford, Butterworth-Heinemann Ltd

Lee, S.-H. *et al.* (2007), 'Evaluation of environmental burdens caused by changes of food waste management systems in Seoul, Korea', *Sci. of the Total Environment* 387, 42–53

LeGood, P. and A. Clarke (2006), 'Smart and active packaging to reduce food waste', Smart Materials, Surfaces and Structures Network (SMART.mat), November 2006, http://www.ktnetworks.co.uk/epicentric_portal/binary/com.epicentric.contentmanage
ment.servlet.ContentDeliveryServlet/AMF/smartmat/Smartand
activepackagingtoreducefoodwaste.pdf

Lewis, L. (2008), Japan is a market pioneer again: the first industrialised nation with no butter', *The Times*, 15 April 2008

Liang, L. *et al.* (1993), 'China's post-harvest grain losses and the means of their reduction and elimination', *Jingji dili* (*Economic Geography*) 1, 92–6

Lillywhite, R. and C. Rahn (2005), 'Nitrogen UK, a Biffaward programme on sustainable resource use', http://www2.warwick.ac.uk/fac/sci/whri/research/nitrogenandenvironment/nmassbal/nitrogen
uk.pdf

Loch, C. (1885), 'Waste food', *The Times*, 17 November 1885, p. 4, Issue 31606, col. F

[Locke, J.] (1690), 'Second Treatise of Government', in [J. Locke], *Two Treatises of Government*, London, A. Churchill

Lok Sabha (2005), *Standing Committee on Agriculture (2004–2005), Fourteenth Lok Sabha, Ministry Of Food Processing Industries, Demands For Grants (2005–2006), Twelfth Report*, April 2005, New Delhi

Lok Sabha (2007), *Standing Committee On Agriculture (2006–07), Four-*

teenth Lok Sabha, Ministry Of Food Processing Industries, Demands For Grants (2007–2008), Thirtieth Report, April 2007, New Delhi

Lundie, S. and G. Peters (2005), 'Life cycle assessment of food waste management options', *J. Cleaner Production* 13:3, 275–86

Lundqvist, J., C. de Fraiture and D. Molden (2008), *Saving Water: From Field to Fork – Curbing Losses and Wastage in the Food Chain*, SIWI Policy Brief, Stockholm, SIWI

Maathai, W. (2005), 'Speech to the Rotarians', Rotarian Conference, Chicago, 22 June 2005, http://greenbeltmovement.org

Maathai, W. (2008), 'On the occasion of TICAD IV Yokohama', Japan, 28–30 May 2008, http://greenbeltmovement.org

Madsen, A. and L. Kromann (2004), 'Alternative possibilities for disposing of food waste from industrial kitchens', in *Food Waste from Catering Centres*, Working Report No. 1, Danish EPA, www.mst.dk/udgiv/publikationer/2004/87-7614-119-5/html/

MAFF (Japan) (2007), '2007 report: outline of survey results on food recycling as renewable resources', MAFF Update, Number 684, 21 December 2007, http://www.maff.go.jp/mud/684.html

MAFF (Japan) (2008), 'Annual report on food, agriculture and rural areas in Japan FY 2007', http://www.maff.go.jp/e/annual_report/2007/pdf/e_all.pdf

MAFF (UK) (1982), *Household Food Consumption and Expenditure 1980 with a Review of 1975–1980*, Annual Report of the National Food Survey Committee (NFS), London, HMSO

Mahro, B. and M. Timm (2007), 'Potential of biowaste from the food industry as a biomass resource', *Engineering in Life Sciences* 7:5, 457–68

[Malthus, T.] (1798), *Principle of Population*, London, J. Johnson

Malthus, T. (1826), *An Essay on the Principle of Population; Or, A View of its Past and Present Effects on Human Happiness; with an inquiry into our prospects respecting the future removal or mitigation of the evils which it occasions*, 2 vols., London, John Murray

Manchester Business School (2006), *Environmental Impacts of Food Production and Consumption*, report for Department for Environment, Food and Rural Affairs, Manchester Business School

Marine Conservation Society (n.d.), 'Fish purchasing guide', http://www.fishonline.org/buying_eating/purchasing_guide.php

Marine Stewardship Council (n.d.), 'Where to buy', http://www.msc.org/where-to-buy

Markham, G. (1676), *Cheap and Good Husbandry*, 13th edn, London, E. H. for George Sawbridge

Marks & Spencer (2008), 'Marks & Spencer updates on the first financial year progress of its "Eco-Plan"™, Plan A', Press Release, 20 May 2008, http://corporate.marksandspencer.com/media/press_releases/RNS/Financial/20052008_MarksSpencerUpdatesOnThe FirstFinancialYearProgressOfItsEcoPlanPlanA (accessed 12 March 2009)

Marks & Spencer (2009), 'Plan A: Waste', http://plana.marksand spencer.com/about/the-plan/waste (accessed 12 March 2009)

Martin, P. (1973), 'The discovery of America', *Science* 179, 969–74

Martin, P. (2005), *Twilight of the Mammoths: Ice Age Extinctions and the Rewilding of America*, Berkeley, University of California Press

Martin, P. and R. Klein (eds.) (1984), *Quaternary Extinctions*, Tucson, University of Arizona Press

Mastoris, S. and R. Malcolmson (2001), *The English Pig: A History*, London, Continuum International Publishing Group

Mata-Alvarez, J. *et al.* (2000), 'Anaerobic digestion of organic solid wastes: an overview of research achievements and perspectives', *Bioresource Technol.* 74:1, 3–16

Matteson, G. and B. Jenkins (2007), 'Food and processing residues in California: resource assessment and potential for power generation', *Bioresource Technol.* 98, 3098–105

Mattson, J. *et al.* (2004), *Analysis of the World Oil Crops Markets*, Agribusiness & Applied Economics Report No. 529, Fargo, Center for Agricultural Policy and Trade Studies, North Dakota State University, http://ageconsearch.umn.edu/bitstream/23621/1/aer529.pdf

Mauss, M. (1990), *The Gift: The Form and Reason for Exchange in Archaic Societies*, London and New York, Routledge

McL. Dryden, G. (2008), *Animal Nutrition Science*, Wallingford, CABI Publishing

MCT (2004), 'Primeiroinventário Brasileiro de emissões antrópicas de gases de efeito estufa', Ministry of Science and Technology (Brazil), www.mct.gov.br/Clima/comunic_old/invent1.htm (accessed 24 February 2006)

Meikle, J. (2008), 'Sales of whiting rise as credit crunch bites', *Guardian*, 11 October 2008

Mellanby, K. (1975), 'Wasteline', *Nature* 257, 639

Mendes, M. *et al.* (2003), 'Assessment of the environmental impact of management measures for the biodegradable fraction of municipal solid waste in São Paulo City', *Waste Management* 23, 403–9

Metro (2007), 'Turning waste into watts', 19 November 2007, http://www.metro.co.uk/news/climatewatch/article.html?in_article_id=76228&in_page_id=59

Millennium Ecosystem Assessment (2005), *Millennium Ecosystem Assessment Synthesis Report*, Washington DC, Island Press

Miller, D. (1979), 'Man's demand for energy', in K. Blaxter (ed.), *Food Chains and Human Nutrition*, London, Applied Science Publishers

Miller, P. and N. Clesceri (2003), *Waste Sites as Biological Reactors – Characterizing and Modeling*, Lewis Publishers

Ministère de la Justice, Luxembourg (2001), *Code Pénal*, last revised 1 August 2001, http://www.legilux.public.lu/leg/textescoordonnes/codes/code_penal/cp_L2T10.pdf

Mitchell, D. (2008), *A Note on Rising Food Prices*, Washington DC, World Bank Development Prospects Group, http://www-wds.worldbank.org/servlet/WDSContentServer/WDSP/IB/2008/07/28/000020439_20080728103002/Rendered/PDF/WP4682.pdf

MNP (Dutch Environmental Assessment Agency) (2008), 'Local and global consequences of the EU renewable directive for biofuels', http://www.mnp.nl/en/publications/2008/Localandglobal consequencesoftheEUrenewabledirectiveforbiofuels.html

Monbiot, G. (2007), *Heat: How to Stop the Planet from Burning*, London, Penguin

Monsterfodder, J. (2006), 'Did Wal-Mart's food donation policy really change?', *Daily Kos*, 8 January 2006, http://www.dailykos.com/story/2006/1/8/154147/0564

Morgan, H. and F. Nuti (1982), *Harvesters and Harvesting, 1840–1900*, London, Croom Helm

Morgan, S. (2006), *Leftover Food (Dealing with Waste)*, London, Franklin Watts

Morley, N. and C. Bartlett (2008), *Mapping Waste in the Food Industry for Defra and the Food and Drink Federation*, Aylesbury, Oakdene Hollins, http://www.fdf.org.uk/publicgeneral/mapping_waste_in_the_food_industry.pdf

Morrisons (2008), *Taking Good Care: Making Good Progress*, Morrisons Corporate Social Responsibility Report 2008, http://www.morrisons.co.uk/Documents/Morrisons_CSR_2008.pdf (accessed 16 January 2009)

M[ortimer.], J. (1707), *The whole art of Husbandry*, London

Mosimann, J. and P. Martin (1975), 'Simulating overkill by Paleo-indians', *American Scientist* 63, 304–13

MSC (2007), 'Birds Eye to launch sustainable fish finger', Marine Stewardship Council, 2 August 2007, www.msc.org

Myer, R. *et al.* (2000), 'Dehydrated restaurant food waste as swine feed', in M. Westendorf (ed.), *Food Waste to Animal Feed*, Ames, Iowa State University

Nation, The (2008), 'Poverty and food crisis', 8 May 2008

National Audit Office (2009), *Department for Environment, Food and Rural Affairs: Managing the Waste PFI Programme*, 14 January 2009, London, National Audit Office, http://www.nao.org.uk/publications/0809/managing_the_waste_pfi_program.aspx

National Grid (2009), *The Potential for Renewable Gas in the UK*, January 2009, http://www.nationalgrid.com/NR/rdonlyres/9122AEBA-5E50-43CA-81E5-8FD98C2CA4EC/31630/renewablegasWPfinal.pdf

National Renderers Association (US) (2006), *Essential Rendering: All about the Animal By-Products Industry*, Alexandria, Va., National Renderers Association

Natural Environment Research Council (2003), *The Royal Commission on Environmental Pollution Study on the Environmental Effects of Marine Fisheries: The Natural Environment Research Council Memorandum*, Swindon, Natural Environment Research Council

Naturvårdsverket (2007), *Åtgärder för minskat svinn i livsmedelskedjan*

Nelson, S. (1998), *Ancestors for the Pigs: Pigs in Prehistory*, Philadelphia, University of Pennsylvania Press

Nemeth, D. (1995), 'On pigs in subsistence agriculture', *Current Anthropology* 36:2, 292–3

Netherlands Statline database (2008), 'Municipal waste quantities', http://statline.cbs.nl/StatWeb/publication/?DM=SLEN&PA= 7467ENG&D1=5,51,76,87-88,113&D2=0&D3=a&LA=EN& HDR=T&STB=G1,G2&VW=T (accessed 12 March 2009)

NHS Estates (2000), *Reducing Food Waste in the NHS*, London, The Stationery Office

Niaz, S. (2008), 'Unaddressed issues in food security', *Dawn Economic and Business Review*, 5–11 May 2008

Nielsen, P. *et al.* (2007), 'LCA food database', last updated March 2007, http://www.lcafood.dk/

Nierenberg, D. (2005), *Happier Meals: Rethinking the Global Meat Industry*, Worldwatch Paper 171, Washington DC, Worldwatch Institute

Nijdam, D. *et al.* (2005), 'Environmental load from Dutch private consumption. How much damage takes place abroad?', *J. Industrial Ecology* 9:1–2, 153

Nilsson, S. and W. Schopfhauser (1995), 'The carbon-sequestration potential of a global afforestation program', *Climatic Change* 30:3, 267–93

Nordberg, Å. and M. Edström (2003), 'Treatment of animal waste in co-digestion biogas plants in Sweden', in T. Al Seadi and J. Holm-Nielsen (eds.), *The Future of Biogas in Europe II*, 2–4 October 2003, University of Southern Denmark

NRC (1983), *Underutilized Resources as Animal Feedstuffs*, US National Research Council, Washington DC, National Academy Press

Nuthall, K. (2005), 'Research programme (Extranat) gets usable compounds from fruit waste', 29 November 2005, www.just-food.com

OECD (2008), 'Policy brief', April 2008, Organization For Economic
    Co-Operation and Development, http://www.oecd.org/dataoecd/
    40/33/40556614.pdf

Ogden, S. (2005), *Food and Drink Waste in the Yorkshire and Humber
    Region*, report prepared for Yorkshire Forward by Safe Waste and
    Proper (SWAP) Ltd, Project No. 204/10, June 2005

Ogino, A. *et al.* (2007), 'Environmental impact evaluation of feeds
    prepared from food residues using life cycle assessment', *J. Environ-
    mental Quality* 36, 1061–8

Olson, W. (2006), 'Wal-Mart ends food donations to charity', *Overlaw-
    yered*, 9 January 2006, http://overlawyered.com/2006/01/wal-mart
    -ends-food-donations-to-charity/

Ophuls, Marcel (1969), *The Sorrow and the Pity* (*Le Chagrin et la pitié*)
    (documentary film)

Opie, I. and P. Opie (1951), *The Oxford Dictionary of Nursery Rhymes*,
    London, Oxford University Press

OPSI (1996), Statutory Instrument 1996 No. 1499, The Food Label-
    ling Regulations 1996, Part II, Food to be delivered as such to
    the ultimate consumer or to caterers, Scope and general labelling
    requirement, London, Office of Public Sector Information, http://
    www.opsi.gov.uk/si/si1996/Uksi_19961499_en_3.htm#mdiv22

Osava, M. (2004), 'Deforestation inflates inventory of gases in Brazil',
    Inter Press Service, 9 December 2004, http://forests.org/articles/
    reader.asp?linkid=37174

Osborne, C. (1916), 'Waste of food', *The Times*, 17 November 1916,
    Issue 41327

Osner, R. (1982), 'Food wastage', *Nutrition and Food Science* 77, 13–17

OSPAR Commission (2008a), *List of Threatened and/or Declining Species
    and Habitats* (Ref: 2008-6), London, OSPAR Commission, www
    .ospar.org/v_measures/get_page.asp?vo=
    08-06eOSPAR%20List%20species%20and%20habitats.doc&v1=5;

OSPAR Commission (2008b), *Case Reports for the OSPAR List of
    Threatened and/or Declining Species and Habitats*, London, OSPAR
    Commission

Ovandoa, P. and A. Caparrós (2009), 'Land use and carbon mitigation in Europe: a survey of the potentials of different alternatives', *Energy Policy* 37:3, 992–1003

PA News (2008), 'Britons ignoring food sell-by dates', 1 January 2008, http://www.channel4.com/news/articles/society/health/britons +ignoring+food+sellby+dates/1252447 (accessed 26 March 2009)

Pagnamenta, R. (2008), 'At Sainsbury's, where's there's muck, there's gas', *The Times*, 3 November 2008, http://business.timesonline.co .uk/tol/business/industry_sectors/retailing/article5068704.ece (accessed 26 March 2009)

Pailliet, J.-B. (1837), *Manuel de droit français*, 9th edn, Le Normant

Parliamentary and Health Service Ombudsman (2007), *The Introduction of the Ban on Swill Feeding*, December 2007, London, Parliamentary and Health Service Ombudsman

Paul, L. (1994), 'The problem of waste: dropouts, ghostfishing and the by-catch', in L. Paul (ed.), *High Seas Driftnetting: The Plunder of the Global Commons. A Compendium*, rev. edn, Kailua, EarthTrust, http://www.earthtrust.org/dnpaper/waste.html

Peacock, T. (1817), *Melincourt*, London, T. Hookham

Peden, D. *et al.* (2007), 'Water and livestock for human development', in D. Molden (ed.), *Water for Food, Water for Life: A Comprehensive Assessment of Water Management in Agriculture*, London/Colombo, IWMI/Earthscan Publications, 485–514

Pengue, W. (2006), 'Increasing roundup ready soybean export from Argentina', in N. Halberg *et al.* (eds.), *Global Development of Organic Agriculture: Challenges and Prospects*, Wallingford, CABI Publishing

Phillips, T. (2008a), 'Police break up riot over logging', *Guardian*, 26 November 2008

Phillips, T. (2008b), 'Brazil announces plan to slash rainforest destruction', *Guardian*, 2 December 2008

Phillips, T. (2008c), 'Rush to buy sends prices soaring', *Guardian*, 22 November 2008

PICME (2006), *Chilled Food Manufacturing Waste Minimisation Study*, Defra Project FT0351, Process Industry Centre for Manufacturing Excellence, http://randd.defra.gov.uk/Default.aspx?Menu=Menu&

Module=More&Location=None&ProjectID=13439&From
Search=Y&Publisher=1&SearchText=Chilled%20Food%20
Manufacturing%20&SortString=ProjectCode&SortOrder=Asc&
Paging=10#Description

Polian, P. (2003), *Against Their Will: The History and Geography of Forced Migrations in the USSR*, tr. A.Yastrzhembska, Budapest, Central European University Press

Pollan, M. (2008), *In Defence of Food: An Eater's Manifesto*, London, Penguin

Poppendieck, J. (1986), *Breadlines Knee-Deep in Wheat: Food Assistance in the Great Depression*, Piscataway, NJ, Rutgers University Press

Poppendieck, J. (1998), *Sweet Charity*, New York, Viking Penguin

Potato Council (Britain) (2008), 'Consumption and processing in GB: potato industry "Redbook" statistics updated to June 2008', http://www.potato.org.uk/media_files/MIS_reports_2006_07/red bookjun08.pdf

Poulter, S. (2008), 'Supermarkets urged to bin BOGOFs', *Daily Mail*, 8 July 2008

Press Association (2004), 'Pile of dead fish gives Greenpeace ammunition against EU policy', *Guardian*, 12 August 2004

Price, C. and J. Harris (2000), 'Potential food loss recovery from farmers markets: a preliminary analysis', *J. Food Distribution Research* 30:1

Proctor, D. (ed.) (1994), *Grain Storage Techniques: Evolution and Trends in Developing Countries*, FAO Agricultural Services Bulletin No. 109, Rome, FAO

Proctor, R. (1999), *The Nazi War on Cancer*, Princeton University Press

Putnam, J. (1999), 'US food supply providing more food and calories', *Food Review* 22:3, 2–12, http://www.ers.usda.gov/publications/foodreview/sep1999/frsept99a.pdf

Rabo India Finance (2005), *Vision, Strategy and Action Plan for Food Processing Industries in India*, prepared by Rabo India Finance Pvt Ltd for Ministry of Food Processing Industries, Government of India, April 2005

Rediff (2007), 'How much food does India waste?', *Rediff India Abroad*, 16 March 2007, http://www.rediff.com/money/2007/mar/16food. htm

Renton, A. (2007), 'No such thing as a free lunch?', *Observer Food Monthly*, 19 August 2007

Renton, A. (2009), 'Our culture of wasting food will one day leave us hungry', *Observer*, 8 February 2009

Reuters (2008), 'Biofuels major driver of food price rise – World Bank', 28 July 2008, http://www.alertnet.org/thenews/newsdesk/ N28615016.htm

RFA (2008), *The Gallagher Review of the Indirect Effects of Biofuels Production*, ed. E. Gallager, July 2008, St Leonards-on-Sea, Renewable Fuels Agency, http://www.dft.gov.uk/rfa/_db/ _documents/Report_of_the_Gallagher_review.pdf

Rice, X. (2008), 'Wildlife and livelihoods at risk in Kenyan wetlands biofuel project', *Guardian*, 24 June 2008

Riches, G. (1986), *Food Banks and the Welfare Crisis*, Ottawa, Canadian Council of Social Development

Riches, G. (ed.) (1997), *First World Hunger*, London, Macmillan

Ristow, E. (2008), 'Waste not, want not' (Associated Press) in D. Stringer, 'British families urged to cut food waste', *International Business Times*, 10 July 2008, http://www.ibtimes.com/articles/ 20080710/british-families-urged-to-cut-food-waste_all.htm

Roberts, C. and J. Hawkins (2000), *Fully-Protected Marine Reserves: A Guide*, Washington DC/York, WWF Endangered Seas Campaign

Rosen, S. *et al.* (2008), *Food Security Assessment, 2007*, Washington DC, Economic Research Service, US Dept of Agriculture

Roy, R. (1976), *Wastage in the UK Food System*, London, Earth Resources Research Publications

Ruddock, J. (2007), 'Speech on food waste and anaerobic digestion to the Environmental Services Association Conference', 16 October 2007, London, http://www.defra.gov.uk/corporate/ministers/ speeches/joan-ruddock/jr071016.htm

Ruiz López, M. *et al.* (2003), 'Pickled vegetable and fruit waste mixtures as an alternative feedstuff', *Food Policy – Economics Planning and Politics of Food and Agriculture* 28, 1–13

Runge, C. and B. Senauer (2007), 'How biofuels could starve the poor', *Foreign Affairs*, May/June 2007

Russ, G. and A. Alcala (2004), 'Marine reserves: long-term protection is required for full recovery of predatory fish populations', *Oecologia* 138:4

Russ, W. and R. Meyer-Pittroff (2004), 'Utilizing waste products from the food production and processing industries', *Critical Reviews in Food Science and Nutrition* 44, 57–62

Russ, W. and M. Schnappinger (2006), 'Waste related to the food industry: a challenge in material loops. 1.1. Food waste in the EC: status, definition & handling', in V. Oreopoulou and W. Russ (eds.), *Utilization of By-Products and Treatment of Waste in the Food Industry*, New York, Springer-Verlag

Safeway (2009), 'Community caring', http://www.safeway.com/ifl/grocery/Community-Caring

Safeway (n.d.), 'Environmental responsibility', http://www.safeway.com/ifl/grocery/Environment-Sustainability (accessed 27 February 2009)

Sainsbury plc, J. (2008a), *Corporate Responsibility Report 2007: Make the Difference*, http://www.j-sainsbury.co.uk/files/reports/cr2007/files/report.pdf (accessed 16 January 2009)

Sainsbury plc, J. (2008b), 'Would you buy great tasting, unusually shaped fruit & veg?', 22 October 2008, http://www.sainsburys.co.uk/yourideas/forums/10606/ShowThread.aspx

Sainsbury plc, J. (n.d.), 'Financials', corporate website, http://www.j-sainsbury.co.uk/index.asp?pageid=206

Sancho, P. *et al.* (2004), 'Microbiological characterization of food residues for animal feeding', *Waste Management* 24, 919–26

Sandars, D. *et al.* (2006), 'An environmental Life Cycle Assessment (LCA) of the centralised anaerobic digester at Holsworthy in North Devon (England)', in S. Petersen (ed.), *12th RAMIRAN International Conference: Technology for Recycling of Manure and Organic Residues in a Whole-Farm Perspective*, DIAS Report No. 122, Vol.

1, 203–7, http://www.manure.dk/ramiran/djfma122.pdf (accessed 30 January 2009)

Sarsby, J. (2001), *Deep Trouble* (BBC documentary film), *Blue Planet* DVD 1089, BBC

Sauer, C. (1969), *Seeds, Spades, Hearths, and Herds: The Domestication of Animals and Foodstuffs*, Cambridge, Mass., MIT Press

Schmidt, J. (2007), 'Life cycle assessment of rapeseed oil and palm oil', Ph.D. thesis, Part 3: 'Life cycle inventory of rapeseed oil and palm oil', http://vbn.aau.dk/fbspretrieve/10388016/inventory_report

Scottish Government (2008), 'The "madness" of dumped fish', 25 September 2008, http://www.scotland.gov.uk/News/Releases/2008/09/25092305

Second Harvest (1997), *Hunger 1997: The Faces and Facts*, Chicago, Second Harvest

Sen, A. (1981), *Poverty and Famines: An Essay on Entitlements and Deprivation*, Oxford, Clarendon Press

Sen, A. (1987), *Hunger and Entitlements*, Helsinki, Wider

Senewiratne, H. (2006), 'Rs. 9 billion worth fruits, vegetables go to waste', *Daily News* (Sri Lanka), 27 June 2006

Serra-Majem, L. *et al.* (2003), 'Comparative analysis of nutrition data from national, household, and individual levels: results from a WHO-CINDI collaborative project in Canada, Finland, Poland, and Spain', *J. Epidemiology and Community Health* 57, 74–80

Sevenster, M. and F. de Jong (2008), 'A sustainable dairy sector: global, regional and life cycle facts and figures on greenhouse-gas emissions', CE Delft, October 2008

Severson, K. (2001), 'The dating game: freshness labels are a manufacturers' free-for-all', *San Francisco Chronicle*, 10 January 2001, http://www.sfgate.com/cgi-bin/article.cgi?f=/c/a/2001/01/10/FD171775.DTL&hw=the+dating+game+freshness&sn=001&sc=1000

Shaler, N. (2007), *Domesticated Animals – Their Relation to Man and to His Advancement* (1895), Read Books

Shanklin, C. and D. Ferris (1995), 'A comparative analysis of quantity and type of waste generated in institutional foodservice operations', *J. Amer. Dietetic Assoc.* 95:9, Supplement A47

Shell, E. Ruppel (2003), 'Are we turning our children into "fat" junkies?', *Observer*, 12 October 2003

Shemeis, A. *et al.* (1994), 'Offal components, body fat partition, carcass composition and carcass tissues distribution in Danish Friesian cull cows of different age and body condition', *Livestock Production Sci.* 40:2, 165–70

Sibrián, R. *et al.* (2008), *Estimating Household and Institutional Food Wastage and Losses in the Context of Measuring Food Deprivation and Food Excess in the Total Population*, FAO Statistics Division Working Paper Series, ESS/ESSA/2005-1, Rome, FAO

Siddique, H. (2007), 'Minister vows to persuade EU to let fishermen catch more cod', *Guardian*, 20 November 2007

Silsoe Research Institute (2004), *Physical Assessment of the Environmental Impacts of Centralised Anaerobic Digestion*, Silsoe Research Institute (BBSRC), Research project CC0240 for Defra, Final Report, http://sciencesearch.defra.gov.uk/Document.aspx?Document=CC0240_3173_FRP.doc

Silver, W. *et al.* (2000), 'The potential for carbon sequestration through reforestation of abandoned tropical agricultural and pasture lands', *Restoration Ecology* 8, 394–407

Singer, D. (1979), 'Food losses in the UK', *Proc. Nutrition Society* 38, 181–6

Smil, V. (2001), *Feeding the World: A Challenge for the Twenty-First Century*, Cambridge, Mass., MIT Press

Smil, V. (2004), 'Improving efficiency and reducing waste in our food system', *Environmental Sciences* 1:1, 17–26

Sobal, J. and M. Nelson (2003), 'Food waste', in S. Katz (ed.), *Encyclopedia of Food and Culture*, Volume I, New York, Charles Schribner's Sons

Society for Bettering the Condition and Increasing the Comforts of the Poor (1802), *The reports of the Society for Bettering the Condition and Increasing the Comforts of the Poor*, Volume 3, London

Sogi, D. *et al.* (2002), 'Effect of tomato seed meal supplementation on the dough and bread characteristics of wheat (Pbw 343) flour', *Int. J. Food Properties* 5:3, 563–71

Sonesson, U. *et al.* (2005), 'Home transport and wastage – environmentally relevant household activities in the life cycle of foods', *Ambio* 34:4, 371–5

Sowell, T. (2007), *Basic Economics: A Common Sense Guide to the Economy*, 3rd edn, New York, Basic Books

Spear, S. (2006), 'Step up the gas', 4 August 2006, http://www.cieh.org/ehp1/article.aspx?id=2070

Spencer, R. (2008), 'Britain agrees deal to sell pigs trotters to China', *Daily Telegraph*, 23 January 2008

Statistisches Bundesamt Deutschland (German National Statistical Office) (2005), 'Waste management: tables', http://www.destatis.de/jetspeed/portal/cms/Sites/destatis/Internet/EN/Navigation/Statistics/Environment/EnvironmentalSurveys/WasteManagement/Tables.psml (accessed 12 March 2009)

Steinfeld, H. *et al.* (2006), *Livestock's Long Shadow: Environmental Issues and Options*, Rome, FAO

Stevenson, B. (1948), *The Home Book of Proverbs*, New York

Stone, T. (1800), *A letter to the Right Honourable Lord Somerville, . . . late president of the Board of Agriculture*, London

Stuart, R. (1995), *Pigs, Goats and Poultry 1580–1660*, Bristol, Stuart Press

Stuart, T. (2006a), *The Bloodless Revolution*, London/New York, HarperCollins/W. W. Norton

Stuart, T. (2006b), 'Malthusians and Anti-Malthusians', in A. Grayling *et al.* (eds.), *The Continuum Encyclopedia of British Philosophy*, London, Continuum/Thoemmes

Sukhatme, P. and S. Margen (1982), 'Autoregulatory homestatic nature of energy balance', *Amer. J. Clinical Nutrition* 35, 355–65

Supervalu (n.d.), 'Environmental stewardship: environmental sustainability', http://www.supervalu.com/sv-webapp/about/envstewardship.jsp (accessed 10 March 2009)

Sustain (2009), 'Who is affected by food poverty?', http://www.sustainweb.org/page.php?id=235 (accessed 14 March 2009)

Sustainable Development Commission (2008), *Green, Healthy and Fair: A Review of Government's Role in Supporting Sustainable Supermarket Food*, February 2008, London, The Sustainable Development Commission, http://www.sd-commission.org.uk/publications/downloads/GreenHealthyAndFair.pdf

Système d'informations juridiques, institutionnelles et politiques (Côte-d'Ivoire) (1969), *Législation 1969*, http://droit.francophonie.org/df-web/publication.do?publicationId=2098

Target Corp. (2008), *2008 Corporate Responsibility Report*, http://sites.target.com/site/en/company/page.jsp?contentId=WCMP 04-034305 (accessed 27 February 2009)

Target Corp. (n.d.), 'Community', http://sites.target.com/site/en/corporate/page.jsp?contentId=PRD03-001820#ash (accessed 10 March 2009)

Taylor, D. and A. Fearne (2006), 'Towards a framework for improvement in the management of demand in agri-food supply chains', *Supply Chain Management: An International Journal* 11:5, 222–36

Taylor, J. (2006), 'Chinese like food but waste it on etiquette', ABC *The World Today*, 3 March 2006, http://www.abc.net.au/world today/content/2006/s1583287.htm

Tekelenburg, T. *et al.* (2003), 'Impact of global land use change on biological diversity: application of the Global Biodiversity Model (GLOBIO3)', paper for the *International Workshop on Transition in Agriculture and Future Land Use Patterns*, 1–3 December 2003, Wageningen

Teresa, Mother (1996), *A Gift for God: Prayers and Meditations*, New York, HarperCollins

Tesco (2008), *Tesco Corporate Responsibility Review 2008*, http://www.tescoreports.com/crreview08/ (accessed 7 March 2009)

Tescopoly (2008), 'Supermarket Code of Practice', http://www.tesco poly.org/index.php?option=content&task=view&id=69 (accessed 12 March 2009)

Tescopoly (n.d.), 'Farmers', http://www.tescopoly.org/index.php?option=com_content&task=view&id=58&Itemid=184 (accessed 12 March 2009)

Themelis, N. and P. Ulloa (2007), 'Methane generation in landfills', *Renewable Energy* 32, 1243–57

Thirsk, J. (1985), *The Agrarian History of England and Wales. V: 1640–1750, Part 2: Agrarian Change*, Cambridge University Press

Thomas, C. *et al.* (2004), 'Extinction risk from climate change', *Nature* 427, 145–8

Thompson, J. (2009), 'Tesco and Asda attack Competition Commission crackdown', *Independent*, 27 February 2009

Times, The (1917), 'Defence of the Realm Regulation', 8 May 1917, p. 8, Issue 41472, col. B

Times, The (1918), 'Fate of British pigs. Finding food from waste', 31 October 1918, p. 3, Issue 41935, col. E

Times, The (1946), 'Food from waste: 2,000,000 tons saved', 18 September 1946, p. 2, Issue 50559, col. D

TNS Retail Forward (2007), 'Retail Knowledgebase: net sales 2007', TNS Retail Forward

Torr, R. (2008), 'Shame as 25pc of all food is binned', *Gulf Daily News*, 13 December 2008

Trivedi, B. (2008), 'Dinner's dirty secret', *New Scientist*, 13 September 2008

Tryon, T. (1700), *Tryon's Letters, Upon Several Occasions*, London

Tudge, C. (2004), *So Shall We Reap*, London, Penguin

Turff, R. (2008), *Potato Review*, 18:1

Uhlin, S.-E. (1997), *Energiflöden i livsmedelskedjan* ('Energy flows in the food chain') Report No. 4732, Stockholm, Swedish Environmental Protection Agency

UN Population Division (2007), *UN 2006 Population Revision*, New York, UN Population Division, http://esa.un.org/unpp/

UN Statistics Division (2007), 'United Nations environmental indicators & selected time series. Composition of municipal waste: latest year, last update June 2007', http://unstats.un.org/unsd/environment/Questionnaires/Website%20tables%20and%20Selected%20Time%20Series/composition_municipal_waste_latestyear.pdf (accessed 7 March 2009)

UNEP (2009), *The Environmental Food Crisis – The Environment's Role* in *Averting Future Food Crises. A UNEP Rapid Response Assessment*, ed. C. Nellemann *et al.*, February 2009, GRID-Arendal, Norway, United Nations Environment Programme, www.grida.no

United Nations Trust Fund for Human Security (2007), 'Timor-Leste: reducing post-harvest losses to improve food security', ochaonline. un.org/LinkClick.aspx?link=2125&tabid=2110

United Nations University Press (1979), *Food and Nutrition Bulletin*, 1:2

USDA (1999), *Food Recovery and Gleaning Initiative, A Citizen's Guide to Food Recovery*, Washington DC, United States Dept of Agriculture

USDA (2004), *The Amazon: Brazil's Final Soybean Frontier*, Washington DC, Production Estimates and Crop Assessment Division, Foreign Agricultural Service, United States Department of Agriculture, http://www.fas.usda.gov/pecad/highlights/2004/01/Amazon/ Amazon_soybeans.htm

USDA (2007), 'Food labeling', United States Dept of Agriculture, http://www.fsis.usda.gov/Fact_Sheets/Food_Product_Dating/ index.asp

USDA (2009), *Foreign Agricultural Service, Oilseeds: World Markets and Trade*, Circular Series FOP 2-09, Washington DC, Foreign Agricultural Service, United States Dept of Agriculture, http://www.fas .usda.gov/psdonline/circulars/oilseeds.pdf

Valdemarsen, J. *et al.* (2007), *Options to Mitigate Bottom Habitat Impact of Dragged Gears*, FAO Fisheries Technical Paper No. 506, Rome, FAO, www.fao.org/docrep/010/a1466e/a1466e00.htm

Van De Reit, S. (1985), 'Food discards: nature, reasons for discard, and relationship to household variables', unpublished PhD thesis, Corvallis, Oregon State University

Vance Haynes, C. (1980), 'The Clovis culture', *Canadian J. Anthropol.* 1, 115–21

Varda, A. (2000), *Les Glaneurs et la glaneuse* (documentary film), Cinétamaris

Veblen, T. (1970), *The Theory of the Leisure Class*, London, Unwin Books

Vidal, J. (2005), 'More than 30% of our food is thrown away – and it's costing billions a year', *Guardian*, 15 April 2005

Vidal, J. (2006), 'The 7,000km journey that links Amazon destruction to fast food', *Guardian*, 6 April 2006

Vidal, J. (2008), 'Film of fishermen dumping catch causes uproar', *Guardian*, 13 August 2008

Waggoner, P. and J. Ausubel (2001), 'How much will feeding more and wealthier people encroach on forests?', *Population and Development Review* 27:2, 239–57

Waitrose (2008), *Waitrose Food Illustrated*, January 2008

Waldron, K. (2007), *Handbook of Waste Management and Co-product Recovery in Food Processing*, Cambridge, Woodhead Publishing

Walker-Palin, J. (n.d.), 'People, prices, planet', CIWM presentation, http://www.ciwm.co.uk/mediastore/FILES/15872.pdf (accessed 16 January 2009)

Wal-Mart (2008a), 'Sustainability progress to date 2007–2008', http://walmartstores.com/FactsNews/FactSheets/#Sustainability (accessed 27 February 2009)

Wal-Mart (2008b), 'The Wal-Mart Foundation: hunger relief fact sheet', www.walmartstores.com/download/3396.pdf (accessed 10 March 2009)

Wal-Mart (2008c), 'Wal-Mart announces new national food donation program', 30 April 2008, http://walmartstores.com/FactsNews/NewsRoom/8237.aspx (accessed 10 March 2009)

Wal-Mart (n.d.), 'Fact sheet', http://walmartstores.com/FactsNews/FactSheets/#Sustainability (accessed 27 February 2009)

Washington, B. (1980), *Booker T. Washington Papers*, eds. L. Harlan and R. Smock, Urbana, University of Illinois Press

Watson, K. (2007), *Retail Perspectives: Food Landscape in Transition*, Columbus, Ohio, TNS Retail Forward, http://www.retailforward.com/marketing/freecontent/foodlandscape.pdf (accessed 26 March 2009)

Webster, J. and P. Cottee (1997), *Waste Not, Want Not – Report on Surplus Fresh Food in the Food Industry*, London, Crisis

Webster, P. (2008), 'Waste not want not, Gordon Brown tells families', *The Times*, 7 July 2008

Weidema, B. *et al.* (2008), *Environmental Improvement Potentials of Meat and Dairy Products* (IMPRO), Luxembourg, European Commission

Weiland, P. (2000), 'Anaerobic waste digestion in Germany – status and recent developments', *Biodegradation* 11, 415–21

Wells, H. and J. Buzby (2007), 'ERS food availability data look at consumption in three ways', *Amber Waves: Behind the Data* 5:3, 40–41, Economic Research Service, US Department of Agriculture, www.ers.usda.gov/AmberWaves/June07/PDF/Datafeature.pdf

Wenlock, R. and D. Buss (1977), 'Wastage of edible food in the home: a preliminary study', *J. Human Nutrition* 31

Westendorf, M. (ed.), (2000), *Food Waste to Animal Feed*, Ames, Iowa State University

Westendorf, M. and R. Meyer (2003), *Feeding Food Wastes to Swine*, Rutgers Cooperative Research & Extension Fact Sheet, December 2003

Westendorf, M. *et al.* (1998), 'Recycled cafeteria food waste as a feed for swine: nutrient content, digestibility, growth, and meat quality', *J. Anim. Sci.* 76, 3250

Westendorf, M. *et al.* (1999), 'Nutritional quality of recycled food plate waste in diets fed to swine', *Prof. Anim. Sci.* 15:2, 106–11

WFP (2009), 'Financial speculation and the food crisis', 6 February 2009, World Food Programme, http://beta.wfp.org/stories/dr-timmer

Whole Foods Market (2009), 'Values overview, community giving, and green action', http://www.wholefoodsmarket.com/values/

Williams, A. *et al.* (2006), *Determining the Environmental Burdens and Resource Use in the Production of Agricultural and Horticultural Commodities*, main report, Defra Research Project IS0205, Bedford, Cranfield University and Department for Environment, Food and Rural Affairs, www.silsoe.cranfield.ac.uk

Winnett, R. (2008), 'G8 summit: Gordon Brown has eight-course dinner before food crisis talks', *Daily Telegraph*, 8 July 2008

Wlcek, S. and W. Zollitsch (2005), 'Sustainable pig nutrition in organic

farming: by-products from food processing as a feed resource',
Renewable Agriculture and Food Systems 19:3, 159–67

Wolf, M. (2008), 'Food crisis is a chance to reform global agriculture',
Financial Times, 29 April 2008

World Bank (2007), *2008 World Development Report: Agriculture for Development*, Washington DC, World Bank

World Bank (2008a), 'Food price crisis imperils 100 million in poor
countries, Zoellick says', 14 April 2008, http://web.worldbank.org/
WBSITE/EXTERNAL/NEWS/0,,contentMDK:21729143
~pagePK:64257043~piPK:437376~theSitePK:4607,00.html
(accessed 26 March 2009)

World Bank (2008b), *Rising Food and Fuel Prices: Addressing the Risks to
Future Generations*, Washington DC, World Bank

World Bank (2008c), *Biofuels: The Promise and the Risks*, Washington
DC, World Bank

World Bank (2009), *Global Economic Prospects: Commodities at the Crossroads*, Washington DC, World Bank

World Resources Institute (1998), 'Disappearing food: how big are
postharvest losses?', World Resources Institute

Worldwatch Institute (2006), *State of the World 2006: Special Focus: China
and India*, Washington DC, Worldwatch Institute

WRAP (2007), *Understanding Food Waste: Research Summary*, Banbury,
Waste & Resources Action Programme, http://www.wrap.org.uk/
downloads/FoodWasteResearchSummaryFINALADP29_3_07
.8a3253d1.3659.pdf (accessed 26 March 2009)

WRAP (2008a), *The Food We Waste*, food waste report v2, project
code: RBC405-0010, Banbury, Waste & Resources Action Programme, http://www.wrap.org.uk/downloads/The_Food_We
_Waste_v2_2_.940ff987.5635.pdf (accessed 26 March 2009)

WRAP (2008b), *Helping Consumers Reduce Fruit and Vegetable Waste:
Interim Report*, project code: RTL044-001, Banbury, Waste & Resources Action Programme, http://www.wrap.org.uk/downloads/
Helping_Consumers_reduce_fruit_veg_waste_Apr
_08.b98bbc85.5286.pdf (accessed 26 March 2009)

WRAP (2008c), *Courtauld Commitment Case Studies*, November 2008,

Banbury, Waste & Resources Action Programme, http://www.wrap
.org.uk/downloads/CC_Case_Studies_29_Jan_09_final
.1568e4da.6249.pdf (accessed 26 March 2009)

WRAP (2008d), *Organics Funding Guide*, November 2008, Banbury,
Waste & Resources Action Programme, http://www.wrap.org.uk/
downloads/Funding_guide_FINAL_printed_.f6251809.6551.pdf
(accessed 26 March 2009)

WRAP (2009a), 'UK grocery sector commits to reduce household food waste', 26 January 2009, http://www.wrap.org.uk/wrap_
corporate/news/uk_grocery_sector.html (accessed 12 March 2009)

WRAP (2009b), 'Courtauld Commitment', http://www.wrap.org
.uk/retail/courtauld_commitment/ (accessed 12 March 2009)

WRAP (2009c), 'Non-household food waste', http://www.wrap.org
.uk/retail/food_waste/nonhousehold_food.html (accessed 12
March 2009)

WRAP (2009d), 'Consumers save £300 million worth of food going
to waste', 14 January 2009, http://www.wrap.org.uk/wrap
corporate/news/consumers_save_300.html (accessed 12 March
2009)

WRAP (2009e), 'Love food, hate waste', www.lovefoodhatewaste
.com

Wright, K. (2008), 'Consuming Passions', *Psychology Today Magazine*,
March–April 2008

WWF (2006), *Species Factsheet: Bycatch*, Gland, Switzerland, World
Wildlife Fund, http://assets.panda.org/downloads/bycatch_apr_
2006.pdf

WWF (Germany) (2008a), 'Meerestiere sind kein Müll!' ('Sea creatures
are not rubbish'), November 2008, World Wildlife Fund, http://
www.panda.org/index.cfm?uNewsID=149401 and http://www.
divephotoguide.com/news/one_million_tons_of_fish
_discarded_in_the_north_sea_annually

WWF (2008b), 'European Fisheries Quota 2009: how can Europe
tackle the problem of discard?', press briefing, 8 December 2008,
World Wildlife Fund, http://assets.panda.org/downloads/wwf
_briefing_fisheries_2009_final.pdf

WWF (2008c), 'Norway forces EU to reduce cod discards', 10 December 2008, World Wildlife Fund, http://www.panda.org/what_we_do/footprint/smart_fishing/bycatch/bycatch_news/?152601 (accessed 12 March 2009)

Yang, S. *et al.* (2006), 'Lactic acid fermentation of food waste for swine feed', *Bioresource Technol.* 97, 1858–64

Zaehringer, M. and J. Early (1976), *Proc. National Food Loss Conference*, Boise, University of Idaho

Zhang, P. *et al.* (2005), 'Banana starch: production, physicochemical properties, and digestibility – a review', *Carbohydrate Polymers* 59, 443–58

Zhang, R. *et al.* (2007), 'Characterization of food waste as feedstock for anaerobic digestion', *Bioresource Technol.* 98, 929–35

Index